CELLULAR BASIS OF MORPHOGENESIS

The Ciba Foundation is an international scientific and educational charity. It was established in 1947 by the Swiss chemical and pharmaceutical company of CIBA Limited—now CIBA-GEIGY Limited. The Foundation operates independently in London under English trust law.

The Ciba Foundation exists to promote international cooperation in biological, medical and chemical research. It organizes about eight international multidisciplinary symposia each year on topics that seem ready for discussion by a small group of research workers. The papers and discussions are published in the Ciba Foundation symposium series. The Foundation also holds many shorter meetings (not published), organized by the Foundation itself or by outside scientific organizations. The staff always welcome suggestions for future meetings.

The Foundation's house at 41 Portland Place, London W1N 4BN, provides facilities for meetings of all kinds. Its Media Resource Service supplies information to journalists on all scientific and technological topics. The library, open seven days a week to any graduate in science or medicine, also provides information on scientific meetings throughout the world and answers general enquiries on biomedical and chemical subjects. Scientists from any part of the world may stay in the house during working visits to London.

Ciba Foundation Symposium 144

CELLULAR BASIS OF MORPHOGENESIS

A Wiley-Interscience Publication

1989

JOHN WILEY & SONS

Chichester · New York · Brisbane · Toronto · Singapore

Wiley Editorial Offices

John Wiley & Sons Ltd, Baffins Lane, Chichester,
West Sussex PO19 1UD, England

John Wiley & Sons, Inc., 605 Third Avenue,
New York, NY 10158-0012, USA

Jacaranda Wiley Ltd, G.P.O. Box 859, Brisbane,
Queensland 4001, Australia

John Wiley & Sons (Canada) Ltd, 22 Worcester Road,
Rexdale, Ontario M9W 1L1, Canada

John Wiley & Sons (SEA) Pte Ltd, 37 Jalan Pemimpin #05-04,
Block B, Union Industrial Building, Singapore 2057

Published in 1989 by John Wiley & Sons Ltd

Suggested series entry for library catalogues:
Ciba Foundation Symposia

Ciba Foundation Symposium 144
viii + 307 pages, 44 figures, 9 tables

Library of Congress Cataloging in Publication Data

Cellular basis of morphogenesis.
p. cm.—(Ciba Foundation symposium; 144)
Based on a conference held in Madrid, Spain in Oct. 18–20, 1988.
Edited by David Evered and Joan Marsh.
'A Wiley–Interscience publication'.
Includes bibliographies and index.
ISBN 0 471 92306 0
1. Morphogenesis. 2. Developmental cytology. I. Evered, David.
II. Marsh, Joan. III. Series.
[DNLM: 1. Morphogenesis—congresses. W3 C161F v. 144/QH 491
C3931 1988]
QH491.C45 1989
591.3′32—dc20
DNLM/DLC
for Library of Congress

89-5831
CIP

British Library Cataloguing in Publication Data

Cellular basis of morphogenesis.
1. Organisms. Morphogenesis
I. Evered, David II. Marsh, Joan III. Series
574.3

IBSN 0 471 92306 0

Phototypeset by Dobbie Typesetting Limited, Devon.
Printed and bound in Great Britain

Contents

Symposium on Cellular Basis of Morphogenesis, held at the Juan March Foundation, Madrid, Spain 18–20 October 1988

Editors: David Evered (Organizer) and Joan Marsh

L. Wolpert Introduction 1

A. Garcia-Bellido The cellular interphase 5
Discussion 10

D. A. Melton, A. Ruiz i Altaba, J. Yisraeli and **S. Sokol** Localization of mRNA and axis formation during *Xenopus* embryogenesis 16
Discussion 29

C. Nüsslein-Volhard and **S. Roth** Axis determination in insect embryos 37
Discussion 55

G. Struhl Morphogen gradients and the control of body pattern in insect embryos 65
Discussion 86

General discussion I 92

J. Heasman, A. Snape, A. Turner and **C. C. Wylie** The establishment of regional identity in the *Xenopus* blastula 99
Discussion 109

D. A. Weisblat and **S. H. Astrow** Factors specifying cell lineages in the leech 113
Discussion 124

P. A. Lawrence Cell lineage and cell states in the *Drosophila* embryo 130
Discussion 140

General discussion II 150

R. R. Franks, R. J. Britten and **E. H. Davidson** *In vivo* competition identifies positive *cis*-regulatory elements required for lineage-specific gene expression in the sea urchin embryo 156
Discussion 166

R. L. Gardner Cell allocation and lineage in the early mouse embryo 172
Discussion 181

J. Cooke Induction and organization of the body plan in *Xenopus* development 187
Discussion 201

General discussion III 208

J. Kimble and **J. Austin** Genetic control of cellular interactions in *Caenorhabditis elegans* development 212
Discussion 221

A. Busturia, J. Casanova, E. Sanchez-Herrero and **G. Morata** Structure and function of the bithorax complex genes of *Drosophila* 227
Discussion 239

M. Takeichi, K. Hatta, A. Nose, A. Nagafuchi and **M. Matsunaga** Cadherin-mediated specific cell adhesion and animal morphogenesis 243
Discussion 250

T. M. Jessell, P. Bovolenta, M. Placzek, M. Tessier-Lavigne and **J. Dodd** Polarity and patterning in the neural tube: the origin and function of the floor plate 255
Discussion 276

A. Tomlinson Cellular communication in the developing *Drosophila* eye 281
Discussion 287

General discussion IV 290

L. Wolpert Summary 296

Index of contributors 298

Subject index 300

Participants

C. V. Cabrera MRC Laboratory of Molecular Biology, Hills Road, Cambridge, CB2 2QH, UK

J. Cooke Laboratory of Embryogenesis, National Institute for Medical Research, The Ridgeway, Mill Hill, London NW7 1AA, UK

E. H. Davidson Division of Biology 156-29, California Institute of Technology, Pasadena, CA 91125, USA

A. Garcia-Bellido Centro de Biologia Molecular, CSIC, Universidad Autonoma de Madrid, Campus de Cantoblanco, 28049 Madrid, Spain

R. L. Gardner Imperial Cancer Research Fund, Developmental Biology Unit, Department of Zoology, University of Oxford, South Parks Road, Oxford, OX1 3PS, UK

J. Heasman Department of Zoology, University of Cambridge, Downing Street, Cambridge, CB2 3EG, UK

T. M. Jessell Neurobiology & Behavior Center, Howard Hughes Medical Institute, Columbia University, 722 West 168th Street, New York, NY 10032, USA

J. Kimble Laboratory of Molecular Biology, University of Wisconsin, 1525 Linden Drive, Madison, WI 53706, USA

T. Kornberg Department of Biochemistry & Biophysics, School of Medicine, University of California, San Francisco, CA 94143-0554, USA

P. A. Lawrence MRC Laboratory of Molecular Biology, Hills Road, Cambridge, CB2 2QH, UK

A. Martinez-Arias Department of Zoology, University of Cambridge, Downing Street, Cambridge, CB2 3EG, UK

M. McKeown Salk Institute for Biological Studies, PO Box 85800, San Diego, CA 92138, USA

H. Meinhardt Molekularbiologische Abteilung, Max Planck Institut für Entwicklungsbiologie, Spemannstrasse 35/IV, D-7400 Tubingen 1, Federal Republic of Germany

D. A. Melton Department of Biochemistry & Molecular Biology, Harvard University, 7 Divinity Avenue, Cambridge, MA 02138, USA

G. Morata Centro de Biologia Molecular, CSIC, Universidad Autonoma de Madrid, Campus de Cantoblanco, 28049 Madrid, Spain

C. Nüsslein-Volhard Max Planck Institut für Entwicklungsbiologie, Spemannstrasse 35, D-7400 Tubingen, Federal Republic of Germany

K. Sander Institut für Biologie I (Zoologie), Albert-Ludwigs-Universitat, Albertstrasse 21A, D-7800 Freiburg, Federal Republic of Germany

M. P. Scott Department of Molecular, Cellular & Developmental Biology, University of Colorado, Porter Biosciences Building, Campus Box 347, Boulder, CO 80309-0347, USA

G. Struhl Neurobiology & Behavior Center, Howard Hughes Medical Institute, Columbia University, 722 West 168th Street, New York, NY 10032, USA

M. Takeichi Department of Biophysics, Faculty of Science, Kyoto University, Sakyo-ku, Kyoto 606, Japan

A. Tomlinson MRC Laboratory of Molecular Biology, Hills Road, Cambridge CB2 2QH, UK

D. A. Weisblat Department of Zoology, University of California at Berkeley, Berkeley, CA 94720, USA

E. Wieschaus Department of Biology, Princeton University, Princeton, NJ 08544, USA

L. Wolpert Department of Anatomy & Developmental Biology, University College & Middlesex School of Medicine, Windeyer Building, Cleveland Street, London W1P 6DB, UK

Introduction[1]

Lewis Wolpert

Department of Anatomy & Developmental Biology, University College & Middlesex School of Medicine, Windeyer Building, Cleveland Street, London W1P 6DB, UK

It is fourteen years and forty-four Ciba volumes since we had a very exciting meeting on cell patterning at which several of us were present. In that time, the field has changed dramatically. The emphasis has switched largely from vertebrates to flies and worms. This is the result of a combination of genetics and molecular biology. Although there have been very important changes over those last fourteen years, I think the basic problems are still with us. Morphogenesis is the most important issue in biology—the working of the brain is important too, but there is nothing more important than going from genes in the egg to the five fingers of the hand or the wing of the fly or to the brain itself.

From the beginning of our subject there has been a tension between the 'left wing' and the 'right wing'. On the right I see those who thought that pattern formation, the organization of spatial differentiation, was something internal to the cell with little influence of cell interactions. Those people believed in cytoplasmic localization and autonomous cell lineages generating diversity between cells, following in the tradition of A. Weismann and E. B. Wilson. What I regard as the left wing were the people who thought that things were much more global, more interactive, in the tradition of Hans Driesch. I think that distinction is still present today—on the one hand are those who pursue the lineage cytoplasmic line and on the other hand are those who prefer gradients and more global interactions. In the middle are those who think in terms of local interactions, such as induction.

If we look for one moment at the left wing, a curious feature is that over the years the distance over which cells interact seems to have become smaller and smaller. The distances for which there is strong evidence that cells interact with each other directly is really very small, may be 10–30 cell diameters or even less. On the other hand, the right wing has problems too. If you look at the evidence for autonomous cell lineages (as distinct from cytoplasmic

[1]1989 Cellular basis of morphogenesis. Wiley, Chichester (Ciba Foundation Symposium 144) p 1–4

localization) then, at least in vertebrates and flies, the evidence I would argue is very poor (Wolpert 1989). I would argue that most of the evidence is for interactions—but I am not saying over what range they occur.

The best evidence for autonomous cell lineages comes from worms and yeast. For yeast mating types there is most persuasive evidence that a cell can generate daughter cells which are different from each other without any influence of the environment. There is almost a stem cell lineage, an unswitchable cell generating at each generation one daughter cell that is switchable. There is evidence from the work of A. J. S. Klar that the molecular basis of that difference between the cells is a difference between the Watson & Crick strands of the DNA. I regard that as the best evidence of a mechanism for autonomous generation of differences between cells at cell division.

One of the best pieces of evidence that I know against lineages comes from the experiment of Antonio Garcia-Bellido, Ripoll and Morata (1973), which demonstrated compartments. They used somatic crossing-over to generate a marked cell together with a slow growing mutant to show that while in the normal situation one marked cell in the early wing disc could give rise to about one-tenth of the wing, it could also give rise to half the wing and yet the pattern of the wing was completely normal.

The global positional type of mechanism originated from Hans Driesch. He rearranged the cells of the early sea urchin embryo in many different ways and argued that one particular cell will behave according to its position in the field, as if there is a system of co-ordinates and the cells can read these co-ordinates. If you are going to think in terms of positional information and gradients, you should recognize that you immediately imply the importance of boundary regions and polarity, as required by a system of co-ordinates.

Those on the 'left wing' think in terms of two main classes of global interactive theories. One is that there is some sort of monotonic concentration gradient and that cells can read thresholds and that determines how they behave—this is a positional information model. The other, from the reaction–diffusion ideas of Alan Turing, is that there is a non-monotonic variation in concentration and there is some relationship between the pattern of the morphogen concentration and the pattern that develops. This is a pre-pattern model (Wolpert & Stein 1984).

Over the last few years there has been some evidence that morphogens may have been identified. I think it would be a great pity if the word morphogen were used for any substance that changes the behaviour of a cell. When we use the word morphogen, we are really using it as Alan Turing did, in the sense that it has a global distribution that leads to the specification of the pattern that develops. The important point is how are we going to decide that we really do have a morphogen. The historical analogue is the problem people had in identifying neural transmitters. It is dangerous to assume that because you can mimic something or you can change a pattern by putting on a chemical, you have really identified the morphogen. Here are some criteria: there must be some relationship between the distribution of the putative morphogen and the final pattern. You have to show that when you change this distribution, the cells respond in the appropriate way. You also have to show that the cells are responding to that particular substance and that if you block that interaction in some way, you block the development of the pattern.

I am personally devoted to the concept of positional information. By positional value, I mean that cells have a parameter which corresponds with their position relative to a particular boundary. It doesn't mean necessarily that there is a diffusion gradient, since position can be specified in a variety of ways. What one is interested in is the nature of the cell parameter that varies with position. Some of the best evidence that cells do have such a parameter comes from regeneration in insects and vertebrates (Bryant et al 1981). In insects particularly, grafts which bring together regions not normally in contact result in intercalation of the missing positional values.

It is curious how few ideas we have in our field. We have a lot of data, but I feel I have already mentioned quite a lot of the ideas, admittedly very superficially. It would be nice if there were some coming together between them, for few of the things that I have spoken about are necessarily mutually exclusive. For example, you can have a positional mechanism which relies on cell lineage, and cell lineages may be involved with local interactions. Ultimately, we want to link cell behaviour to genes. One of the big issues is to what extent is the specification of cell fate, or cell position, or cell differentiation, combinatorial or not. By combinatorial I mean this: if we assume that a cell state is specified by genes, then a mechanism is combinatorial if the number of genes involved is quite small compared to the number of cell states specified. If they were about the same order of magnitude, I would say it was not combinatorial.

I have spoken mainly about specifying pattern. I've hardly said anything about change in form, which is so important in morphogenesis. I would argue that in all changes in form, whether it is the folding of the neural tube or gastrulation, the mechanical changes are preceded by a patterning mechanism which specifies the cells that generate the forces. This can be, for example, the specification of adhesive molecules or the concentration of microfilaments.

Finally, I would hope that the 'left' and the 'right' and the 'middle' will ultimately come together. I would be very sad if it turned out that there were lots of very different mechanisms of morphogenesis that bear no relation to each other whatsoever. I hope that general mechanisms for going from genes to flesh and blood are going to emerge. But I may just have to face the fact that I am going to be disappointed.

References

1975 Cell patterning. Elsevier Science Publishers, Amsterdam (Ciba Foundation Symposium 29)

Bryant SV, French V, Bryant PJ 1981 Distal regeneration and symmetry. Science (Wash DC) 212:993–998

Garcia-Bellido A, Ripoll P, Morata G 1973 Developmental compartmentalization in the wing disc of drosophila. Nat New Biol 245:251–253

Wolpert L 1989 Stem cells: a problem in asymmetry. In: Lord BI (ed) Stem cells. J Cell Sci Suppl 10:1–10

Wolpert L, Stirn WD 1984 Positional information and pattern formation. In: Malacinski GM, Bryant SV (eds) Pattern formation. Macmillans, New York, p 2–21

The cellular interphase

Antonio Garcia-Bellido

Centro de Biologia Molecular, CSIC, Universidad Autonoma de Madrid, Campus de Cantoblanco, 28049 Madrid, Spain

Abstract. A brief outline is given of the impact of molecular genetics on the study of morphogenesis. Two approaches have been used, one beginning at the molecular level and applying knowledge gained in this way to observations of whole organisms; the other starting with descriptions of the organism and tracing the causes of developmental changes to cellular and then molecular events. The availability of experimentally induced mutants has facilitated the study of development, but such work must be combined with detailed information on the normal process. Comparative studies have exploited evolutionary conservation to identify functionally important gene activities, and allowed knowledge gained from one organism to be rapidly applied to other species. The successes achieved at the molecular level are impressive, but the time has come to return to studying the cells that are the real mediators of the transmission of the genetic information.

1989 Cellular basis of morphogenesis. Wiley, Chichester (Ciba Foundation Symposium 144) p 5–15

This is the second Ciba Foundation symposium dealing with the problem of morphogenesis. The first took place in 1974, under the general title of 'Cell patterning'. The problem addressed then is still the same today, as it has been for almost one hundred years, from Roux's Manifesto of 1894 onwards: how do eggs generate, through cell division, organisms of constant, species-specific size and shape? We have made some advances in our understanding of this problem since 1894, adding to the role of the egg cytoplasm the role of genes in the generative function. Since the genes carry morphogenetic information, the problem can now be stated in other terms, namely, how does the linearly encoded information of the DNA interact with the cytoplasm to generate the morphogenetic space? This is a problem of transformation between different levels of complexity. In confronting this transformational problem, we are in the company of sociologists, ethologists and linguists, as well as that of chemists, cosmologists and physicists. The problem has two aspects, one epistemological (what constitutes an explanation?) and the other logical (which explanations are not merely historical descriptions?).

At the symposium on cell patterning, Sydney Brenner, an eminent representative of the successful discipline of molecular biology, proposed a pragmatic approach. He suggested that the description of molecular interactions in the transformation of genetic information be extended from prokaryotes to eukaryotes and multicellular organisms. Brenner's strategic proposal of proceeding 'from the bottom up' seemed to be both clear and feasible. The bottom was the level of molecules and the top was the level of organisms. In terms of a linguistic analogy, he was proposing to study a phonological grammar, expecting that there was probably little, if any, syntax or semantic grammar in morphological evolution. According to the then still reigning notions of the New Synthesis, evolution was the result of the accumulation of multiple variants with small phenotypic effects, mainly in genes coding for enzymes, that happened to lead to morphological changes with a higher adaptive value. There was not much for the student of morphogenesis to do but to describe contingencies.

Meanwhile a 'top-down' research programme had begun, whose extremely successful descent from the organismal to the cellular level was rapid and surprising. The detailed description of development uncovered *quasi*-invariant developmental cell lineages in the fruit fly, the roundworm, the leech and other taxa, including the paradigmatic organism of regulative development, the sea urchin. Organisms appeared to be constructed from modules, by reiterative operations associated with stereotyped cell divisions. The specific perturbation of stereotyped cell lineages by gene mutations, and subsequent genetic analysis of the mutated loci, uncovered the existence of genuine morphogenetic genes, operating specifically to generate morphological space. Further genetic analysis uncovered hierarchies of morphogenetic genes, acting as teams or syntagmas, which carried out specific morphogenetic operations. Diverse gene teams were found to be working independently of each other in the same cell, thus combinatorially defining cell specificity and characteristic behaviour.

The descent to the cellular level was followed by an even more rapid and spectacular descent to the molecular level. The molecular analysis of morphogenetic genes uncovered examples of a diversity of products arising from a single coding region. Moreover, *cis*-regulatory DNA sequences were discovered in or near morphogenetic genes, which are the targets for the *trans*-regulation among genes proposed from the results of genetic analysis. These interactions effect the temporal, spatial and quantitative control of gene expression. Finally, the DNA sequences of some genes that perform a certain morphogenetic function in one organism were found to be remarkably similar to DNA sequences in other, phylogenetically far removed, organisms. This high degree of evolutionary conservation of morphogenetic genes reveals strong functional and operational constraints, suggesting an invariant semantic grammar in the generation of forms in evolution.

This analytical approach, aimed at elucidating the molecular elements and biophysical mechanisms of the transformation of morphogenetic information between levels of organization in development, uncovered a new world that transcends the comfortable and naive one of historical contingency in morphogenesis. We found that there is more for the developmental biologist to do than merely to describe: there is, at least, a morphogenetic syntax to be ascertained. After taking apart our four-dimensional toy from the top down, we are now left with the task of finding out, from the bottom up, the logical rules by which it was put together. This may prove to be a difficult task, especially if, working at the bottom level, we forget what it was at the top that we wanted to explain.

History repeats itself. As has always been the case, new conceptual fashions and new research tools continue to redefine the problems that are to be solved. Scores of researchers are now engaged in working out the structure of novel genes, apparently interesting because of their normal products or because of the phenotypic effect of their mutations. They follow a standard protocol that ends in the deposition of a nucleotide sequence in a gene bank. This is followed by cataloguing the sequence, by comparing it with other sequences to identify canonical regions and by matching it with putative homologous sequences in other organisms.

The possibility of describing patterns of gene expression *in situ*, in intact embryos, has carried morphogenetic analysis one step further up, to the description of spatiotemporal changes in the pattern of gene expression during the course of development. Moreover, whenever mutant alleles of morphogenetic genes are available, one can describe the mutant pattern of expression, or the pattern of expression of the normal gene on a mutant background, and relate this pattern to the morphology of the mutant phenotype. It should be remembered, however, that whenever this approach has been interpreted successfully, it was carried out in combination with a detailed cellular and genetic description of the normal morphogenetic process (as, for example, in the analysis of embryonic segmentation in *Drosophila*). This approach is a synchronic analysis, but we are now in a position to compare similar morphogenetic processes among more or less related taxa in a diachronic analysis. These interphyletic comparisons will be made without necessarily submitting each of the taxa to a classical (transmission) genetic analysis. Instead, the spatiotemporal patterns of expression of homologous genes will be ascertained by using molecular probes developed in just one of the organisms surveyed (e.g. *Drosophila*). It is hoped that such a comparison will provide us with the semantic aspect or meaning of gene syntagmas, i.e. with the functional logic of morphogenesis. It is to be expected that the conservation through evolution of the specific molecular product of a given gene occurs together with conservation of the developmental or genetic operation in which it participates. These genetic operations correspond in structuralist transformational analysis to the 'deep structure' of morphogenesis.

If development, like evolution, is not a serial phenomenon, its understanding will not come through exhaustive reiterative description of a particular sequence of steps at the lowest level of analysis. It is doubtful that by merely articulating molecular reactions and genetic elements we shall elucidate how form is generated and evolves via complex interactions of multiple elements. Rather, we shall have to shift our focus of attention to higher levels of diachronic integration, in search of phenomena to be investigated in an approach from the top down. Three historical examples illustrate this idea. The founding Padres of the California missions moved from Mexico to California in the eighteenth century. The order in which they founded the missions did not, however, follow a south-to-north sequence but represented a process of intercalation, progressively reducing the average distances between missions. Thus, the founding Padres mapped California top-down before settling. Another example: Gregor Mendel necessarily preceded James Watson and Francis Crick, because it is only in the higher level context of genetics that DNA structures can be said to carry information. Niels Bohr proposed to study living organisms because he thought that they might be governed by physical laws that would pass undetected by studies at the level of atoms.

Many classical phenomena of developmental biology remain unexplored. For example, the notion of entelechy—a term Hans Driesch borrowed from Aristotle—denotes a developmental tendency to completeness or perfection, both in normal and in experimentally perturbed development. Unfortunately, Driesch used entelechy to explain observed biological phenomena, instead of regarding entelechy itself as a subject of investigation, thus, provoking the long-lasting disdain of his (then generally positivistic) fellow embryologists. However, entelechy is a real phenomenon and remains to be understood. One cannot expect, of course, that the explanation of entelechy will emerge from a detailed knowledge of the molecular interactions that are involved. Entelechy is an aspect of the more general phenomenon of biological homeostasis, but classifying it as such does not help to determine its essence. We simply have to search for the most appropriate level of analysis on which to deal with it—a level probably intermediate between molecules and whole organisms.

For the sake of epistemological pragmatism, I propose that the time has come to reascend from the molecular to the cellular level in the study of morphogenesis. Cells correspond to the first level of integration of genetic, molecular and structural interactions. At this level (to stay with our linguistic metaphor), we shall encounter utterances and sentences susceptible to phonological, syntactic and semantic analysis. This is the level that W. Bateson and T. M. Morgan wanted to reach, as developmental biologists, before they made their detour into genetics. Organisms are not only made of cells, they arise from them by self-organization. Phenomena such as cell morphology, the specification of cell fate, differentiation, locomotion, migration, polarity and intercellular communication are waiting to be understood in genetic and molecular terms. What is more important, they correspond to the very properties of cells that underlie morphogenesis.

Mark Bretscher recently pointed out to me that cell biology is in its infancy. We know about molecular elements, structures and cell decorations, but we can neither sort them out functionally, nor account for their origins. They are actors looking for a plot. Cells are not 'bricks' to be used to fill morphogenetic space; they themselves generate that space. In so doing, they engage in complex patterns of cell behaviour, defined by specific genetic operations. Cells possess an integrative machinery that decides when and how the cell should divide and differentiate. Cells talk to each other in very specific ways, possibly holding several conversations simultaneously, in languages about which we know almost nothing. Certainly the study of morphogenesis at the cellular level will not be easy, because identifying morphogenetically relevant operations in the overall cell behaviour will require identification of the individual operational components, i.e. a dissection of molecular functions interrelated in unknown ways. The relationship between intracellularly processed morphogenetic information, its transmission to neighbouring cells and its transformation inside the target cell is a new research agenda for approaching the problem of morphogenesis at the appropriate level. It is to be hoped that in dealing with the complex interactions among genes we shall be helped by the evolutionary conservation of genetic functions and structures. If these are conserved, then the operations in which they participate are likely to be conserved as well. For that reason, comparative studies of the products of morphogenetic genes and of their spatiotemporal pattern of cellular expression will become a fundamental method of morphogenetic analysis.

The proposition that the study of morphogenesis now requires exploration at the cellular level is also of epistemological significance. The ever-increasing complexity of known gene–gene interactions will make developmental phenomena increasingly less intelligible at the molecular level. We shall need a higher level perspective from which to evaluate the lexical value of genes in their semantic transactions and to sort out the levels at which structural details are morphogenetically relevant. In the end, understanding morphogenesis will have become merely the finer and finer mapping of nets of gene interactions. Just as molecular biology blossomed only after it focused on searching for the rules of the storage and transformation of genetic information, ignoring the problem of the energy supply for these transactions, so developmental biology has to learn to leave molecular mechanisms to one side and search for the transformational logic of morphogenesis that connects genes with the spatiotemporal development of organismic form. The formulation of this agenda seems to me to be the main justification for the present symposium. Let us meet fifteen years hence to reconsider our strategies once again.

Reference

1975 Cell patterning. Elsevier Science Publishers, Amsterdam (Ciba Foundation Symposium 29)

DISCUSSION

Garcia-Bellido: Our genetic analysis of the generation of patterns, such as that of the wing veins in *Drosophila*, is showing that patterns are closely related to cell proliferation. It seems that in wing morphogenesis there are two fundamental genetic operations: 1) defining spatially periodic variations in gene activity (amplitudes in wave functions), and 2) communicating these amplitude values between neighbouring cells. Differences in these values between cells elicits cell proliferation. This process stops (attains the condition of entelechy) when the maximal amplitude values are reached and the differences between neighbours become minimal.

There are candidates for amplitude genes (classical growth genes) and cell communication genes (those related to epidermal growth factors, their receptors and other signal transducers). Growth and pattern in this way are controlled from within, by local cell interactions, rather than by external morphogenetic signals diffusing over long distances. This model accounts for classical phenomena such as regeneration (both morphallactic and epimorphic); patterns formed by proliferation of single cells (which are spatially incomplete but internally perfect); and local non-autonomy ('accommodation') in genetic mosaics. In this model, cells are of central importance; they must communicate with each other, holding simultaneous conversations mediated by molecular interactions of specific gene products.

Cooke: I would like to take up this comparison that you are drawing between synchronic and diachronic types of analysis. People are taking the ensemble of genes that can be cross-identified between evolutionarily distant organisms as having homeobox-like regions, and cross probing, looking for clues to the role that particular genes play in different organisms. For the very few genes that have been identified as the closest homologue in evolutionary terms to a gene of known function in *Drosophila*, when one asks what they are doing in vertebrate development, for example, the answer is the uncomfortable one that the gene has been co-opted for an early developmental function that may well be important, but is different from its function in *Drosophila*. The family tree of molecular homologies does not correspond to a family tree of homologous functions.

Garcia-Bellido: This is an important point. The initial reaction to finding a homeobox homologue of segmentation genes of *Drosophila* in vertebrates was to say, here is a vertebrate segmentation gene. This was obviously naïve. The homeobox is simply a domain of a protein that binds to DNA and regulates its transcription: it is a syntactic signal and may not be associated with the same *word* and the same semantic meaning in distantly related organisms. We cannot deduce its meaning (function in development) from its DNA sequence. In fact, expression of particular homeobox genes in vertebrates may vary between different histiotypes as well as between embryonic regions. The understanding of segmentation, or periodicities in vertebrates, will require knowledge of more genes—regulatory or not—and their patterns of spatial expression. Moreover, we will have to visualize the process at the level of individual cells and through time, if we want to make relevant comparisons between different organisms.

Davidson: Many of us in this field who are extremely rigorous in discussion of the things that we know about, i.e. the molecular biology of development, are not so rigorous in using evolutionary considerations. For example, remarks such as *abrupt changes in evolution* are disturbing if you consider the length of evolutionary time and the variation seen within given taxa. We don't really understand how functionally important changes in genomes occur in evolution or at what rate these things happen. It seems dangerous to say one mechanism must be close to another or that because two animals are in the same taxon they must do things in a fundamentally similar way, when the taxon has been around since the Carboniferous period and includes organisms which may do things in very different ways. In the example of the colonization of California, understanding whether the Padres settled by intercalation or not does not help us understand California, because this phase of its development has little to do with its ultimate morphogenesis.

Wieschaus: Antonio, presumably one of the reasons that the Padres settled in California by intercalation is that they defined an area in which they wanted to establish a pattern of settlements. Intercalation works well when you start at one end and choose something far away, then set up a pattern in between. If you look at development, there are alternative approaches for establishing patterns in fields, the equivalent of which for the Padres would be to start at the border of Mexico and add little stations progressively. To what extent do you think that the underlying developmental mechanisms—how genes work and how cells interact with each other—are going to depend on whether you set up pattern by intercalation between end points or by sequential steps?

Garcia-Bellido: I used this analogy as an epistemological scheme for understanding or analysing complexities. It may be used by developing organisms to specify patterns. It is possible that organisms also use other approaches, such as sequential specifications. But even here I would like to propose that both overtly different operations may have common internal mechanisms. Thus, whereas *Drosophila* embryonic segmentation seems to occur by intercalation, more primitive insects, and annelid derivatives in general, segment by budding sequentially. I suspect that such dramatic differences maintain profound similarities at the genetic level, if what is important is, for example, to generate periodicities with different wavelengths in register in certain cells, as references for subsequent intercalation. This would account for the mechanism of defining constant numbers of segments in arthropods and insects.

Meinhardt: Concerning simultaneous versus sequential segment formation, I have proposed that segmentation results from the reiteration of at least three cell states, which locally exclude each other but activate each other over long distances. According to this model, a periodic pattern can be formed either gradually during outgrowth or more or less simultaneously under the control of a gradient (Meinhardt 1982a, 1986). Since the modifications required in the mechanism are minute, it is not too surprising that different insect species use one or the other of these mechanisms; or even that transitions exist within one species in that the more anterior segments are formed simultaneously while the more posterior ones are added by budding.

Let me come back to cell lineage and irreversibility. Clonal inheritance of a particular determined state becomes possible, if genes are involved which exert positive feedback on their own activation. At the same time, alternative genes must be suppressed. If this feedback is strictly confined to cells, daughter cells will obtain the same determination as the mother cell and the probability that a cell will be in a particular cell state is independent of the cell state of neighbouring cells. In contrast, if the substances involved in auto-stimulation are slightly diffusible, the border between two cell states can be smooth and can be shifted, enabling, for instance, size regulation. Therefore, the underlying molecular mechanisms for determination may be very similar, independent of whether there is cell lineage or not. The activation of such a cell state will usually be irreversible, except when the self-enhancing feedback is perturbed such that an alternative feedback loop wins the competition.

Garcia-Bellido: I agree, lineage doesn't mean the cell is irreversibly committed, only that it is following one developmental pathway. However, it is like hormonal systems, where a single shot of hormone is not enough, the signal has to be maintained. In highly interactive genetic systems, the decision has to be maintained as well, especially because there are so many genetic connections with other genes. So the notion of differentiation (as irreversibility) may be wrong, at the cellular level. Differentiation may be something which is simply related to the setting of specific genetic interactions, but it is still capable of changing or reversing, if we disrupt these interactions.

Martinez-Arias: Antonio, do you think that intercalation is the universal way of establishing pattern? Do you reject Eric's (Wieschaus) argument for stable sequential addition?

Garcia-Bellido: I repeat, this metaphor is related to understanding, not to how development actually proceeds. We have to be prepared to encounter a series of developmental operations generating morphological space and pattern, of which this would be one.

Martinez-Arias: A point often forgotten in these discussions is that genes code for proteins and proteins can do many more things than genes. We know very little about proteins. So by focusing on genes you might be missing something very important.

Gardner: Antonio, you were very dismissive about mechanical explanations, could you say why. It seems to me they provide an important intermediate stage of analysis in relating gene activity to cytoskeletal changes.

Garcia-Bellido: Again, I wanted to be provocative. The more we know about the specification of the different regions of the amphibian gastrula, the chord-mesoderm for example, the more we realize it occurs in the blastula. The same is true for other systems. In this case, gastrulation becomes a mechanism for implementing a genetic signal of cell migration or sorting. Gastrulation is an overt expression of a phenomenon that can be described, but not understood, in mechanical terms. When I criticize mechanical explanations of development, I do so against the spirit of W. His in the 19th century or in the spirit of C. H. Waddington in ours. Waddington would say that if you take an embryonic epithelium from *Drosophila* imaginal disc and push it through a tube, you obtain a leg. This notion is absurd because leg properties are implemented at the cellular level, as shown by genetic mosaics.

The problem is that it is difficult to demonstrate that this statement is not true. Exactly one hundred years ago, Boveri showed that segregation of chromosomes in the anaphase is not due to pulling by the spindle, which was the obvious explanation. Mechanically, it was very difficult to imagine a system in which the central mass pushed the chromosomes away. I would say that, in general, mechanical explanations are just a short-hand description of the macro-phenomenon. The more we know about cells, the more we can show that simple mechanical explanations are wrong, simply because the system has to be much more precisely specified. In defining cell behaviour, cells are becoming the important actors in the story.

Wolpert: In spite of your passion for genes, cells have to obey Newtonian mechanics and I am not prepared to abandon that at this stage. However much you don't like mechanical explanations, when anything changes shape it has to follow the principals of mechanics and something has to generate the forces. Antonio has been pleading that the important thing in cells is gene action—that proteins are the really important actors. They can provide the necessary forces, thus linking gene action to mechanics.

Lawrence: Antonio, you said two things had hindered our progress over a long time. One, with which I agree absolutely, is a damaging obsession with how reversible cells are when they are put into different experimental or ectopic situations. The other one you said was size invariance, which I didn't quite understand.

Garcia-Bellido: I think size invariance has been misleading because it has obligated us to think that the system is controlled as a whole. Size invariance means that there is no size control, but what then computes the actual size? We have to look for mechanisms which produce dimensions not properties, and size invariance is a property without dimensions. In the real world, the system has finite shape, size and structure. The system has to know when it is complete. If you analyse pattern components, for example in *Drosophila* imaginal epidermis, the number and position of elements of pattern vary when the size of the anlage changes. That means that the system is not computing borders and then subdividing space, but that the cells are checking only with their neighbours. The same applies to the bone patterns in vertebrate limbs (Oster et al 1988). Size invariance has damaged our thinking because it has placed the emphasis on global qualities or properties, and detracted us from ascertaining the role that the cells are playing.

Cooke: Size invariance cannot so easily be dismissed. It is possible to look only at aspects of development where thinking about size invariance has been misleading. But various sorts of experiments reveal aspects of pattern which do show size invariance. In development of each type of embryo, there seems to be one mode of establishing pattern in which cells talk only to their neighbours, and which doesn't take account of global dimension, and another mode that takes that into account in setting the scales of territories within patterns. There are many classical experimental manipulations of insect blastoderms where such a principle is operating.

Garcia-Bellido: It is true that in *Drosophila*, a normal larva can develop from a blastula with a reduced number of cells. How does it compensate for the low number of cells? I believe that the system solves equations at an interstitial differential level and then grows and proliferates to reach the normal condition. If you take embryos that have been subjected to global perturbations, like cutting, or study morphogenetic mutants as individuals or in mosaics, you see that the system is obviously not size invariant.

Meinhardt: In the anteroposterior dimension of insects, there is no positive size regulation, there is even a negative size regulation. For instance, after ligation in the posterior fragment, less structures are formed compared with the unperturbed situation. But in the dorsoventral dimension there is excellent size regulation. Fragmentation can lead to identical twins. One should not generalize more than nature does. Different mechanisms with very different size-regulating properties may be operating within a single organism, for instance, in two orthogonal axes (for experiments see Sander 1976, for models see Meinhardt 1982b).

Garcia-Bellido: I am only saying that size invariance may be an epiphenomenon that does not directly reflect what is happening between cells generating forms.

Wolpert: Do you believe the size invariance exists—that over a different size range the basic pattern remains invariant, as for example in sea urchins and hydra?

Garcia-Bellido: Within certain physical limits there is size invariance. The question is whether this can be used as an argument that the normal specification of position is due to global properties. The moment you go beyond these limits within which there is regulation, the system shows that it has internal mechanisms that will not generate size invariance. Thus, there seem to be mechanisms that generate distributions of heterogeneities computing only small cell intervals. When you push the system too far experimentally, the issue of size invariance breaks down and the internal mechanisms become apparent.

Wolpert: So are you saying that there is size invariance but that the models being used to explain it are not the right models?

Garcia-Bellido: On the contrary, models taking that notion to explain morphogenesis may be wrong. Because by extrapolating size invariance, by a factor of two, to the general mechanism of shape formation, you may lose the perspective of what you want to explain.

Davidson: The classical experiments in which cleavage stage sea urchins are fused provide a good example of regulative development. The result, which is a normal large embryo, can be explained either by a classical double gradient system or by assuming inductive interactions from cell layer to cell layer. So the phenomenon does not imply a particular kind of explanation. However, it can be concluded that whatever mechanism applies to make cells talk to each other has to be involved in size change.

References

Meinhardt H 1982a The role of compartmentalization in the activation of particular control genes and in the generation of proximo-distal positional information in appendages. Amer Zool 22:209–220

Meinhardt H 1982b Models of biological pattern formation. Academic Press, London

Meinhardt H 1986 Hierarchical inductions of cell states: a model for segmentation in Drosophila. J Cell Sci, Suppl 4:357–381

Oster GF, Shubin N, Murray JD, Alberch P 1988 Evolution and morphogenetic rules: the shape of the vertebrate limb in ontogeny and phylogeny. Evolution 42:862–884

Sander K 1976 Specification of the basic body pattern in insect embryogenesis. Adv Ins Physiol 12:125–238

Localization of mRNA and axis formation during *Xenopus* embryogenesis

D. A. Melton, A. Ruiz i Altaba, J. Yisraeli and S. Sokol

Department of Biochemistry and Molecular Biology, Harvard University, 7 Divinity Avenue, Cambridge, Massachusetts 02138, USA

Abstract. This paper summarizes our recent work concerned with the developmental polarity of the frog egg and the patterning of the embryonic body plan. In two separate projects, we are studying genes involved in setting up basic embryonic axes. One of these genes, *Vg1*, codes for a maternal mRNA that is localized in the frog egg. The *Vg1* gene is used in studies on the induction of mesoderm and as a probe to understand how the polarity of an egg is established. A second gene, *Xhox3*, contains a homeodomain and is differentially expressed in the axial mesoderm. Our studies suggest that this homeobox gene is critically involved in setting up different positional values along the anteroposterior axis.

1989 Cellular basis of morphogenesis. Wiley, Chichester (Ciba Foundation Symposium 144) p 16–36

The role of the localized maternal mRNA, *Vg1*

Xenopus eggs have an easily distinguished top and bottom that define the animal–vegetal axis. The top or animal hemisphere is dark brown because of pigment granules and the bottom or vegetal hemisphere is a yolky yellow. With respect to the development of the fertilized egg, the visible polarity of the animal–vegetal axis foretells an important difference in developmental capacity. The top of the egg gives rise to cells that form ectoderm (skin and neural tissue), whereas the bottom of the egg gives rise to endodermal derivatives, primarily the gut. Thus, these two cytoplasmic regions have the information needed to form different parts of the embryo and, therefore, to direct the synthesis of different gene products. In addition, the vegetal end of the egg produces a signal that is responsible for the formation or induction of embryonic mesoderm. An intercellular signal that emanates from cells at the vegetal pole interacts with overlying cells in the middle region of the cleaving egg to specify the fate of the latter as mesoderm (Nieuwkoop 1973).

The different developmental capacities that are stored along the animal–vegetal axis in the egg cytoplasm are most probably in the form of different maternal mRNAs and/or proteins. This inference is based on the fact that the embryo's genome is not transcriptionally active until about eight hours after fertilization, hours after decisions about cell fates (such as mesoderm induction) have transpired. With this in mind, we examined the animal–vegetal distribution of maternal *Xenopus* mRNAs and concluded that (1) most mRNAs are uniformly distributed in the animal–vegetal axis, and (2) there does exist a rare class of localized maternal mRNAs (Rebagliati et al 1985). Some RNAs are localized to the top or animal hemisphere, e.g. *An1,2,&3* (Weeks & Melton 1987a). One mRNA localized to the vegetal end, *Vg1*, is of special interest because of the pattern of its localization and the protein it encodes.

Vg1 mRNA encodes a 40 kDa protein which contains a putative signal sequence at its amino terminus for secretion through the endoplasmic reticulum. Most intriguing are the 115 amino acids at the carboxy terminus which are very similar to the human transforming growth factor-β (TGF-β) (Weeks & Melton 1987b). On the basis of this sequence information and the intracellular location of the mRNA (Fig. 1), it is reasonable to suggest that the *Vg1* protein will be synthesized by and secreted from cells at the vegetal end of the embryo (the presumptive endoderm).

In separate experiments, it has been shown that heterologous growth factors are capable of inducing mesoderm in isolated animal caps taken from *Xenopus* blastulae. Human TGF-β2 (Rosa et al 1988) and bovine fibroblast growth factor (FGF) (Slack et al 1987, Kimelman & Kirschner 1987) have some inducing activity, as does a homologous growth factor (XTC-MIF) secreted from *Xenopus* tissue culture cells (Smith 1987). As noted above, Nieuwkoop's studies clearly demonstrated that the vegetal end of the cleavage embryo normally produces a mesoderm-inducing signal (Nieuwkoop 1973). These facts, combined with the knowledge that the *Vg1* protein is a growth factor-like molecule synthesized by the inducing tissue (endoderm), point to the possible involvement of *Vg1* in mesoderm induction. However, the evidence is still circumstantial. We are presently trying a number of experiments to demonstrate directly the function of the *Vg1* protein. These include the injection of synthetic *Vg1* mRNA into the animal hemisphere (cells that do not normally have the mRNA) to test for a change in their fate from ectoderm to mesoderm. Additionally, the injection of antisense nucleic acids to block the translation of the endogenous *Vg1* protein may help demonstrate the function of this embyronic protein.

In all, there is a growing body of information which suggests that the synthesis of growth factor-like molecules in the endoderm can account for the mesoderm-inducing capacity of that end of the developing egg. Certainly, a final proof of which molecules are involved and what each does is still missing and the details of how these signals are produced and received will probably provide several

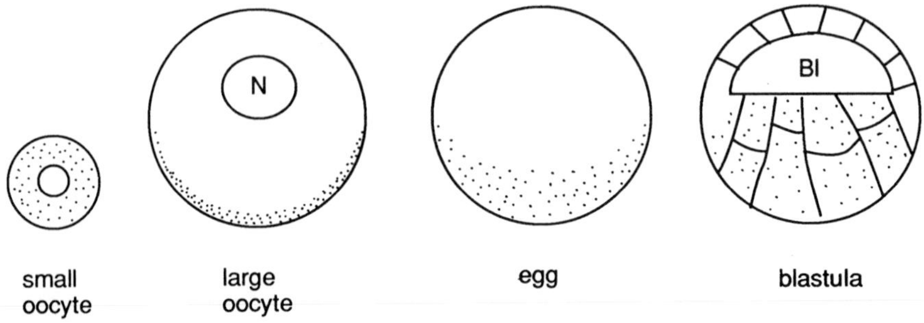

FIG. 1. Localization of the maternal mRNA, *Vg1*. The diagram summarizes results of several *in situ* hybridization experiments using antisense *Vg1* RNA as a hybridization probe (Melton 1985, Weeks & Melton 1987). The animal–vegetal axis is top to bottom in all cases. The small dots represent the location of *Vg1* RNA at various developmental stages. N, nucleus or germinal vesicle in oocytes; Bl, blastocoel.

interesting surprises. Nonetheless, it is reasonable to assume that we have the correct outline of the molecular mechanism used to induce mesoderm. A critical part of the mechanism is the localization of the inducing signal to one end of the egg. In essence, it is the polarity of the signal that underlies the mechanism by which one end of the egg has a fate or an inductive capacity different from that of the other end. With this view, we have focused much of our efforts on understanding how mRNAs are localized, using *Vg1* as a probe.

Vg1 RNA is translocated to the vegetal pole during oogenesis

Vg1 mRNA is synthesized early during oogenesis and the level of RNA remains constant through oogenesis and up to the early gastrula stage. The amount of spliced and polyadenylated RNA present in small pre-vitellogenic oocytes (StI) is equal to that found in large fully grown (StVI) oocytes and early gastrula (Melton 1987). It is not known whether the RNA is synthesized continually and turned over, or whether the RNA is made early on and is stable. Interestingly, *Vg1* RNA is not localized in young pre-vitellogenic oocytes, but becomes localized during oogenesis (Melton 1987), as diagrammed in Fig. 1. The simplest interpretation of the data on *Vg1* RNA synthesis and *in situ* hybridizations is that *Vg1* RNA is moved or translocated to the vegetal pole. *In situ* hybridizations show that *Vg1* RNA is found in a tight crescent at the vegetal hemisphere in fully grown oocytes. The RNA is apparently in a subcortical shell between the large yolk platelets and the cell membrane.

The tight cortical location of *Vg1* RNA changes when oocytes mature into unfertilized eggs. The RNA behaves as if it were released and seems to diffuse toward the animal pole during the first few hours of post-fertilization development. However, because cell cleavage is occurring rapidly during this period, the RNA is confined to cells at the vegetal end of the cleaving embryo (see Fig. 1). It should be emphasized that the localization and movement of *Vg1* is highly unusual; most maternal RNAs are uniformly distributed along the animal–vegetal axis.

Synthetic *Vg1* RNA is correctly localized after injection into oocytes

We should like to understand how *Vg1* is translocated in the cytoplasm of this single cell, the frog oocyte. To this end, we have adapted an *in vitro* culture system (Wallace et al 1980) for the growth of young oocytes outside the female's ovary. Using tissue culture media supplemented with frog serum that contains vitellogenin (a yolk platelet precursor), we are able to obtain continued growth and development of young (StIII) oocytes for at least 10 days in culture. In addition to the normal morphological changes, *in situ* hybridizations show that the endogenous *Vg1* is localized in cultured oocytes during this time. *Vg1* RNA is not localized in the oocytes that are removed at the beginning of the culture period, but the RNA is translocated within 3–5 days in culture. These oocytes cultured *in vitro* form a test system to study the mechanism of RNA localization (Yisraeli & Melton 1988)

Radiolabelled *Vg1* or globin mRNA (control) are synthesized *in vitro* using a bacteriophage transcription system (Melton et al 1984). This radioactive RNA is injected into StIII oocytes and the recipients are cultured for 3–6 days. At the end of the culture period, injected oocytes are fixed, sectioned and exposed to autoradiographic emulsion. The results (Fig. 2) show that injected *Vg1* RNA is localized to the vegetal pole in a manner analogous to endogenous *Vg1* mRNA: in contrast, globin RNA diffuses freely in the cytoplasm. The main conclusion is that *Vg1* RNA as naked RNA contains a signal that is sufficient for its correct localization within the oocyte (Yisraeli & Melton 1988). The localization of injected *Vg1* RNA is not observed if the RNA is injected into fully grown (StVI) oocytes. In this case, the *Vg1* RNA stays primarily in the vegetal hemisphere, but does not accumulate in a tight subcortical crescent. This result suggests that the localization machinery is less active or absent in the fully grown oocytes.

Deleting various portions of the *Vg1* RNA and adding pieces of *Vg1* RNA to other non-localized RNAs before injection should enable us to identify the RNA signal responsible for intracellular localization. The first experiments of this type have been aimed at the 5′ end of the mRNA. The 5′ end of *Vg1* RNA was deleted, including the translation initiation codon and the bases encoding the first 25 amino acids of the protein. Nevertheless, synthetic transcripts with this 5′ deletion are still correctly localized following injection. This result strongly suggests that the localization machinery does not depend on translation of the *Vg1* RNA and that nascent *Vg1* protein is not involved in localizing the mRNA (Yisraeli & Melton 1988).

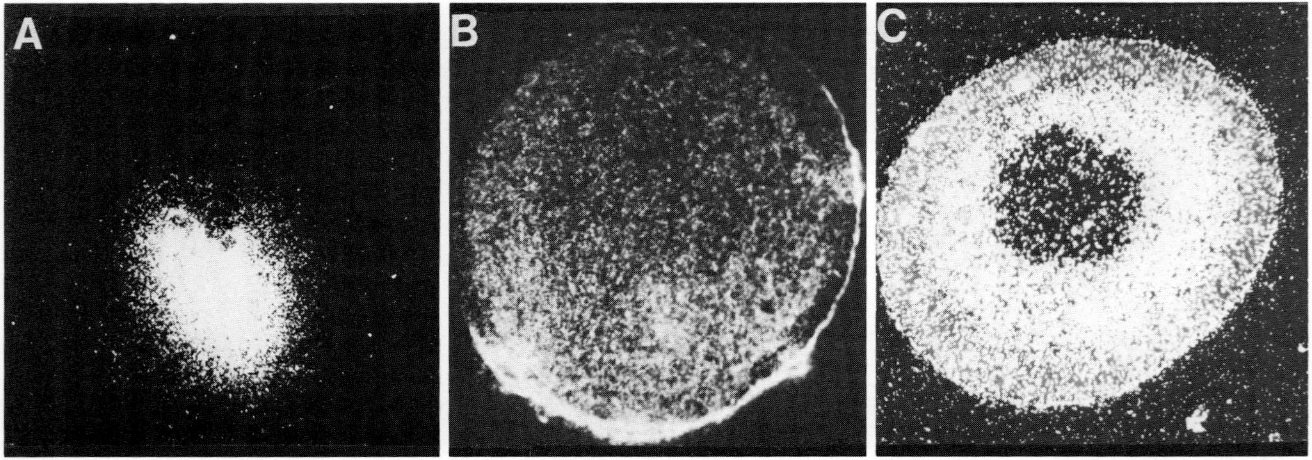

FIG. 2. Synthetic *Vg1* is localized following injection into growing oocytes. [^{32}P]-labelled RNA was injected into the middle of stage III oocytes and the injected cells were cultured *in vitro* for five days, after which the oocytes were fixed, sectioned, and exposed to autoradiographic emulsion. Panel A shows the position of the radioactive RNA at time 0, immediately after injection. Panel B shows the localization of injected *Vg1* RNA and panel C shows the uniform distribution obtained by injected globin RNA. Extraction of the injected RNAs at Day 5 shows that the autoradiographic signals represent *Vg1* and globin RNAs, respectively, and are not due to reincorporation of nucleotides from degraded RNAs.

Vg1 RNA preferentially associates with the cytoskeleton

The cellular machinery involved in localizing mRNAs is not known. There are a few other cases in which RNAs are known to be differentially distributed in cells (Lawrence & Singer 1986, Nusslein-Volhard et al 1987, Fontaine et al 1988), but in no case is it understood how this is accomplished. We believe an understanding of the RNA localization machinery is an important issue because it speaks directly to the question of how developmental polarity is established and maintained. Indeed, the mechanism by which oocytes make one end different from the other may well be used later in development, for example when stem cells undergo a polar division wherein one daughter differentiates and the other remains a stem cell.

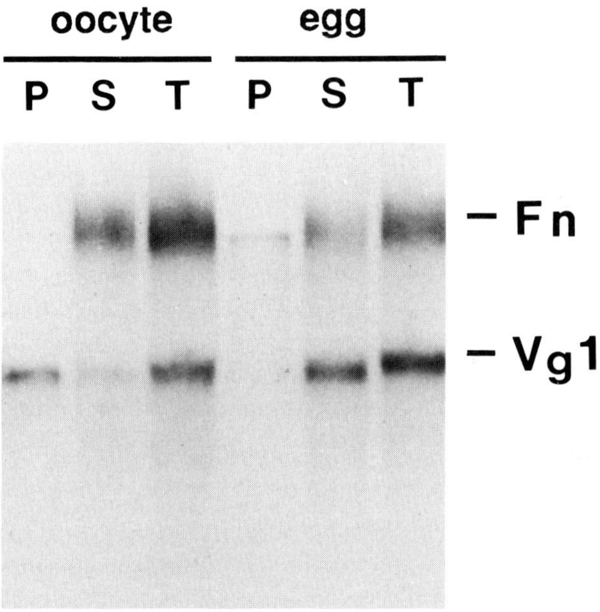

FIG. 3. Detergent-insoluble pellets preferentially contain *Vg1* RNA. Fully grown (StVI) oocytes or fertilized eggs were homogenized in a buffer containing 0.5% Triton X-100 detergent, a buffer used to prepare cytoskeletal fractions modified from Jeffrey (1984). This extraction buffer does not solubilize the cytoskeleton. The homogenates were spun at $10\,000 \times g$ and the pellet (P) and supernatant (S) were assayed for the presence of *Vg1* or fibronectin (Fn) mRNA by hybridization in a Northern blot to the appropriate antisense RNA probes. A control of total (T) or unfractionated RNA is shown in each case to test for RNA recovery. Note that the *Vg1* RNA is preferentially found in the cytoskeletal pellet from oocytes, but not from eggs.

The cytoskeleton warrants consideration as a structure potentially involved in localization because it could, in principle, act as a scaffold or track upon which the *Vg1* RNA could be moved or tethered. To begin to assess the role of the cytoskeleton in *Vg1* localization, we have adapted biochemical methods (Jeffery 1984) for preparing cytoskeletal extracts and assayed for the presence of *Vg1* RNA. As a control, we use a non-localized maternal mRNA, one coding for fibronectin. The results of such an extraction show that *Vg1* RNA is preferentially associated with a cytoskeletal extract in fully grown (StVI) oocytes (Fig. 3) (S. Sokol, J. Yisraeli, D. Melton, unpublished observations) . More than half of the *Vg1* mRNA is found in the detergent-insoluble pellet (the 'cytoskeleton'), whereas only a small proportion of fibronectin or histone (not shown) mRNAs are found in the pellet. *In situ* hybridizations have shown that the *Vg1* RNA is in a tight cortical crescent at this stage of oogenesis (see Fig. 1). Thus, the association with the detergent-insoluble fraction may reflect an association with cytoskeletal elements and/or some type of anchoring protein. Interestingly, when oocytes have matured into eggs, *in situ* hybridizations show that *Vg1* RNA is no longer held in a tight cortical crescent but spreads through the vegetal hemisphere (Fig. 1). Biochemical fractionation shows that *Vg1* RNA in eggs is no longer preferentially associated with the cytoskeletal pellet (Fig. 3).

The cytoskeletal pellet contains only a small portion of total cellular protein and very little ($<2\%$) of the total polyA$^+$RNA. Thus, this extraction method offers an effective initial step in identifying cellular proteins with which localized *Vg1* RNA associates. These experiments, coupled with those which map the portion of the *Vg1* mRNA containing the localization signal, may allow us to identify the proteins and other cellular elements that accomplish the localization (S. Sokol, J. Yisraeli, D. Melton, unpublished observations).

The role of *Xhox3* in specifying fates along the anteroposterior axis later in development

Up to this point we have considered the egg's animal–vegetal polarity and how the localization of maternal RNAs may be involved in mesoderm induction. These processes set up important initial differences along one axis of the embryo, namely that which gives rise to the three germ layers: ecto-, meso- and endoderm. However, gastrulation rearranges the relative positions of the germ layers and this rearrangement sets up new axes, which in turn form the tissue patterns of the swimming tadpole. A variety of grafting and explantation studies have shown that the mesoderm assumes the role of an inducing tissue in gastrula and early neurula. The mesoderm instructs overlying ectoderm as to which tissues and cell types will form. Significantly, the mesoderm along the anteroposterior (A-P) axis has different positional values and these are reflected in the types of neural and other ectodermal tissues that are induced.

Perhaps the clearest evidence that the mesoderm has different inductive capacities along the A-P axis comes from the transplantation studies of O. Mangold (1933). When anterior mesoderm is transplanted to a host embryo, it induces ectoderm to form anterior neural and ectodermal structures, such as the forebrain, sucker and eye. In contrast, posterior mesoderm induces posterior or trunk neural tissue and fin. In both cases, the ectoderm is equally competent to form anterior or posterior structures and the different results are due to different inducing capacities of the anterior and posterior mesoderm. What then is the nature of this difference in the A-P positional information in the mesoderm? Recent experiments suggest that the differential expression of a homeobox gene, *Xhox3*, plays a controlling role.

Xhox3 is a *Xenopus laevis* gene that was isolated by screening a cDNA library with the *Drosophila* pair-rule gene, *even-skipped* (Ruiz i Altaba & Melton 1989a). The *even-skipped* gene was chosen as a probe because the other homeobox genes that we and others have investigated (Carrasco et al 1984, Harvey et al 1986, Condie & Harland 1987) were isolated with homeobox probes of the *Antennapedia-Ultrabithorax* class, whose sequences are distinctly different from the *even-skipped* homeodomain. Transcripts from the *Xhox3* gene first accumulate at the mid-blastula transition and are expressed predominantly, if not exclusively, in the mesoderm. An important aspect of *Xhox3* expression is that it forms a gradient during gastrulation with the highest concentration in the posterior mesoderm (Fig. 4). While the levels of only *Xhox3* RNA have been measured, these are presumed to correspond to the levels of protein. In any case, the gradient of *Xhox3* RNA which is observed is probably established by differential transcriptional activity. The consequence is that the anterior mesoderm has low levels of *Xhox3* RNA, whereas the posterior end expresses the gene at high levels. Thus, the anterior neural plate, from which the head and brain will form, is in contact with mesoderm that has very low levels of *Xhox3* RNA.

Xhox3 expression in the mesoderm is intriguing because of the idea that homeobox genes control cell fates and because this particular gene is differentially expressed along the A-P axis. To test whether this gradient plays a role in A-P patterning, we have manipulated the embryo by injecting *Xhox3* mRNA and antisense *Xhox3* DNA to alter the endogenous *Xhox3* gradient.

In the first set of experiments, synthetic *Xhox3* mRNA was prepared *in vitro* and injected into fertilized eggs. Our previous studies had shown that this is an effective way to oversupply a specific gene product (Harvey & Melton 1988). As controls, we injected two other mRNAs, globin and *Xhox1A* mRNA, which codes for a different homeobox protein, one of the *Antennapedia-Ultrabithorax* class and which may have a role in somite segmentation (Harvey & Melton 1988). Dissection of injected embryos, coupled with assays for the presence of the injected mRNAs, showed that the endogenous *Xhox3* gradient of RNA is

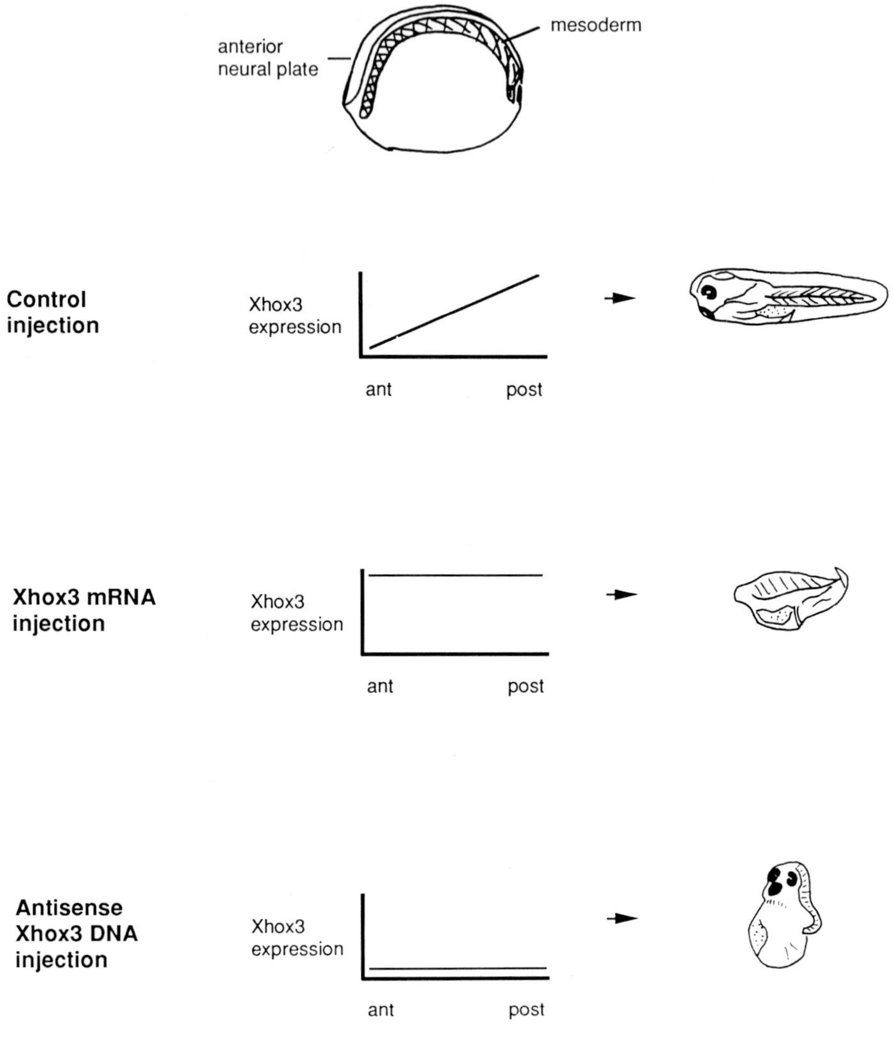

FIG. 4. The effect on the embryonic body pattern of altering *Xhox3* expression. The diagram at the top shows a section of a neurula embryo with the mesoderm (cross-hatched) underlying the ectoderm. In normal or control injected embryos the endogenous gradient of *Xhox3* RNA gives rise to a tadpole, as shown at the top right. The level of *Xhox3* expression is altered by injecting sense mRNA into fertilized eggs (*Xhox3* mRNA injection). This produces a uniform expression of *Xhox3* along the anteroposterior axis and gives rise to anterior-deficient (acephalic) embryos, as shown at the right. In the converse experiment (Antisense *Xhox3* DNA injection), the endogenous level of *Xhox3* is presumed to be destroyed by the action of an antisense RNA (see text). This results in a posterior deficiency (short axis with little or no tail), as shown at the right.

overwhelmed by the injected RNA and the gradient is destroyed (Fig. 4). In effect, there is a high and even level of *Xhox3* RNA across the A-P axis. The result of this injection is that embryos fail to develop normal anterior structures, most notably, the entire head is absent. While there is a graded response, probably related to the stability and location of the injected RNA, the one consistent effect is the loss of anterior structures. Embryos injected with control RNAs rarely show this specific defect (Ruiz i Altaba & Melton 1989b).

Headless embryos can be formed by at least two other techniques. For example, any disruption of gastrulation in which mesoderm is prevented from migrating all the way to the presumptive anterior end will prevent formation of normal anterior structures. This can be accomplished by the injection of trypan blue into embryos at various stages of gastrulation (Gerhart et al 1984). Alternatively, headless (and dorsal-deficient) embryos are formed following UV irradiation of fertilized eggs. In this case and that of trypan blue injection, the cause of the defect is that anterior ectoderm is not supported by or in contact with anterior mesoderm. We have carefully examined *Xhox3*-injected embryos for defects in the time at which gastrulation begins and the extent of mesodermal migration. Unlike the case for UV irradiation or trypan blue injection, *Xhox3*-injected embryos gastrulate normally. Histological sections and visual inspection reveal that the morphogenetic movements associated with gastrulation are indistinguishable from those in control embryos. Embryos injected with *Xhox3* mRNA show a normal invagination of mesoderm in all respects. We, therefore, conclude that the failure to form normal anterior structures is the result of increased levels of *Xhox3* expression in the mesoderm present in the anterior portion of the late gastrula.

The converse experiment, removing or inactivating expression of the endogenous *Xhox3* gene, has been attempted by injecting a DNA that codes for an antisense *Xhox3* RNA. The idea of this experiment is that the antisense RNA will hybridize specifically to endogenous *Xhox3* transcripts and block their translation (Melton 1985, Giebelhaus et al 1987). Controls include injection of DNAs that do not encode an antisense *Xhox3* RNA, but rather encode a small portion of the cDNA in the sense orientation. When a DNA directing the synthesis of antisense *Xhox3* RNA is injected into fertilized eggs, 1–5% of the injected embryos fail to develop normal posterior structures and, in essence, the posterior trunk and tail are missing or greatly reduced. The anterior structures are normal or enlarged. These posterior-deficient embryos are not observed in controls injected with DNA that directs synthesis of a portion of the *Xhox3* mRNA in the sense orientation.

The results of this experiment are complicated by the fact that the injected DNA is not distributed to or transcribed in all tissues of the embryo. The mosaic expression of the injected DNA may account for the observation that only a small portion (1–5%) of the injected embryos show specific posterior defects.

Moreover, because we have not yet assayed for RNA hybrid formation between the endogenous *Xhox3* RNA and transcripts from the injected DNA, we must be cautious in interpreting this antisense DNA injection result. Nevertheless, these preliminary experiments are encouraging because the phenotype of the antisense DNA-injected embryos is, to a first approximation, the opposite of that obtained following overexpression by injection of sense (messenger) RNA. In the former case, injected embryos fail to form or induce normal posterior structures and in the latter case the anterior end is defective.

Consistent with these observations are assays of *Xhox3* RNA levels in embryos that have been treated so as to form anterior/dorsal-deficient or posterior/ventral-deficient patterns. UV irradiation will produce embryos lacking anterior/dorsal structures (Gerhart et al 1984) and our tests show that these embryos have elevated levels (about five times the normal level) of *Xhox3* RNA. Conversely, treatment with lithium produces embryos lacking posterior/ventral structures (Kao et al 1986) and these embryos have depressed levels (about 0.2 times the normal level) of *Xhox3* RNA (Ruiz i Altaba & Melton 1989a).

The sense and antisense injection experiments require more molecular analyses, in particular the use of antibodies to test for altered levels of *Xhox3* proteins. Nevertheless, the results of those injection experiments and results from UV and lithium-treated embryos allow us to suggest the following role for the *Xhox3* protein. Low levels of *Xhox3* in the mesoderm are required for normal anterior development and high levels are required for or lead to posterior development. A gradient in the level of *Xhox3* along the mesoderm will lead to or define different positional values along the A-P axis.

This simple hypothesis begs the question of how different levels of *Xhox3* are set. How is it that during gastrulation the level of *Xhox3* RNA in anterior mesoderm is set low and the posterior mesoderm has high levels? We don't yet know the answer, but one intriguing possibility is that the levels of *Xhox3* are set by growth factors. The evidence for this comes from induction experiments using isolated animal caps. Animal caps derived from early blastulae will not express *Xhox3* RNA in isolation. This is an expected result because *Xhox3* is normally expressed in mesoderm and isolated animal caps form only ectodermal derivatives. When animal caps are treated with XTC factor or FGF, both of which induce mesoderm, these factors induce the expression of *Xhox3*. Interestingly, in these two cases, the levels and kinetics of *Xhox3* expression are distinctly different. A conclusion which can be drawn from these experiments is that one of the first responses following mesoderm induction is the transcriptional activation of the *Xhox3* gene. Secondarily, these experiments suggest that the level of *Xhox3* may be set by different combinations or concentrations of growth factors which act during gastrulation. This raises the possibility that the Spemann organizer acts as a source or sink for diffusible factors, e.g. for growth factors, which in turn set the level of *Xhox3* expression (Ruiz i Altalba & Melton 1989a).

Obviously, the details of how the level of *Xhox3* effects A-P pattern are not clear. We do not know which genes are controlled by *Xhox3* expression nor how these differences in expression are related to induction and pattern formation. Nevertheless, the present experimental results offer a simple starting hypothesis for explaining A-P patterning. Further manipulations of the expression of *Xhox3* should clarify its role in specifying cell fates in the A-P axis.

Summary

In this paper we discuss results which relate to the formation of two axes in the frog embryo. With respect to the animal–vegetal axis, our work suggests that the developmental polarity of the egg is embodied in the localization of maternal mRNAs. We suggest that localized mRNAs like *Vg1* (and perhaps those encoding FGF, TGF-β2 or the XTC factor) code for intercellular inducing signals and thereby generate the embryonic mesoderm. The mesoderm is formed in one region and not all over because the inducing signal(s) is localized. Thus, the problem of how different cell types are specified early in development is reduced, in one sense, to the problem of how informational molecules such as mRNAs are localized. Our results show that the *Vg1* RNA contains a signal that specifies its position within the cell and reinjection experiments should enable us to determine what cellular components are involved in its translocation.

It should be noted that the question of how dorsal/ventral differences in the mesoderm are established has not been directly considered here. The establishment of the embryonic dorsoventral axis is largely understood from the work of Gerhart and his colleagues and is considered elsewhere (Gerhart et al 1984).

Later in development, when the body pattern is set up by inductions during the gastrula and early neurula stages, our injection results suggest that the *Xhox3* gene has an important role in controlling the pattern along the anteroposterior axis. Because the levels of *Xhox3* expression are correlated with or perhaps define different positional values in the A-P axis, it is important to determine how the level of *Xhox3* expression is set during gastrulation. In this regard, the finding that growth factors can induce *Xhox3* expression may provide an important clue in understanding the molecular biology of the patterning of this vertebrate.

Acknowledgement

This work was supported by a grant from the NIH.

References

Carrasco AE, McGinnis W, Gehring W, DeRobertis EM 1984 Cloning of an Xenopus laevis gene expressed during early embryogenesis coding for a peptide region homologous to Drosophila homeotic genes. Cell 37: 409–414

Condie BG, Harland RM 1987 Posterior expression of a homeobox gene in early Xenopus embryos. Development 101:93–105

Fontaine B, Sasson D, Buckingham M, Changeux JP 1988 Detection of the nicotinic acetylcholine receptor alpha-subunit mRNA by in situ hybridization at neuromuscular junctions of 15-day-old chick striated muscles. EMBO (Eur Mol Biol Organ) J 7:603–609

Gerhart J, Vincent JP, Scharf SR, Black SD, Gimlich RL, Danilchik M 1984 Localization and induction in early development of Xenopus. Philos Trans R Soc Lond Biol Sci 307:319–330

Giebelhaus DH, Eib DW, Moon RT 1987 Antisense inhibition of expression of membrane skeleton protein 4.1 during embryonic development of Xenopus. Cell 53:601–609

Jeffrey W 1984 Spatial distribution of messenger RNA in the cytoskeletal framework of ascidian eggs. Dev Biol 103:482–492

Harvey RP, Melton DA 1988 Microinjection of synthetic Xhox-1A homeobox mRNA disrupts somite formation in developing Xenopus embryos. Cell 53:687–697

Harvey RP, Tabin C, Melton DA 1986 Embryonic expression and nuclearlocalization of Xenopus homeobox (Xhox) gene products. EMBO (Eur Mol Biol Organ) J 5:1244–1267

Kao K, Masui YI, Elinson R 1986 Lithium-induced respecification of pattern in Xenopus laevis embryos. Nature (Lond) 322:371–373

Kimelman D, Kirschner M 1987 Synergistic induction of mesoderm by FGF and TGF-β and the identification of an mRNA coding for FGF in the early Xenopus embryo. Cell 51:869–877

Lawrence JB, Singer RH 1986 Intracellular localization of messenger RNAs for cytoskeletal proteins. Cell 45:407–415

Mangold O 1933 Uber die Induktionsfahigkeit der verschieden Bezirke der Neurula von Urodelen. Naturwissenshaften 21:761–777

Melton DA 1985 Injected anti-sense RNAs specifically block messenger RNA translation *in vivo*. Proc Natl Acad Sci USA 82:144–148

Melton DA 1987 Translocation of a localized maternal mRNA to the vegetal pole of Xenopus oocytes. Nature (Lond) 328:80–82

Melton DA, Krieg PA, Rebagliati MR, Maniatis T, Zinn K, Green MR 1984 Efficient *in vitro* synthesis of biologically active RNA and RNA hybridization probes from plasmids containing a bacteriophage SP6 promoter. Nucleic Acids Research 12:7035–7056

Nieuwkoop PD 1973 The organization center of the amphibian embryo: Its origin, spatial organization and morphogenetic action. Adv Morphol 10:1–39

Nusslein-Volhard C, Fronhofer HG, Lehmann R 1987 Determination of anteror-posterior polarity in Drosophila. Science (Wash DC) 238:1675–1681

Rebagliati MR, Weeks DL, Harvey RP, Melton DA 1985 Identification and cloning of localized maternal RNAs from Xenopus eggs. Cell 42:769–777

Ruiz i Altaba A, Melton D 1989a Bimodal and polar expression of the Xenopus homeobox gene Xhox 3. Development 106:173–183

Ruiz i Altaba A, Melton D 1989b Involvement of the Xenopus homeobox gene Xhox3 in embryonic pattern formation along the anterior-posterior axis. Cell, in press.

Rosa F, Roberts AB, Daneilpour D, Dart LL, Sporn MB, Dawid IB 1988 Mesoderm induction in amphibians: The role of TGF β2-like factors. Science (Wash DC) 239:783–785

Slack JMW, Darlington BG, Heath JK, Godsave SF 1987 Mesoderm induction in early Xenopus embryos by heparin-binding growth factor. Nature (Lond) 326:197–200

Smith JC 1987 A mesoderm inducing factor is produced from a Xenopus cell line. Development 99:3–14

Wallace RA, Misulovin Z, Wiley HS 1980 Growth of anuran oocytes in serum-supplemented medium. Reprod Nutr Dev 20:699–708

Weeks DL, Melton DA 1987a A maternal RNA localized to the animal pole of Xenopus eggs encodes a subunit of mitochondrial ATPase. Proc Natl Acad Sci USA 84:2798–2802

Weeks DL, Melton DA 1987b A maternal messenger RNA localized to the vegetal hemisphere in Xenopus eggs codes for a growth factor related to TGF-β. Cell 51:861–867

Yisraeli J, Melton DA 1988 The maternal mRNA *Vg1* is correctly localized following injection into Xenopus oocytes. Nature (Lond) 336:592–595

DISCUSSION

Davidson: What's the evidence that the mRNA synthesized in early oogenesis is stable as opposed to being present throughout oogenesis in a steady state, the rate of synthesis equalling the rate of degradation?

Melton: You can't ever prove that completely; you can argue that the gene is always on and the transcript is immediately degraded.

Davidson: I calculated that for a typical message in a *Xenopus* oocyte, if it had a half-life of 40 days, the gene would have to be transcribed at the maximum rate to maintain a steady state, because of the large size of the cell and the many copies of each transcript (typically about 10^6 per oocyte, even for normally rare message species). So, to say the RNA is stable may be erroneous.

Melton: It may be, but one experiment argues against that view. We have injected radiolabelled ribonucleotides into oocytes and tried to detect labelled (newly synthesized) *Vg1* transcripts by hybridization to *Vg1* cDNA. It is not a good experiment because the label is diluted on injection and the newly synthesized RNA has a rather low specific activity. In any case, we found no evidence for *Vg1* synthesis in fully grown oocytes.

Davidson: The same calculation would apply because of the size of the system, so you might not see a significant quantitative change within a few days. Do you know what the RNA target sequence is that is responsible for *Vg1* RNA localization?

Melton: No, but we are narrowing it down by deletion mapping. At present, we know that it is not contained entirely in the 5′ untranslated region.

Meinhardt: Do you have any idea how the animal–vegetal polarity is set up? Is there a difference between the cytoskeleton at the animal pole and that at the vegetal pole?

Melton: Janet Heasman may talk more about that in detail. In my view, a simple idea would be that the nucleus and the mitochondrial cloud set up, at random, what will be the animal–vegetal axis. That axis is then 'seen' (I don't know what we mean by that) by cytoskeletal elements which align themselves between the nucleus and the mitochondrial cloud. Microtubules have polarity and there is no reason to rule out the possibility that a *Vg1* RNA–protein complex recognizes that polarity—when it bumps into the cytoskeleton it knows which way to move.

McKeown: It looked like there was perinuclear nuclear localization prior to movement to the vegetal cap, is that true?

Melton: It is almost always true. Sometimes we don't see that, so it may be an artifact of the *in situ* hybridization.

McKeown: Does your *in vitro* culture system accumulate yolk?

Melton: Yes.

McKeown: So it's localizing something else to the vegetal pole, as well?

Melton: Yes. If you culture oocytes in saline without vitellogenin in the medium, they don't localize the RNA. But the *Vg1* RNA is localized long before you can see any difference in the animal–vegetal axis of yolk platelets. Both yolk platelet localization and *Vg1* localization recognize some fundamental axis but we don't know its cellular basis yet.

Heasman: At what stage oocytes is the asymmetry of *Vg1* mRNA established? Is it always at a particular size?

Melton: Vg1 RNA is evenly distributed in young oocytes (stage I and II) and begins to localize shortly thereafter. The first reliable signs arise when oocytes are about 0.6 mm in diameter.

Heasman: Which is the same time as the mitochondrial cloud breaks down. That time in oogenesis might be as important in establishing localization patterns as the time in maturation when the nuclear membrane, the Golgi and the endoplasmic reticulum break down and cell surface changes occur.

Melton: There is a difference between the way that the primary axis is established in the frog egg and the way it's done in *Drosophila*. The *Xenopus* oocyte doesn't grow out of a case with attached nurse cells that squirt in RNAs and proteins at one end, as is the case for *Drosophila*. In flies, the anteroposterior axis is set up by a kind of template mechanism that depends on the position of the nurse cells.

Weisblat: Did you say that the *Vg1* RNA is released from the cytoskeleton by the blastula stage?

Melton: Before blastula, when an oocyte matures into an egg. In an unfertilized egg the RNA is not held in a tight crescent but behaves as if it has been released and is floating or migrating towards the animal pole.

Weisblat: So it is cellularization that keeps it localized in the vegetal hemisphere?

Melton: That's right. I would imagine that if you block cytokinesis, the RNA may eventually reach the animal pole. A teleological explanation for how you localize the signal would be that the only reason it is restricted to such a tight sub-cortical crescent is so that as it begins to diffuse following maturation it doesn't have time to get to what will become the ectoderm.

Lawrence: Do you have a functional but unlocalizable *Vg1* RNA that you can inject?

Melton: We think we do, but we haven't done the experiment yet. One way forward is to inject synthetic message *Vg1* at the top of the egg and see if it affects the fates of the cells at the animal pole. Alternatively, one can try to remove *Vg1* message from the bottom of the egg by using an antisense message.

McKeown: Given that some things are spatially localized and then become unlocalized, like *Vg1*, what do you know about the releasing signal? Not just fertilization but any information about how that's released?

Melton: All we know is that the hormone-induced maturation of the oocyte, which brings it through its meiotic division, is coincident with the release of *Vg1* from this tight cortical crescent.

Wolpert: After you injected *Xhox3* mRNA, is the level of *Xhox3* mRNA higher than the control in all regions of the embryo?

Melton: That's right. But we don't have an antibody so we don't know what that means in terms of the protein.

Davidson: You commented that no part of the cleavage stage *Xenopus* embryo is capable of differentiating into mesoderm by itself non-inductively. One cannot exclude that isolated equatorial cells interact with each other and thereby produce mesoderm, but is it really clear that the alternative is untrue—do you really know that no mesodermal precursor cells can differentiate autonomously?

Melton: No, you cannot rule that out. You can say that there is no reason to invoke a mechanism other than induction. There is no convincing experiment which shows that there is a localized factor which is inherited by a cell and causes that cell to become mesoderm.

Martinez-Arias: You isolated *Xhox3* on the basis of its homology to *even-skipped*. How homologous is it to *even-skipped*?

Melton: The *Drosophila* gene was kindly provided by Gary Struhl and Paul Macdonald. The sequence of *Xhox3* is more similar in the homeobox region to *even-skipped* than it is to members of the *Ant-Ubx* class of homeotic genes, but outside the homeobox there is no obvious similarity. The only other similarity is that, like *even-skipped*, the homeobox is near the amino end of the protein.

Wolpert: Have you any speculations why *even-skipped* should be the one that shows the developmental effects?

Melton: We have been trying these types of experiments with other homeobox genes, in particular, Richard Harvey and I reported experiments altering the expression of *Xhox1A*. However, the effects observed after altering *Xhox3* expression are quite different and more dramatic. I should like to mention that others, notably Eddy DeRobertis and Richard Harland, have also been working on *Xenopus* homeobox genes, but to my knowledge those genes are all members of the *Ant-Ubx* class. It is to Ariel Ruiz i Altaba's credit that he decided to look farther afield at this other class, of which *Xhox3* is a member.

Meinhardt: Your gradient notation suggests that *Xhox3* has something to do with positional information, but it could be that it is more analogous to the gap genes in *Drosophila*. It could be activated in response to a positional signal.

Melton: Yes, I was very hesitant to use the word gradient. The graphs I showed of *Xhox3* mRNA levels are intended primarily as a way to remember that there are low levels of expression in the anterior part of the embryo and high levels in the posterior part. We can only presume that the levels of RNA reflect the level of protein. Until we have an antibody and can do a more detailed description of where the protein is, perhaps we shouldn't use the word gradient.

Meinhardt: The observation that *Xhox3* is zygotically and not maternally expressed is another indication that it is more comparable to the gap genes in *Drosophila* than to the positional information genes.

Garcia-Bellido: If you inject *Xhox3* RNA after the blastula stage, into blastomeres, what happens?

Melton: We haven't done that because the cells are so small at that stage—you would have to inject several hundred cells to get a significant result.

Jessell: Xhox3 seemed to be still expressed at high levels later in development. Do you know whether after somite differentiation, for example, its expression looks more like that of *even-skipped* or whether it is uniform?

Melton: Later in development, by the swimming tadpole stage, the gene is on, but like many *Drosophila* genes it is being expressed in a different place. In the nervous system it is expressed only in the brain. It is also switched on in the posterior tip of the tail, which continues to grow and induce more tail as development proceeds.

Cooke: I would like to ask about the earliest time point at which you have examined the endogenous distribution of *Xhox3*. Early on during gastrulation, several complicated movements occur. How far have you got in finding out where the *Xhox3* transcript is in the anatomy of the embryo at stage 11?

Melton: We have not got very far at all, the levels are too low to detect it easily by *in situ* hybridization. We are now concentrating on making an antibody. If you use an RNAse protection assay to ask when does the gene come on, it is one of the very first genes to be transcribed. Its transcription is activated at the time of the mid-blastula transition. So the evidence shows that when the zygotic genome is activated, *Xhox3* comes on.

Cooke: So the implication would be that in what we call pre-involution mesoderm—mesoderm which is laid out in the reverse dimension in the embryo to the way in which it will be distributed after gastrulation—the gene is already being activated?

Melton: Yes. We have no information about whether or not it's differentially activated in the mesoderm at that point. It could be that it is initially uniformly distributed and this changes during gastrula movements.

Cooke: In these embryos which look like mild versions of UV embryos, which are missing their anterior structures, that you produce by over-expressing *Xhox3*, do they have notochord or not? Embryos with the same external morphology produced by other disturbances to pattern formation come in two forms—those that have the notochord of the posterior part of the body and those that don't.

Melton: They do have notochord and their anterior deficiencies are not caused by any delay or change in the migration of the mesoderm. In UV-irradiated embryos the mesoderm involutes late and uniformly, in these embryos that does not happen. Careful histological examination shows that embryos injected with *Xhox3* have anterior mesoderm as defined by its position in the embryo. However, this 'anterior mesoderm' is evidently not capable of inducing anterior structures.

Cooke: That information about timing is very interesting because it suggests that the spread of positional information in the mesoderm is not abnormal in these embryos. In nearly all other experiments, the absolute timing of movement in groups of mesoderm cells is predictive of the pattern region they are going to form. But in this case, because of the abnormal level of RNA caused by the exogenous RNA, the embryos have an abnormal spectrum of interpretation of positional information. Mesoderm which still has anterior position value (i.e. timing of movement) is nevertheless unable to differentiate or induce anterior structure.

Melton: To speak very loosely, you could say that the Spemann organizer is the source or sink of a growth factor and that as the mesoderm moves past the blastopore lip its level of *Xhox3* is set high or low, depending on the concentration of the growth factor. I don't want to imply that it is all that simple, but there is no reason not to test that type of simple hypothesis.

Struhl: I was wondering about the level of *Xhox3* protein in terms of the type of pattern that is produced. Your evidence seems to suggest that the amount of *Xhox3* protein is not instructive for a particular type of pattern, but rather permissive. In your experiments you change the global level of *Xhox3* mRNA by making it high everywhere or low everywhere. Under these circumstances, some regions of the embryo develop normally, whereas others do nothing, even though they all receive the same signal. In that sense it is rather different from the way we think the homeotic genes, for example, may work in *Drosophila*.

Melton: I agree. One could say that low levels of *Xhox3* are necessary but not sufficient to produce anterior structures, and high levels are necessary but not sufficient to produce posterior structures.

Struhl: The fact that it is permissive in that sense implies that there must be other information which actually specifies the pattern. Another graded cue of some kind is needed which plays an instructive role.

Melton: It is reasonable to assume that other genes, perhaps homeobox-containing genes, are involved in the process. I was only trying to make the point that when you change the level of expression of this one gene (*Xhox3*), and not that of other homeobox genes that we have tested, there is this dramatic effect on anteroposterior patterning.

Struhl: But you appear to have changed only the ability of cells along the prospective anteroposterior axis to respond to something else that must be there and must be non-uniform.

Wieschaus: In these cases where you inject *Xhox3* RNA and get abnormal development, do you know how the anterior deficiencies arise? Is it that cells fall off and die or are they transformed or shifted in fate?

Melton: We don't know the answer, but we have two relevant pieces of information. The first is that we can inject RNA into one of two blastomeres so that only half of the embryo will have the excess RNA. When we do that, the first indication of a defect is that the anterior neural plate fails to form, there is no anterior neural fold. The embryo has a half-sucker and neural fold on one side but nothing on the injected side.

Wieschaus: Even at a stage when the posterior part of the neural plate would have formed?

Melton: That's right. We haven't done enough of those experiments to be sure, but in the 10–20 cases we have done, it seems that the effects of *Xhox3* RNA injection are restricted to one half of the embryo, the half that received the *Xhox3* RNA. The point of that experiment was to try and see, with a contralateral control, what is the first indication of a defect, and it is a deficiency in the formation of the anterior neural plate.

The second fact is that we have done histological sections and know that cells which look to be mesoderm histologically are in the anterior end of the embryo. We conclude from that that those cells don't know what to induce because they have the wrong level of *Xhox3* RNA. In other words, high levels of *Xhox3* RNA do not lead to death of all those cells. What we don't know is what happens to anterior ectoderm when it is underlaid by mesoderm which has high levels of *Xhox3*. All we see is the absence of these structures; we have no reason to believe its fate has been transformed to a more posterior one.

Wieschaus: A more relevant question is what happens to anterior/posterior fate in the mesoderm.

Melton: If you take a mesodermal cell and ask whether it is anterior or posterior, the only way we can tell is to test its inductive capacity.

Cooke: The issue keeps arising that a spatial distribution of this one zygotically transcribed gene product can't be used as an explanatory principle by itself. It needs to be responding to something else. Whenever that consideration comes up, the automatic response seems to be that there may be other homeobox-containing genes which have already set up a situation to which this gene is responding. There are intrinsic problems with that argument, because our main model at the moment is the *Drosophila* blastoderm. In the *Drosophila* blastoderm the crucial regionalization with respect to gene activities occurs largely when it is a syncytium. The *Xenopus* embryo at the equivalent stage is not a syncytium and hasn't been for a long time, it is a true tissue of separate cells. So there are problems in supposing that gene products which essentially act intracellularly and are localized within the cells are going to 'bootstrap' themselves into a spatial organization without using other principles. The kind of molecules like that which could be encoded by *Vg1*, which are probably growth factor molecules that naturally are secreted and are intercellular signalling molecules, are much better candidates for setting up the preliminary map to which all these things respond. There is a dynamic organization that isn't itself composed of regionally segregated gene activities at that stage, but to which the first regionally activated zygotic genes respond.

Melton: In the same regard, there is no reason to rule out the possibility that the homeobox-containing genes like *Xhox3* control the expression of intercellular signals. In general, in *Drosophila* one thinks that homeobox genes control the expression of other homeobox genes or other intracellularly acting processes, but there is no reason to think that *Xhox3* expression couldn't control the level of some growth factor and then set up another signalling process. It just emphasizes our ignorance of what genes are involved.

Heasman: Concerning the *Drosophila* syncytium not being comparable with the *Xenopus* situation, would it be possible to go back to the *Xenopus* oocyte and make the comparison, because that is the stage when the asymmetry of *Vg1* is set up. Is that more comparable to the syncytial stage of *Drosophila*?

*Scott:*As far as the zygotically active homeobox genes are concerned, the syncytium could be a red herring in some cases. There are certainly molecules that are anchored in a way that we don't understand in the *Drosophila* embryo, as apparently there are in *Xenopus*. To what extent the gene systems we are talking about take advantage of the apparent openness of the embryo to movement and diffusion really isn't clear yet. In the case of the maternally provided *bicoid* products that Janni Nüsslein-Volhard and her co-workers have studied, the protein does appear to diffuse (or at least move) through the syncytium.

Jonathan (Cooke) made the point in an earlier discussion that in cases where we are beginning to learn something about possible functions of vertebrate homeobox genes, it appears that they may have very different functions than the closest relatives among the *Drosophila* genes. This gene may be the best example of that—even though we can't say this is the *even-skipped* equivalent because you don't know what the rest of the protein looks like. I would not really expect the vertebrate homeodomain proteins, most of which differ greatly outside of the homeodomain, to have exactly the same sorts of functions, or that mutations in the vertebrate genes would result in the same phenotypes as those in the *Drosophila* genes. But you might predict that the proteins would have some related target genes on which they act. If homeodomains indicate that a protein is a transcription factor, you might argue that related homeodomains bind to similar DNA sequences and therefore regulate a similar array of target genes. Do you have any ideas what the targets might be in this case?

Melton: No, the only way to do that in frogs would be a rather tedious biochemical approach. I don't find that as interesting as what sets up the graded *Xhox3* distribution. At present, I am more interested in what you might call the upstream activities than in what happens downstream.

Martinez-Arias: To return to the idea that other molecules might be involved in establishing this pattern, you mentioned that they should be asymmetrically distributed. It is equally likely that these molecules can be modulating a product that is everywhere and that would be the active principle of the pattern.

Wolpert: I think it is very exciting, it is the first molecular approach to the anteroposterior axis in a vertebrate.

Axis determination in insect embryos

Christiane Nüsslein-Volhard and Siegfried Roth

Max-Planck-Institut für Entwicklungsbiologie, Spemannstrasse 35, D-7400 Tübingen, FRG

Dedicated to Professor Dr. Klaus Sander on the occasion of his 60th birthday

Abstract. The principles of pattern formation in insects have been studied extensively using classical experimental approaches. In *Drosophila*, a powerful combination of genetics and transplantation experiments, as well as molecular biology, has helped to elucidate the mechanisms that operate during oogenesis to establish a set of positional cues required for axis determination in the early embryo. These studies suggest the following model: for the anteroposterior axis of the embryo, three groups of maternal genes define three largely independent systems that determine(1) the *anterior* segmented region of head and thorax,(2) the *posterior* segmented region of the abdomen, and(3) the *terminal* non-segmented regions of acron and telson. In contrast, the dorsoventral egg axis appears to require only one system. In each of the four systems, one key gene has an active product that is unequally distributed in the egg. This product provides the spatial signal for the region-specific activation of the transcription of at least one zygotic target gene. The other members within each group serve accessory functions such as determining the correct spatial distribution of the key gene products or controlling their localized activation. The unique expression patterns of the individual zygotic target genes provide a coarse spatial framework which is then refined by the action and interaction of zygotic genes.

The notion of three independent systems determining the anteroposterior axis is at variance with a previous model (Meinhardt 1977, Nüsslein-Volhard 1979) of only one gradient, with a high point at the posterior pole, determining a series of states in a concentration-dependent manner. Concentration-dependent determination of more than one quality is likely to occur in the anterior and the dorsoventral system. In contrast, position and polarity within the posterior pattern appear to depend largely on the interaction between gap genes expressed in neighbouring regions rather than on the concentration of the posterior signal.

1989 Cellular basis of morphogenesis. Wiley, Chichester (Ciba Foundation Symposium 144) pp 37–64

The body plan of insects and its development

Insects display a clear axial organization that is already apparent at gastrulation and which persists throughout development. The majority of the anteroposterior pattern, head, thorax and abdomen, is composed of repeating units, the segments. At the anterior and posterior ends of the insect body there are non-segmented, terminal regions that contain unique structures such as mouth, foregut and brain at the anterior, and the hindgut, malphigian tubules and anus at the posterior. (These terminal regions, acron and telson, may include, in addition to the unique structures just mentioned, rudiments of segments that have been lost during evolution.) Thus, from anterior to posterior, the insect body is composed of the acron, head, thorax, abdomen and telson.

In insects, superficial cleavage results in a blastoderm of a nearly uniform cell layer surrounding a central yolk mass. During gastrulation this sheet of cells is condensed ventrally and reorganized to form the germ band which already displays the basic organization into acron, head, thorax and telson. A complete abdomen anlage is also present at this stage in dipterans, while in more primitive insects the segments of the abdomen develop later by intercalary growth.

The first asymmetry along the dorsoventral axis is the gastrulation furrow which in insects forms along the ventral side. By invagination of a longitudinal band of cells, the mesodermal anlage, along the ventral midline, the embryo establishes its three-dimensional, multilayered organization. During gastrulation, large portions of the acronal and telsonal regions of the embryo also invaginate to give rise to the anlagen of the gut.

In this article, we concentrate on those mechanisms that establish a stable axial organization of the embryo. Although many of the properties of the system have been worked out in elegant transplantation and isolation experiments in a variety of insect species, recent research has focused almost exclusively on *Drosophila*. This system provides the unique advantage of an established genetics. Thus, complex biological mechanisms can be analysed and dissected using mutations as tools both to alter and block developmental decisions. In addition to the more conventional techniques of transplantation experiments, the powerful genetics and molecular biology of *Drosophila* have allowed much progress in our understanding of the basic principles of development in this organism.

The anteroposterior axis: experimental evidence

Experiments of classical embryology have elucidated basic principles of axis determination in insects. It was demonstrated that there are two organizing centres located at the anterior and posterior egg poles, which exert long range influences on neighbouring regions. For the posterior centre, the organizing and polarizing effect was demonstrated in transplantation experiments (Sander 1975), while the best evidence for an anterior centre came from experiments in which abdominal mirror-image duplications were induced by the destruction of an 'anterior determinant' (Kalthoff 1983). Recent experiments in *Drosophila* support the existence of two organizing centres. In transplantation experiments, both the anterior and the posterior cytoplasm were shown to harbour activities that could induce 'anteriorness' or 'posteriorness', respectively, at ectopic positions. Moreover, the activities have been shown to have long range effects and to exert mutual inhibition (Frohnhöfer et al 1986, Nüsslein-Volhard et al 1987).

The anteroposterior axis: genetic analysis

The identification of maternal genes that specifically affect axis determination in the embryo (coordinate genes) has revealed that there are not two but three pattern-forming systems for the anteroposterior axis. While the existence of the anterior and the posterior organizing centres finds excellent support in the phenotypes of two classes of coordinate genes (the anterior and posterior groups), the third class, the terminal group, is new and was not predicted by the classical experiments. This third group of genes is required for the formation of the unsegmented terminal regions of the larva, the acron and telson (Nüsslein-Volhard et al 1987).

The three systems are defined by three groups of coordinate genes with similar or identical mutant phenotypes (Table 1). Further analysis shows that within each group at least one gene functions as a 'key' gene, the active product of which provides a spatial cue in the developing egg. The other genes in each group may serve accessory functions required for the transport, localization or local activation of the key gene products. For each system, there is at least one zygotic gene of the gap class of segmentation genes that is locally expressed in response to the presence of the key gene product. Tentatively, *bicoid, nanos* and *torso* have been identified as key genes of the three systems, while the responding zygotic target genes may be *hunchback, knirps* and *tailless*, respectively (Table 1) (Frohnhöfer & Nüsslein-Volhard 1986, Lehmann & Nüsslein-Volhard 1989, Klingler et al 1988).

The independence of the three pattern-forming systems is apparent from single and double mutant phenotypes (Figs. 1,2). In the absence of function of any of the systems, a partial pattern forms. These partial patterns, in the case of *bicoid* lack head and thorax, in *nanos* the abdomen, and in *torso* the terminal structures of the acron and telson. Double mutant combinations develop even smaller portions of the pattern, such that *nanos torso* embryos have head and thorax, in *bicoid torso* just an abdomen develops, while *nanos bicoid* embryos consist of nothing but two telsons (in the absence of *bicoid* the anterior acron is transformed into a telson). Only in the case of the triple mutant is no discernible pattern formed. It is noteworthy that the phenotypes in double and triple mutant combinations are almost perfectly additive. Thus the partial patterns observed in the appropriate single and corresponding double mutants roughly add up to the pattern of a complete larva (Figs. 1,2). This means that, to a first

TABLE 1 Genes involved in the establishment of the anteroposterior axis of the *Drosophila* embryo[a]

	Anterior	*Posterior*	*Terminal*
Maternal accessory functions[b]	exuperantia swallow staufen	vasa valois tudor oskar staufen pumilio	trunk torsolike fs(1)ph fs(1)N
Maternal key genes[c]	bicoid	nanos	torso
Zygotic target genes[d]	gene x hunchback	knirps	tailless

[a]Most of the genes have been identified in systematic mutagenesis experiments (Nüsslein-Volhard & Wieschaus 1980, Schüpbach & Wieschaus 1986. See Nüsslein-Volhard et al 1987 for further references). [b]Functions in localization of sources of morphogen, distribution or activation of the morphogenetic substances; see text. [c]Gene product unequally distributed in egg or locally activated. Distribution determines spatial pattern of expression of zygotic target gene. [d]Only those genes that are thought to be positively regulated by the maternal gene products.

approximation, there is little interaction between the components of the three systems and that large regions of the embryo are determined independently of each other (Nüsslein-Volhard et al 1987).

Localization, distribution and function of the maternal signals

In addition to the genetic analysis, two experimental approaches have greatly aided our understanding of the mode of action of the three systems: cytoplasmic transplantation and molecular biology. Our understanding is most complete for the anterior system, whereas the molecular analysis for posterior and terminal genes has just begun. Cytoplasmic transplantations between mutant and wild-type embryos have revealed the localization of organizing centres at the anterior and posterior poles, the activity of which depends on the respective genes of the anterior and posterior group (Frohnhöfer & Nüsslein-Volhard 1986, Lehmann & Nüsslein-Volhard 1986). No localization of *torso*-dependent activity can be detected, although the *torso* product appears to be required only at the ends of the egg (Kingler et al 1988). These experiments raise the following important questions:

1) How are the maternal products localized in the egg during oogenesis?
2) How are the pattern-forming substances distributed in the early embryo?
3) How is the pattern of response of the zygotic target genes related to the maternal pre-pattern?

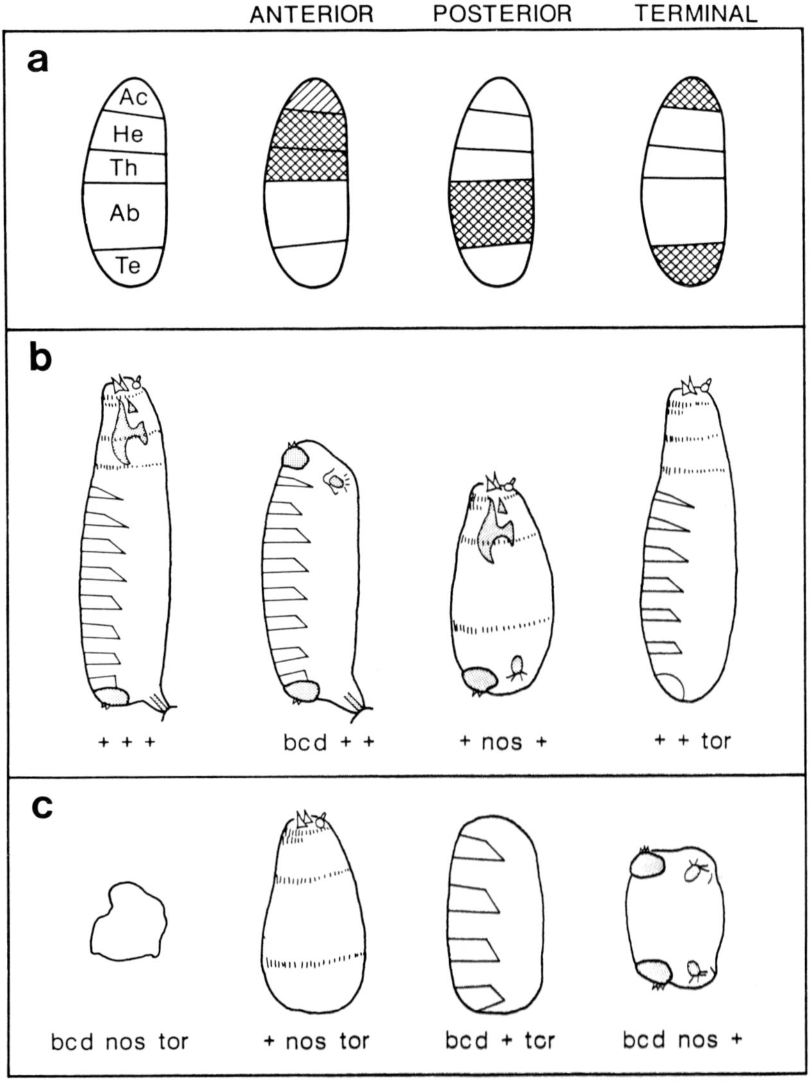

FIG. 1. The regions affected in mutants of the three systems determining the anteroposterior embryonic axis. Panel a: Schematic drawings of the fate map of the *Drosophila* embryo indicating, from anterior to posterior, the origin of the body regions acron (Ac), head (He), thorax (Th), abdomen (Ab) and telson (Te). In mutants of the key genes of the three systems, the tagmata deriving from the shaded regions do not form or are transformed (acron into telson in the case of *bicoid*). Panels b and c: Schematic drawings of the phenotypes of double and triple mutant combinations, displaying patterns that are approximately complementary to those of the wild-type or respective single mutants. The derivatives of the terminal regions, acron and telson, are shaded. bcd, *bicoid*; nos, *nanos*; tor, *torso*.

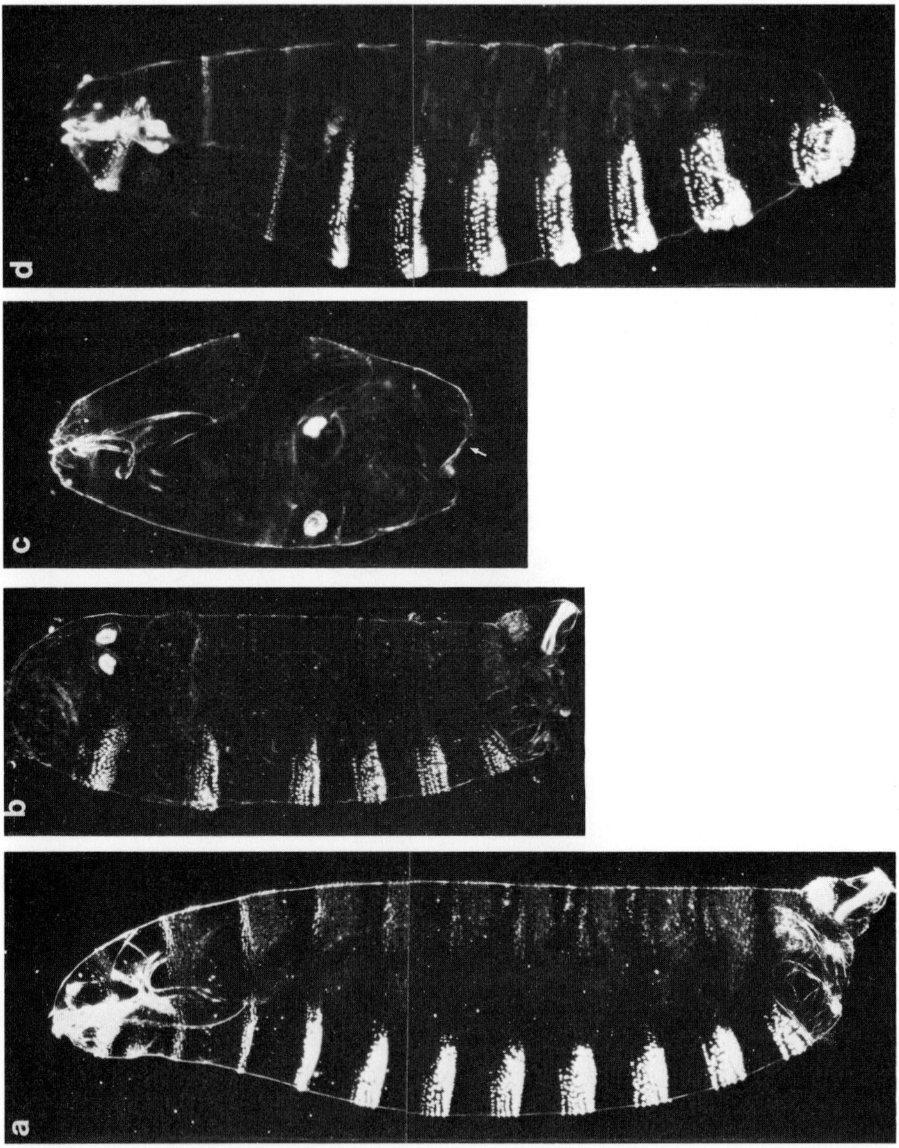
d
c
b
a

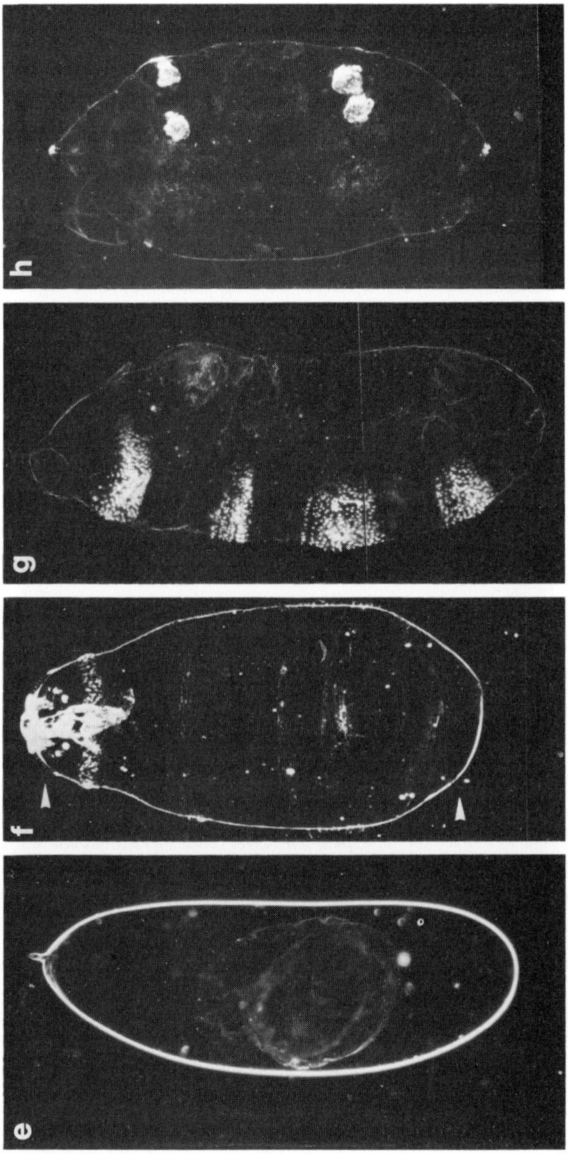

FIG. 2. The cuticle patterns of wild-type and mutant embryos. Dark field photographs. a) wild-type, b) *bicoid*, c) *oskar*, d) *torso*, e) *bicoid oskar torsolike*, f) *oskar torsolike*, g) *bicoid torsolike*, h) *oskar bicoid*.

Transplantable activities and their localization

During oogenesis, the oocyte is located posteriorly to a cluster of 15 nurse cells. All 16 sister cells are derived from a common ancestor germline cell and are interconnected by cytoplasmic bridges. Except for a few nutrients that enter the oocyte via the somatic follicle cells, most of the molecules present in the freshly laid egg are synthesized in the nurse cells and enter the oocyte through cytoplasmic bridges. In general, these molecules (nucleic acids or proteins) quickly distribute within the growing oocyte.

Given the mode of growth of the oocyte, it is easy to visualize how the product of a maternal gene like *bicoid* becomes localized at the anterior (Fig. 3). Using molecular probes, it has been shown that *bicoid* RNA is synthesized in the nurse cells and then trapped immediately upon entering the oocyte (Frigerio et al 1986). For this trapping process, the products of the genes *exuperantia* (*exu*) and *swallow* (*swa*) are required (Frohnhöfer & Nüsslein-Volhard 1987, Berleth et al 1988). A simple mechanism would involve the formation of a specific *exu* protein–*bicoid* mRNA complex, which is in turn attached to the cytoskeleton of the oocyte by the *swa* protein (Berleth et al 1988). If one assumes equal distribution of an excess amount of both the *swa* and *exu* protein in the oocyte, anterior localization of the *bicoid* mRNA would be the immediate consequence of the polar arrangement of nurse cells and oocyte and no other assymmetries or pre-localizations are required. A similar model, however, cannot account for the localization of the posterior activity. In this case, the molecular probes for the gene products involved are not yet available. However, results from transplantation experiments reveal activity present initially in the nurse cell cytoplasm and then in the posterior pole plasm (Sander & Lehmann 1988). This activity is most likely the product (RNA?) of the gene *nanos*, while the functions of most of the other genes of the posterior group are required for the transport or localization of the *nanos* product to the posterior pole plasm (Lehmann & Nüsslein-Volhard 1989). The RNA product of the *torso* gene of the terminal group is not localized in the egg (Sprenger et al 1989).

Distribution of morphogenetic substances in the egg

The localization of the primary products of the three key genes does not coincide with the pattern of their requirement (Fig. 3). We assume that the protein products of *bicoid* and *nanos* spread from the poles. In the case of *bicoid*, the protein product is synthesized from the localized mRNA soon after the egg is laid and then spreads posteriorly. In early development, the protein is distributed in an exponential concentration gradient with a maximum at the anterior tip. It can be detected down to 30% egg length. The distribution pattern is probably the result of diffusion from the anterior source, coupled with dispersed decay

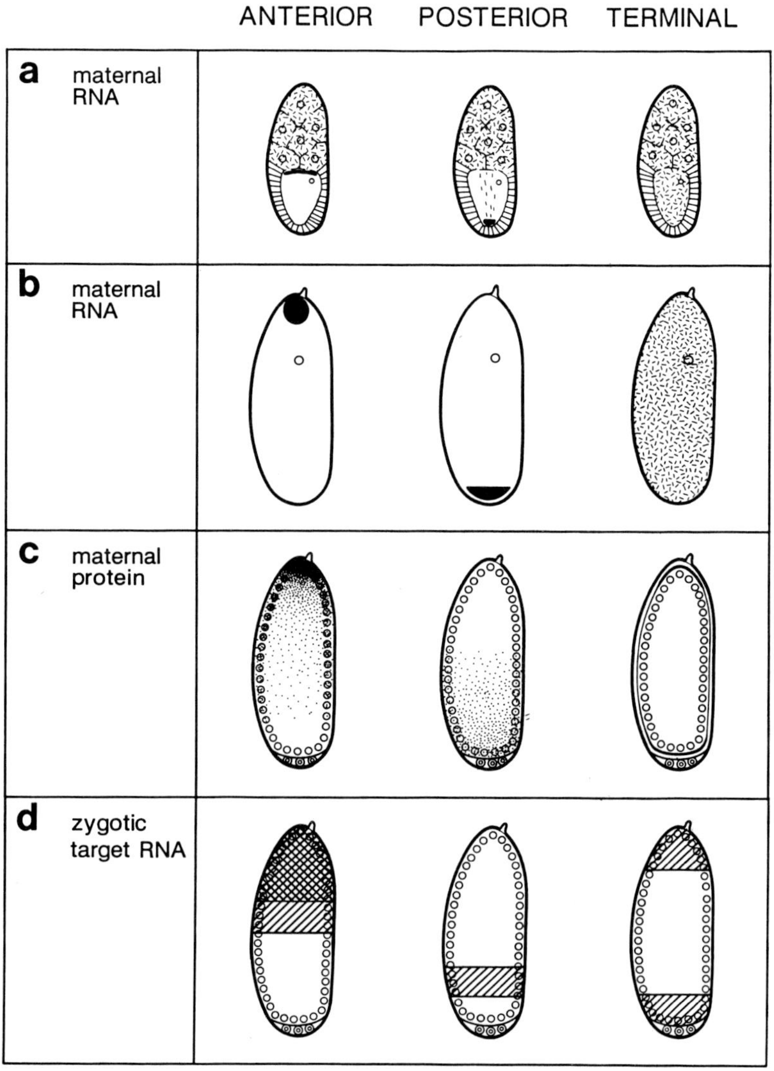

FIG. 3. Spatial distribution of the products of the key genes of the anteroposterior systems in oogenesis and embryogenesis and the pattern of expression of the zygotic target genes. The maternal genes illustrated are, for the anterior system, *bicoid*; for the posterior system, the putative key gene *nanos*; for the terminal system, *torso*. The zygotic genes illustrated are, for the anterior system, gene x and *hunchback*; for the posterior system, *knirps*; for the terminal system, *tailless*. a) oogenesis, stage 10 follicle. b,c,d) early embryogenesis. The distributions of *nanos* mRNA and *nanos* and *torso* proteins are hypothetical. The expression patterns of all the zygotic genes except *hunchback* are hypothetical.

(Driever & Nüsslein-Volhard 1988a). However, more active spreading mechanisms cannot be excluded. A model of a local source and a spreading signal has been postulated for the posterior centre on the basis of transplantation experiments (Lehmann & Nüsslein-Volhard 1986). In this case, at least one gene, *pumilio*, is specifically involved in the transport of the posterior signal to more anterior egg regions (Lehmann & Nüsslein-Volhard 1987). In contrast, in the terminal system, there is no indication of localization during oogenesis. Genetic and molecular evidence suggests that the region-specificity in the terminal system is created by local activation of the *torso* protein, which appears to be a membrane-spanning tyrosine kinase. The *torso* protein may function as a receptor that is responding to a local signal, perhaps provided by the follicle cells that are adjacent to the oocyte at its anterior and posterior ends. Other genes of the terminal group of genes may play a role in this local activation (Klingler et al 1988, Sprenger et al 1989).

Local response of the zygotic target genes

The spatial distribution of the (active) products of the maternal key genes is responsible for the region-specific expression of zygotic segmentation genes. It is believed that the direct targets of the maternal genes are segmentation genes of the gap class, required in large unique regions along the anteroposterior axis (Nüsslein-Volhard & Wieschaus 1980). Based on the similarity in phenotypes of maternal and zygotic genes, one likely target for *bicoid* is *hunchback*. It has been shown with molecular probes that the expression of *hunchback* is eliminated in *bicoid* embryos (Tautz 1988). In addition, at least one other target gene expressed more anteriorly must be postulated because *hunchback* is required for thorax development, while the *bicoid* product can also induce head structures at high concentrations. The formation of an anterior-terminal acron requires *bicoid* in addition to the terminal group of genes, whereas in the absence of *bicoid* a posterior-terminal telson is formed. The target for *nanos* appears to be *knirps*, and for *torso*, *tailless*. The expression of one gap gene, *Krüppel*, is independent of an activating influence of the maternal coordinate genes. *Krüppel* is subject to repression by the maternal gene products or those of their targets and is expressed in the middle region of the embryo (Gaul & Jäckle 1987).

The distribution patterns of the maternal protein products are much more diffuse than the patterns of expression of the zygotic target genes (Fig. 3). The transition from the graded information provided by the maternal products to the sharp regionalized expression of the zygotic target genes is one of the crucial steps in axis determination in the embryo.

Activation of transcription of a zygotic target gene may occur above a certain threshold concentration of a maternal protein that has DNA-binding properties. An example for such a mechanism is provided by *bicoid*, which contains a homeobox (Frigerio et al 1986) and has been shown to bind to the *hunchback* promoter. Cooperative binding of *bicoid* protein to several sites in the *hunchback* promoter could lead to a sharp transition between an on or off state at a particular position along the anteroposterior axis (Driever & Nüsslein-Volhard 1989). Other target genes may respond with lower affinity and would therefore be expressed only more anteriorly. The existence of at least two target genes with different affinities would permit the transmission of the polarity of the maternal gradient to a serially ordered expression of segmentation genes.

In the present model, the spatial pattern of expression of the zygotic target genes specifies a minimum of six bands. They are defined, from anterior to posterior, by the expression of *tailless*, gene x, *hunchback*, *Krüppel, knirps* and *tailless*. Whether or not their domains overlap and how their spatial patterns of expression are transmitted to the segmentation genes following in the hierarchy, is the subject of intensive research (see Ingham 1988 for review).

The dorsoventral axis

The principles involved in establishing the dorsoventral axis in insects appear to be very different from those of the anteroposterior axis. This difference is apparent in the outcome of simple ligation experiments performed in a number of different species (for review see Sander 1976, 1984). The separation of anterior from posterior egg portions by transverse ligature results in partial anterior and posterior embryos, the patterns of which may add up to *less* than a complete pattern. Pattern elements close to the ligation point fail to be specified if the separation is performed before the blastoderm, whereas in later stages the embryo behaves as a mosaic. In contrast, longitudinal separation of the embryo into a right and a left portion leads to the formation of *more* than the normal number of pattern elements and in extreme cases the two partial embryos each reorganize to form two complete patterns such that identical twins result. Depending on the taxon to which the insect species belongs, twinning can be induced until quite late in development, well after gastrulation. The ability to produce two complete sets of dorsoventral patterns in one egg indicates that the mechanisms operating in the dorsoventral axis have great self-regulative properties, while pre-localization appears to play a rather subordinate role, if any (reviewed in Sander 1976).

In higher insect species, like *Drosophila*, longitudinal ligation experiments cannot be done. Transplantation and pricking experiments in the wild-type failed to reveal information concerning the distribution or nature of the pattern-forming signals.

Genetic analysis reveals that in contrast to the anteroposterior axis, the establishment of the dorsoventral axis appears to rely on a single integrated system. Mutations in 11 maternal genes (the *dorsal*-group genes) cause a dorsalization and loss of polarity along the entire dorsoventral axis (Anderson & Nüsslein-Volhard 1986, Schüpbach & Wieschaus 1988). Partial ventralization is observed in mutants at the *cactus* locus (Schüpbach & Wieschaus 1989), as well as dominant alleles of the *Toll* locus (Anderson et al 1985a). For most of the *dorsal*-group genes, partial or complete normality can be restored in mutant embryos by the injection of cytoplasm from wild-type embryos (or from embryos mutant for any of the other genes of the *dorsal*-group). The response to injection, however, varies considerably among the different genes (Anderson & Nüsslein-Volhard 1986). In contrast to the anteroposterior axis, in no case could a strong pre-localization of rescuing activity be detected in early embryos. The results of the transplantation experiments do, however, single out two genes: *dorsal* and *Toll*. In *dorsal*, the activity is equally distributed in young wild-type embryos, but in blastoderm embryos cytoplasm from the ventral side is more active than that from the dorsal side (Santamaria & Nüsslein-Volhard 1983) (Table 2). *dorsal* shows a strong dosage dependence both in transplantation experiments and in the phenotypic analysis, and it has been proposed as a candidate gene for a morphogen that determines pattern along the dorsoventral axis in a gradient mechanism (Nüsslein-Volhard 1979). In double mutants with ventralizing mutants (*cactus* or *Toll*), *dorsal* is the only one of the *dorsal*-group genes that still causes a dorsalized phenotype; in all other cases an intermediate, lateralized, phenotype is produced (Anderson et al 1985a, Roth & Nüsslein-Volhard, unpublished). This result indicates that the *dorsal* gene product is absolutely required for the formation of lateral or ventral structures and supports its role as a key gene. Use of molecular probes confirms this, showing that the protein product of the *dorsal* gene is equally distributed in the early embryo, but the distribution in the blastoderm embryo is that of a steep dorsoventral gradient

TABLE 2 Ventral concentration of dorsal rescuing activity

Age of donor embryos	*Position in donor embryo*	*Embryos with rescued structures*		
		Filzkörper (%)	*Ventral denticles (%)*	*n*
stage 11–13	ventral	87	21	62
	dorsal	64	3	66
stage 14	ventral	94	31	16
	dorsal	75	0	24

Cytoplasm was taken from the ventral or dorsal region (as indicated) of wild-type embryos of the stages shown. Stages refer to nuclear cycles, 11–13 being syncytial blastoderm, 14 after the last nuclear division. The cytoplasm (about 5% of the egg volume) was injected into a posterior ventral position in recipient embryos of stage 11–13 from *dorsal/dorsal* females. The resulting embryos were scored for the presence of structures not normally found in *dorsal* embryos, like Filzkörper and ventral denticle bands. All embryos that developed ventral denticle bands also had Filzkörper. In the case of stage 14 embryos as donors, care was taken to avoid nuclei in the transplanted cytoplasm. In control experiments including nuclei, essentially the same results were obtained.

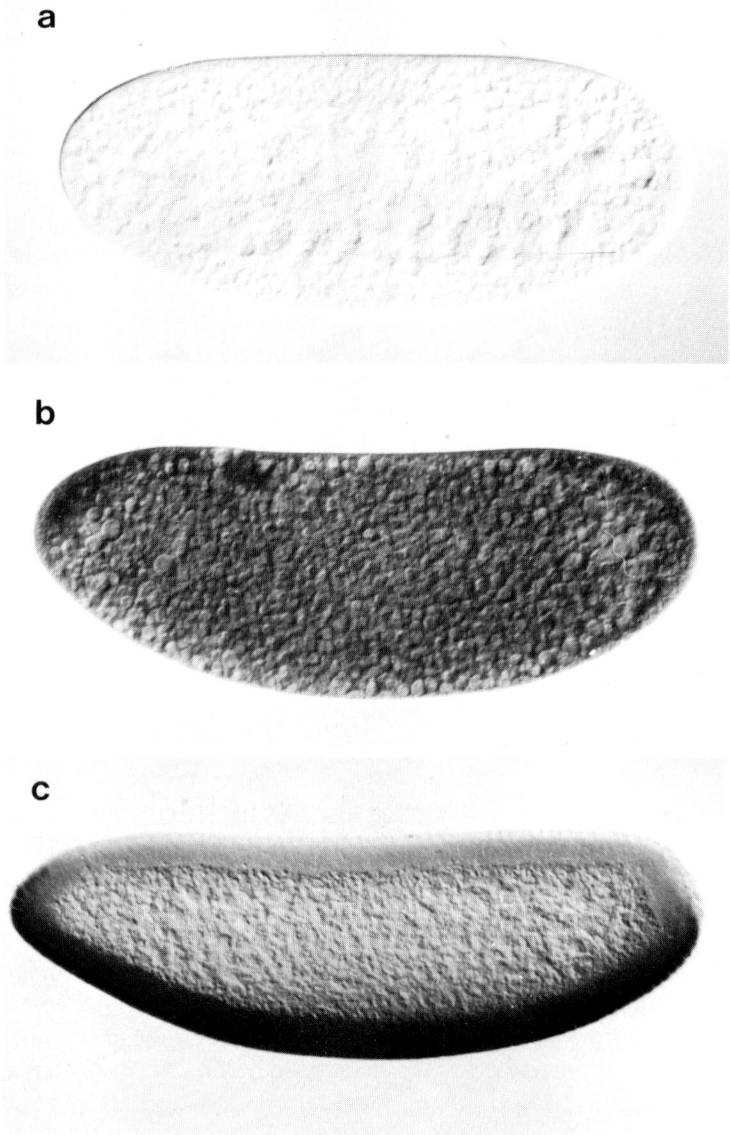

FIG. 4. *dorsal* protein distribution in whole mounts of embryos. Staining with polyclonal antibody against a *dorsal*-β-galactosidase fusion protein (Roth & Nüsslein-Volhard, unpublished). a) control embryo from *dorsal*$^{-}$ female. b) wild-type embryo showing specific uniform *anti-dorsal* staining. Both embryos are in the early cleavage stage. c) wild-type embryo in the cellular blastoderm stage, displaying darkly stained nuclei at the ventral side (bottom), while the nuclei at the dorsal side of the blastoderm are unstained.

with a maximum at the ventral side (Steward et al 1988, Roth & Nüsslein-Volhard, unpublished, Fig. 4).

Although the *dorsal* gene product has many of the properties of a morphogen, in transplantation experiments with *dorsal* embryos as recipients the *dorsal* product is required at the ventral position and the normalized patterns always form in the normal orientation (as indicated by the egg curvature) (Santamaria & Nüsslein-Volhard 1983). It has not been possible to reverse the polarity of the embryo or induce ventral structures at ectopic positions in mutants of *dorsal* or any other *dorsal*-group genes, with the exception of *Toll*. In embryos lacking the *Toll* product, a complete dorsoventral pattern can be induced in any orientation by injection of wild-type cytoplasm. In this case, the site of injection determines the ventral side of the embryo. The induced pattern is complete and includes all pattern elements from the ventral to the dorsal midline, independent of the amount of injected material (Anderson et al 1985b). Molecular analysis shows that it is the mRNA of the *Toll* gene that is the active principle in the transplantation experiments into *Toll* embryos. *Toll* mRNA is equally distributed in the wild-type embryo and codes for a membrane protein (Hashimoto et al 1988). A model explaining these and other findings is that the establishment of the *dorsal* gradient is mediated via local activation of the *Toll* gene product. This activation in normal development would take place only at the ventral side of the egg, whereas in the artificial situation of transplantation it may occur anywhere. Other genes of the *dorsal*-group may play individual roles in this local activation process (Anderson et al 1985b). The finding that the product of the *dorsal*-group gene *snake* appears to be a serine protease suggests that an activation cascade is involved (DeLotto & Spierer 1986).

Whereas the active *Toll* product may promote the accumulation of the *dorsal* product at the ventral side, the gene product of *cactus* appears to exert a negative effect on the *dorsal* product. In the ventralized *cactus* embryos, the *dorsal* product is detectable at the dorsal, as well as the ventral side of the embryo (Roth & Nüsslein-Volhard, unpublished). The molecular basis for the positive and negative influences of *Toll* and *cactus* on the formation of the *dorsal* gradient is being investigated.

The product of the *dorsal* gene is a nuclear protein (Steward et al 1988) and therefore it is likely that *dorsal*, by analogy to *bicoid*, functions as a positive regulator of transcription of zygotic target genes. A candidate target gene is *twist*. *twist* embryos lack the ventral furrow and later the entire mesoderm. The gene product is expressed in a band that encompasses 20% of the egg circumference along the ventral midline and is absent in *dorsal* embryos (Thisse et al 1988). *Dorsal* may repress *zerknüllt* and *decapentaplegic*, zygotic genes that are normally expressed in a mid-dorsal region (Rushlow et al 1987, St. Johnston & Gelbart 1987). A tentative model for the determination of the dorsoventral axis is given in Fig. 5.

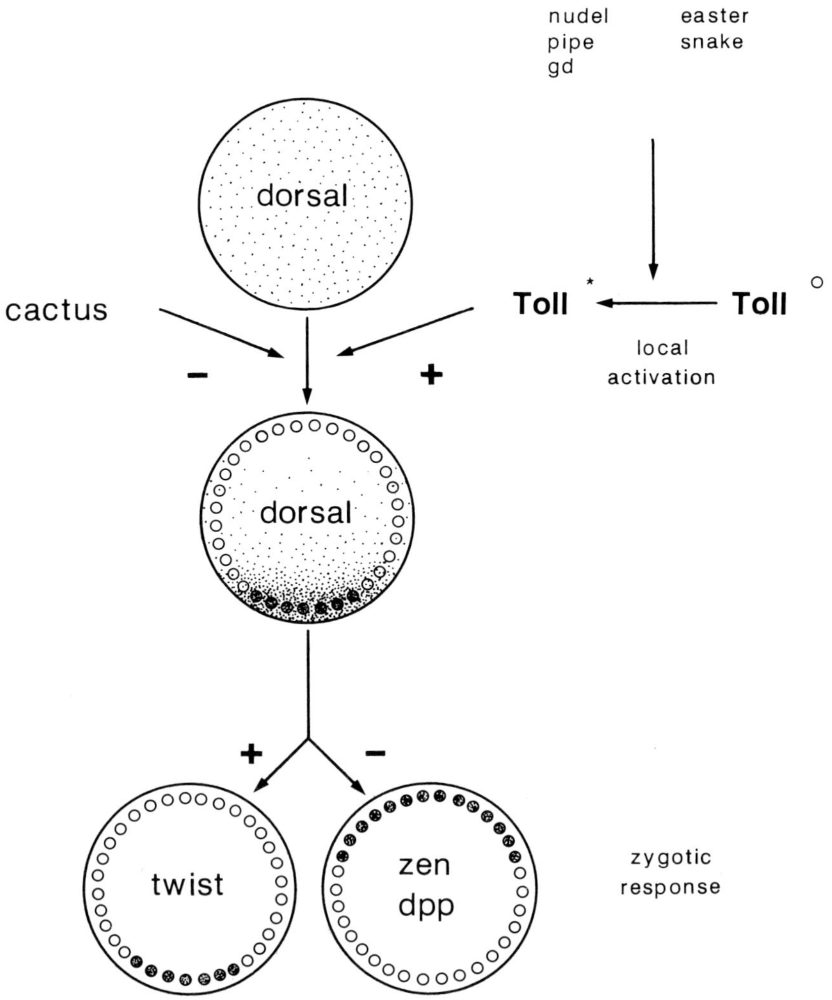

FIG. 5. Model of the establishment of pattern along the dorsoventral axis of the egg.

Gradients in axis determination

In insects, gradient mechanisms have frequently been postulated for axis determination. Sander proposed a double gradient hypothesis, in which each position along the anteroposterior egg axis is determined by the ratio of two graded values from anterior and posterior (Sander 1975, 1976). Using a gradient model based on lateral inhibition, Meinhardt could simulate most of the experimental data by postulating only one gradient with a high point at the posterior pole (Meinhardt 1977, 1986). This model found apparent support in the phenotypes of the *Drosophila* mutant *bicaudal*, and, by analogy, a similar model was postulated for the dorsoventral axis (Nüsslein-Volhard 1979). With the first molecular probes for potential gradient molecules available, it is perhaps appropriate to review critically the old concepts.

A gradient mechanism has to fulfill two requirements: (1) A substance (morphogen) must be distributed within a morphogenetic field in a concentration gradient. (2) Different concentrations of the substance should determine different qualities of structures within the morphogenetic field. Although it is not necessary for the concept that minute concentration differences are sensed in the process of gradient interpretation, at least two different qualities have to be determined by two different ranges of concentration of one morphogen or it would not be appropriate to refer to a gradient mechanism.

In *Drosophila*, in several instances the graded distribution of a gene product has been demonstrated. The first case was that of the *caudal* protein, which is distributed in a smooth linear gradient with a high point at the posterior pole. The role of this gradient in pattern determination is indirect, however (McDonald & Struhl 1986). The *bicoid* product is distributed in an exponential concentration gradient with its maximum at the anterior pole. In this case, a clear dependence of anterior pattern on the *bicoid* protein concentration has been demonstrated (Driever & Nüsslein-Volhard 1988a,b). One target gene for *bicoid, hunchback*, has been tentatively identified. At least one other target gene, gene x, has to be postulated. At present, the anterior system in *Drosophila* best fulfills the requirements of a gradient mechanism. It is noteworthy that in the case of *bicoid*, a gradient mechanism was predicted on the basis of phenotypic, genetic and transplantation studies well before the molecular probes were available (Frohnhöfer & Nüsslein-Volhard 1986, 1987). As outlined above, although the detailed mechanisms are not yet worked out for the dorsoventral system, all evidence supports the notion of *dorsal* as a morphogen determining position along the dorsoventral axis in a concentration-dependent manner.

For the posterior pattern, the case for gradients, although long postulated and seemingly best supported, is weakening, however. As the source of the posterior product, *nanos*, is located in the posterior pole plasm while the product is required more anteriorly in the abdominal region, it is probable that the *nanos* product distribution follows that of a concentration gradient. However, it is unlikely that the local *nanos* concentration determines position within the abdomen. In contrast to the anterior or dorsoventral system, there is no evidence for a dosage dependence of segmental quality, neither in the phenotypes of weak alleles nor in transplantation experiments (Lehmann & Nüsslein-Volhard 1986). These results are more consistent with a one step response, i.e. the determination of 'abdomen' as one quality only. Position and polarity within the abdomen appear to require an interaction with the neighbouring regions, which are defined by the terminal system of the telson and the anterior system. In particular, the *bicaudal*-type mirror-image duplications are perhaps best explained by an interaction of the terminal system with the abdominal system on the level of the zygotic target genes, *tailless* and *knirps* (Lehmann & Frohnhöfer 1989). Why the mirror-image duplications observed in *bicaudal* embryos are of such perfect symmetry and how, in molecular terms, the interaction between the zygotic target genes determines polarity of the pattern is not yet understood.

The problem of establishing a stable concentration gradient in an almost uniform egg cell appears to be solved in a unique manner in the four systems. The anterior system simply uses the orientation and polarization of oocyte and nurse cell cluster, occurring early in oogenesis, while in the posterior system a more elaborate mechanism of transport and establishment of the pole plasm as source for the posterior signal is required. The local stimulus for the terminal system is probably provided by specialized follicle cells that are located at the anterior and posterior of the somatic envelope of the oocyte–nurse cell complex. In the dorsoventral system, finally, a gradient is established from an initially equal distribution triggered by local activation mechanisms.

Acknowledgements

We would like to thank Kavita Arora, Wolfgang Driever, Martin Klingler, Ruth Lehmann, Klaus Sander, Vivian Siegel, Leslie Stevens, Daniel St. Johnston and Frank Sprenger for stimulating and constructive discussion and criticism of the manuscript. The photographs of Figure 2 have been provided by Hans Georg Frohnhöfer, Martin Klingler and Ruth Lehmann.

References

Anderson KV, Nüsslein-Volhard C 1986 Dorsal-group genes of *Drosophila*. In: Gall J (ed) Gametogenesis and the early embryo. Symp Soc Dev Biol, Alan R Liss Inc, 43:177–194

Anderson KV, Jürgens G, Nüsslein-Volhard C 1985a Establishment of dorsal-ventral polarity in the *Drosophila* embryo: genetic studies on the role of the *Toll* gene product. Cell 42:779–789

Anderson KV, Bokla L, Nüsslein-Volhard C 1985b Establishment of dorsal-ventral polarity in the *Drosophila* embryo: the induction of polarity by the *Toll* gene product. Cell 42:791–798

Berleth T, Burri M, Thoma G et al 1988 The role of localization of *bicoid* RNA in organizing the anterior pattern of the *Drosophila* embryo. EMBO (Eur Mol Biol Organ) J 7:1749–1756

Delotto R, Spierer P 1986 A gene required for the specification of the dorso-ventral pattern in *Drosophila* appears to encode a serine protease. Nature (Lond) 323:688–692

Driever W, Nüsslein-Volhard C 1988a A gradient of *bicoid* in *Drosophila* embryos. Cell 54:83–93

Driever W, Nüsslein-Volhard C 1988b The *bicoid* protein determines position in the *Drosophila* embryo in a concentration dependent manner. Cell 54:95–104

Driever W, Nüsslein-Volhard C 1989 The *bicoid* protein is a positive regulator of *hunchback* transcription in the early *Drosophila* embryo. Nature (Lond) 337:138–143

Frigerio G, Burri M, Bopp D, Baumgartner S, Noll M 1986 Structure of the segmentation gene *paired* and the *Drosophila* PRD gene set as part of a gene network. Cell 47:735–746

Frohnhöfer HG, Nüsslein-Volhard C 1986 Organization of anterior pattern in the *Drosophila* embryo by the maternal gene *bicoid*. Nature (Lond) 324:120–125

Frohnhöfer HG, Lehmann R, Nüsslein-Volhard C 1986 Manipulating the anteroposterior pattern of the *Drosophila* embryo. J Embryol Exp Morphol 97:169–179

Frohnhöfer HG, Nüsslein-Volhard C 1987 Maternal genes required for the anterior localization of *bicoid* activity in the embryo of *Drosophila*. Genes & Dev 1:880–890

Gaul U, Jäckle H 1987 Pole region-dependent repression of the *Drosophila* gap gene *Krüppel* by maternal gene products. Cell 51:549–555

Hashimoto C, Hudson K, Anderson KV 1988 The *Toll* gene of *Drosophila*, required for dorso-ventral embryonic polarity, appears to encode a transmembrane protein. Cell 52:269–279

Ingham PW 1988 The molecular genetics of embryonic pattern formation in *Drosophila*. Nature (Lond) 335:25–34

Kalthoff K 1983 Cytoplasmic Determinants in Dipteran Eggs. In: Time, space, and pattern in embryonic development. Alan R Liss Inc, p 313–348

Klingler M, Erdelyi M, Szabad J, Nüsslein-Volhard C 1988 Function of *torso* in determining the terminal anlagen of the *Drosophila* embryo. Nature (Lond) 335:275–277

Lehmann R, Nüsslein Volhard C 1986 Abdominal segmentation, pole cell formation, and embryonic polarity require the localized activity of *oskar*, a maternal gene, in *Drosophila*. Cell 47:141–152

Lehmann R, Nüsslein-Volhard C 1987 Involvement of the *pumilio* gene in the transport of an abdominal signal in the *Drosophila* embryo. Nature (Lond) 329:167–170

Macdonald P, Struhl G 1986 A molecular gradient in early *Drosophila* embryos and its role in specifying the body pattern. Nature (Lond) 324:537–545

Meinhardt H 1977 A model of pattern formation in insect embryogenesis. J Cell Sci 23:117–139

Meinhardt H 1986 Hierarchical inductions of cell states: a model for segmentation in *Drosophila*. J Cell Sci Suppl 4:357–381

Nüsslein-Volhard C 1979 Maternal effect mutations that alter the spatial coordinates of the embryo of *Drosophila melanogaster*. In: Konigsberg I, Subtelney S (eds) Determinants of Spatial Organization. Academic Press Inc, p 185–211

Nüsslein-Volhard C, Frohnhöfer HG, Lehmann R 1987 Determination of posterior polarity in *Drosophila*. Science (Wash DC) 238:1675–1681

Nüsslein-Volhard C, Wieschaus E 1980 Mutations affecting segment number and polarity in *Drosophila*. Nature (Lond) 287:795–801

Rushlow C, Frasch M, Doyle H, Levine M 1987 Maternal regulations of *zerknüllt*, a homeobox gene controlling differentiation of dorsal tissue in *Drosophila*. Nature (Lond) 330:583–586

Sander K 1975 Pattern specification in the insect embryo. In: Cell Patterning. Elsevier, Amsterdam (Ciba Found Symp 29) p 241–264

Sander K 1976 Specification of the basic body pattern in insect embryogenesis. Adv Insect Physiol 12:125–238

Sander K 1984 Embryonic pattern formation in insects: basic concepts and their experimental foundations. In: Malacinski GM, Bryant SV (eds) Pattern Formation. A Primer in Developmental Biology. Macmillan, p 245–268
Sander K, Lehmann R 1988 *Drosophila* nurse cells produce a posterior signal required for embryonic segmentation and polarity. Nature (Lond) 335:68–70
Santamaria P, Nüsslein-Volhard C 1983 Partial rescue of *dorsal*, a maternal effect mutation affecting the dorso-ventral pattern of the *Drosophila* embryo, by the injection of wild-type cytoplasm. EMBO (Eur Mol Biol Organ) J 2:1695–1699
Schüpbach T, Wieschaus E 1986 Maternal-effect mutations altering the anterior-posterior pattern of the *Drosophila* embryo. Roux's Arch Dev Biol 195:302–317
Schüpbach T, Wieschaus E 1989 Female sterile mutations on the second chromosome of *Drosophila melanogaster*. I. Maternal effect mutations. Genetics 121:101–119
Sprenger F, Stevens L, Nüsslein-Volhard C 1989 The *Drosophila* gene *torso* encodes a putative receptor tyrosine kinase. Nature (Lond) 339:61
Steward R, Zusmann SB, Huang LH, Schedl P 1988 The *dorsal* protein is distributed in a gradient in early *Drosophila* embryos. Cell 55:487–495
St. Johnston RD, Gelbart W 1987 *Decapentaplegic* transcripts are localized along the dorsal-ventral axis of the *Drosophila* embryo. EMBO (Eur Mol Biol Organ) J 6:2785–2791
Tautz D 1988 Regulation of the *Drosophila* segmentation gene *hunchback* by two maternal morphogenetic centres. Nature (Lond) 332:281–284
Thisse B, Stoetzel C, Gorostiza-Thisse C, Perrin-Schmidt F 1988 Sequence of the *twist* gene and nuclear localization of its protein in endomesodermal cells of early *Drosophila* embryos. EMBO (Eur Mol Biol Organ) J 7:2175–2183

DISCUSSION

Davidson: I am having trouble, taking your ideas literally, with the theory that *bicoid* binds to, say, *hunchback* to different extents depending on its concentration. Considering what we know about these DNA–protein interactions *in vitro*, how would such small differences in concentration, as for example in the dosage changes, have the all-or-nothing effects that individual cells along the anteroposterior axis would have to interpret? The concentration changes that you need to get something to go from 10% to 90% bound in equilibrium reactions are way beyond the minor changes in concentration that you described over these distances in your experiments. Something else has to happen, the protein has to be modified or there has to be an ancillary protein or some additional causal mechanism.

Nüsslein-Volhard: In the case of *hunchback*, we have evidence for three clustered binding sites and there might be cooperativity of binding (Driever & Nüsslein-Volhard 1989).

Davidson: But something else then has to be changed along the axis?

Nüsslein-Volhard: We don't know how far away this gene x has to be or the range of concentration differences responsible for turning it on. *bicoid* has to have a slightly higher affinity for the *hunchback* promoter than for the gene x promoter.

Davidson: In those dosage experiments, when the concentration of *bicoid* protein is a little higher further back, then cells are doing something that they are not doing when the concentration is a little lower. Literally, this means that particular parameter causes the state of binding of that gene to go from zero to one, although there is only a very small change in concentration. If you do that in a nuclear extract, you never get results like that.

Nüsslein-Volhard: We can do the experiment. We will change the *hunchback* promoter and the *bicoid* protein and see whether we can shift this threshold.

Martinez-Arias: When you increase the dosage of *bicoid*, is there no obvious aberration in the final larva?

Nüsslein-Volhard: Not until the dosage is increased by fourfold.

Meinhardt: One of the reasons why a system like the *Drosophila* embryo may be more robust to changes in concentrations of these gene products could be that the different gradients are doing much less than we originally thought. For example, the *nanos* gradient may be doing nothing other than activating the *knirps* gene, not a sequence of several genes.

For the *bicoid* gradient, you expect that higher concentrations activate gene x and lower concentrations activate the gene *hunchback*. Another possibility would be that there is a symmetrical distribution of the *torso* gene product, and a combination of *torso* and *bicoid* activates x, whereas if torso is below a certain threshold, only *hunchback* is activated. That would also distinguish between the most anterior structure (*torso* and *bicoid*) and the most posterior structure (*torso* without *bicoid*).

Nüsslein-Volhard: There is one very important piece of evidence which argues against that—the independence of the systems. A head can be formed without *torso* but not without *bicoid*, and *hunchback* is only responsible for the thorax. When you remove *torso*, the embryos still have a head and a thorax, they don't have the acron. There might be some influence of the terminal system on *bicoid* function, but I think there still has to be another gene x which is independent of *torso*.

I don't think it is correct to say that low concentrations of *bicoid* protein activate *hunchback* and high concentrations activate gene x. Above a certain concentration *hunchback* is activated, then it is on throughout the anterior of the egg. Above another, higher, concentration gene x is activated.

Meinhardt: Would you agree that there is not only gene x but also gene y which is activated anterior to gene x but for this you need the *torso* gene?

Nüsslein-Volhard: That gene is probably *tailless*.

Meinhardt: To make the correct anterior pattern, you need a combination of *tailless* plus *bicoid*, because if *bicoid* is missing you get a telson, the most posterior structure.

Nüsslein-Volhard: This is true, for the anterior terminal region we need a combination of the two.

Cooke: I wanted to ask about the *bicoid nanos torso* embryo that has nothing left to play with. These kinds of pictures always show the pattern, or non-pattern, at the almost due-to-hatch larval stage. These embryos are always very small. The picture is not often shown of the blastoderm stage. In the last few years we have come to think of a sharp distinction between morphallactic kinds of pattern formation, of which this would have been considered an example, where there's a landscape that cells take notice of in making choices but which does not affect the cell cycle or growth, and other systems which are very different. It looks as if the real situation is in between. When do blastoderms, which have been manipulated to be lacking most of the machinery that is going to do this early partitioning, start producing fewer cells because of cell cycle retardation?

Nüsslein-Volhard: In the triple mutant we haven't looked, but in the single or double mutants there are regionalized effects on cell proliferation or size regulation. In some cases the remaining pattern takes over, it is expanded in the blastoderm fate map; later it is more or less normal in size, there is strong size regulation at later stages of development.

Cooke: But if *torso*, which is going to interfere with the extremes of the normal pattern, is missing, it seems as if cell division and growth go haywire very early.

Nüsslein-Volhard: The first round of cell division after blastoderm formation is not impaired, but later on cells may start to die. There is no major difference between the effect of *torso* and the other genes.

Cooke: Where I work, at the National Institute for Medical Research in London, if you hear '*membrane-spanning tyrosine kinase*', you are taught to think immediately of a growth controlling protein responding to intercellular signals.

Cabrera: If one is to evaluate the importance of the *bicoid* gradient and its overall effect in the embryo (the fate map undoubtedly shifts), one should bear in mind that changing the dosage of the gene from half to four doses is something with which the embryo can cope. Janni referred to this as the remarkable ability of the embryo to correct a transient disarray produced in the *bicoid* gradient. I think the most important feature of the expression of *bicoid* is the localization of its RNA in the anterior pole, in contrast to the graded distribution of its protein product. When one disrupts that localization by mutation, the embryo cannot develop.

Nüsslein-Volhard: It is probably because the amount of *bicoid* protein at the anterior tip falls below a certain value. The embryo can tolerate half the normal amount, but if the level falls slightly lower, the embryo can no longer cope. When the RNA is not localized, you don't get the highest protein concentration at the anterior. When you increase the dose of *bicoid* in a genetic background which prevents localization, the embryos may have a complete anterior pattern but a distorted fate map, and the sizes of the anlagen are out of proportion but there is a lot of size regulation going on later. The earlier models were too precise. It was a surprise that in *Drosophila* you can get away with so much sloppiness, a lot of perturbation can be corrected quite late in development. Embryos which start with a very small anlage of the abdomen can make normal looking abdomens.

Wolpert: You mean your embryos which seem abnormal early on regulate and give normal flies?

Nüsslein-Volhard: We haven't looked at when they size regulate. We can make embryos for which in the fate map the thorax is three times as large as in the wild-type and in the final differentiated larva the thorax is 1.2 times as large as wild-type.

McKeown: If you moved the cephalic furrow from those embryos so that the stripes of the fate map would be five cells wide or three cells wide, do you know when they would regulate back up to proper size?

Nüsslein-Volhard: No, we haven't investigated that yet.

Wieschaus: When you are talking about size regulation, you are probably talking about events that occur later in embryonic development, after the blastoderm stage, presumably involving cells talking to each other and setting up some final stable pattern. You said that you can start with a thorax that is three times larger than wild-type and it will be reduced to 1.2 times—not to wild-type size but to 1.2 times. Do you have any idea how large a thorax with the normal pattern you could make using these maternal systems to manipulate initial cell identities? Could you make a thorax that's 1.5 times wild-type? Is the pattern-regulating system such that you could make a segment that's twice or three times as big? And, is that a relevant question? Is it a sensible question for us ask?

Nüsslein-Volhard: You could argue for the posterior group genes that they have a third thoracic segment which is really very big and it comes from the entire abdominal region.

Wieschaus: Normally patterned?

Nüsslein-Volhard: Yes, more or less, and it is certainly oversized. The embryo cannot regulate everything down to wild-type size. This has to be investigated in more detail.

Sander: In the leaf hopper you can get heads which are about half of the entire body length of the normal larva, at least twice the normal size, so apparently there is no stringent size regulation.

Struhl: There are other gradient systems in the egg that you didn't discuss, some clearly associated with primary determinants as defined genetically, others not so clearly associated with them. Some of those gradients may define only single thresholds, for example the posterior system, whereas there is good evidence, from your own work, that the anterior system provides several thresholds. The conventional view of a morphogen gradient is that it provides several thresholds which have several distinct responses.

Nüsslein-Volhard: At least two thresholds, in order to confer polarity on the system.

Struhl: To me, in a more general sense, what a gradient provides is at least one threshold which has essential information relevant to the patterning of the system. Do you want to say that the criterion is the number of thresholds? Do you think that a morphogen gradient is only something which provides more than one threshold?

Nüsslein-Volhard: I think it has to have more than one target gene, otherwise you don't get polarity.

Struhl: But there could be many target genes.

Nüsslein-Volhard: I think a gradient has to determine at least two different qualities by two different quantities of the same molecule. If your gradient is used only to specify the localization of one product, I would not call it a gradient mechanism. For example, in the posterior system you can say that the source of the product is the posterior pole plasm and the *nanos* product which is spread anteriorly in the form of a gradient, which is simply a way of placing a substance in the posterior half of the embryo. It is there where it is required, but the amount is not important. Further, the polarity of the pattern it determines need not be dependent on the distribution of *nanos*—that's why I think the posterior system is not a gradient mechanism.

Sander: When interpreting my leaf hopper data, I was unable to develop a model specifying the individual segments in the abdomen by the overall gradient(s). The gradient ratio for the abdomen was given as a single value because nothing else would fit the complete set of data (Sander 1960, p700).

Melton: I was struck by the fact that different axes—the anteroposterior and dorsoventral axes—seem to be controlled by strikingly different molecular mechanisms. In the one case, it is transcriptional regulation via homeobox genes and in the dorsoventral case, it is the expression of proteins either by their movement or degradation. Should we attach any significance to the fact that such different molecular mechanisms are used for these two different axes? Is it significant in either a developmental or an evolutionary sense?

Nüsslein-Volhard: I find it very significant. The notion that the dorsoventral pattern-forming system is very different from the anteroposterior one is an old idea. You can make longitudinal ligatures in other insect embryos and you get twinning. This is reminiscent of *Xenopus*. There is no pre-localization in this case, which is also very significant. The problem of how the orientation of the egg case is passed on to the orientation of the embryo becomes more significant than in the anteroposterior system where this follows from the geometry of the nurse cell–oocyte pattern.

Martinez-Arias: Do you think that the difference between the anterior and the posterior systems is related to the phylogeny of the arthropods—the differences between short and long germ band insects and the wide spectrum in between?

Nüsslein-Volhard: There is a striking similarity between this posterior group phenotype and that of the short germ band insects in the very early stage of embryonic development. This phenotype is reminiscent of the way that the more primitive insects make their abdomen. It is very different from how the anterior pattern is laid down. The *bicoid* system is present initially and the posterior system intercalates the abdominal segments between the thorax and the telson that is also there initially.

Sander: I think that the anterior system is more advanced. You can see this in insects such as the cricket or leaf hopper, where the segment pattern in the anterior part of the body is laid out more or less simultaneously, as in *Drosophila*, but the posterior part of the body is formed by some kind of budding, presumably based on intercalation (cf Sander 1988).

Morata: You sub-divide the embryo into five regions, two of which, the acron and the telson, are unsegmented, and the other three are segmented. This seems to be a fundamental difference between these regions, which I suppose has implications for the genetic construction of these body parts. What is the criterion for the telson, for example?

Nüsslein-Volhard: The criterion by which acron and telson are defined is that they are the regions which are not affected or duplicated in particular maternal mutants of this class. For example, the telson that is present in posterior group mutants is duplicated in the *bicoid* mutant phenotype, and absent from the posterior end of *torso* embryos. The acron is more difficult to define, this is possible essentially only by the *torso* phenotype. Everyone agrees that the telson and acron are not segmented but there are some structures that are apparently rudiments of lost segments which are present in other insects.

Morata: But you say *torso* is responsible for telson development and the phenotypic limit of *torso* appears to be the seventh abdominal segment. Would that mean that anything posterior to segment A7 is unsegmented?

Nüsslein-Volhard: It depends which phenotypes you mean. There may be an interaction between *torso* and the posterior group genes. This would account for the fact that in *torso* mutants the eighth abdominal segment is usually also missing: you can find it in some alleles. But in *bicoid* mutants, and in mutants of the posterior group genes, the telson is very clearly defined. This is the definition which is easiest.

Morata: There are mutants which put an extra segment in the telson, such as *spalt* or *forked head.*

Nüsslein-Volhard: To call anything which you can transform into something which has structures usually found in segments a segment is probably not a good definition.

Morata: But there is a pre-existing segmental primordium there.

Nüsslein-Volhard: This is your interpretation. We could discuss this for ever! It's a very clear definition if you take these mutant phenotypes and if we take these three systems. Your notions about abdominal segmentation are much more detailed than our notions, because we look at a much simpler level. On the other hand, it's interesting that you include in these mutants that the telson contains these remnants of the abdominal segments 9,10,11 or whatever and not everything which is visible now. This is an evolutionarily significant observation.

Melton: Is it known whether the follicle cells around the dorsoventral boundary of the egg case give any information to the dorsoventral axis?

Nüsslein-Volhard: I think they probably do. The local activation in *Toll* embryos, where you can induce polarity at all positions along the dorsoventral axis, in normal development always occurs at the ventral side. There are mutants—this is an entire new class of genes which I haven't talked about at all—which probably affect earlier steps in pattern formation (Schüpbach 1987). The genes affect both the polarity and pattern of the follicle cells and the egg cell. Among that class of mutants, one should look for those which give the spatial cues for this local activation process. This is probably also the case for *torso*, that there is local activation at the anterior and posterior tips.

Kornberg: This very pretty evidence that you presented, showing that the RNA is so well localized and the protein exists in a gradient, could be ascribed to a difference in sensitivity between the two methods. Can you comment on the probability that the *bicoid* RNA might also define a gradient pattern? Secondly, has anyone done similar kinds of experiments to those Doug (Melton) described, injecting molecules into the egg and asking whether proteins diffuse and is there any specificity to the diffusion?

Nüsslein-Volhard: We don't think it is a difference in the sensitivity of the two methods. It is clear from the *in situ* hybridizations that there is pre-localization because we find localized RNA. There is a slight issue whether the RNA moves posteriorly or whether the RNA stays and the protein diffuses. I think it is more likely that the protein diffuses and not the RNA. When you look at the *in situ* patterns carefully in embryos with overdoses of *bicoid*, the RNA stays localized where it is put. We should do the experiment to see whether it remains in the same condition. There is a strong translational control on *bicoid*, so this could be some change of the tagging there, but we have no evidence for or against that.

Garcia-Bellido: One thing that is missing from descriptions of phenotypes of this type is what is happening at the cellular level. For example, when and where do cells die? If you start with stripes of gene expression of different sizes, what happens later on? You are disturbing the equilibrium of the system and the stripes must compensate somehow. It would be very interesting if genes like these could be cloned from closely related insects, for example *Hymenoptera*.

In *Hymenoptera*, the process of cellularization with respect to gastrulation is different to that in *Drosophila* (Fleig & Sander 1985). We place a lot of emphasis on diffusion because we think that the system is a syncytium. A gene may not be working at the syncytial stage in other insects, it may not be operationally active until after cellularization. Then instead of simple diffusion of proteins, we have to consider passage across cell membranes and intercellular signalling.

Nüsslein-Volhard: You always tell me that I should look at the later level—people are looking there, that is their business, I am looking at my system!

McKeown: Knowing that the homeobox is a DNA-binding protein, how do you know that a particular homeobox-containing protein directly activates a particular target? With the *bicoid* system, you can come pretty close but it is not a direct assay. For other things, like some of the homeodomain proteins, you can show that a protein is capable of activating a given gene, but you can't show that it is doing so biologically.

Scott: One possibility is an experiment where you change the DNA sequence of the binding site and change the protein, so that it can bind to the new site. Then they are both novel, and could be tested within the organism. If you could show that the same interaction occurs between a protein and a target gene in that novel situation, that would be a very tight demonstration.

Garcia-Bellido: Such experiments are being done in Walter Gehring's lab. They are taking the *ftz* gene and substituting the neighbouring flanking regions with the corresponding regions of other genes with larger or smaller introns. Apparently, the specificity of the autoactivation function is in the homeobox domain.

Struhl: A lot is known about how the products of yeast and bacterial genes which contain homeoboxes, such as the mating-types *a* and α and lamda repressor, regulate transcription. In these cases, it is now quite certain that the proteins directly recognize and bind distinct sequences in the DNA via their homeobox domain, and that other domains of the protein are involved in transcriptional activation or repression. That seems to be a good precedent for the molecular biologists who want proof. Do you doubt that this is the mechanism by which things are occurring?

McKeown: No, I don't doubt the mechanism. I am worried more about the circuitry—specifically, which molecules are inducing or repressing which other ones. It is experimentally difficult, because homeoboxes are promiscuous; *in vitro* one homeobox will bind to many different targets and many of those targets are found upstream of different genes. So if you do any specific pair-wise test, you can say that a particular homeobox-containing gene product binds to a given sequence that is normally upstream of a certain gene and even transcriptionally activates it, without knowing if it is relevant.

Garcia-Bellido: Promiscuous means that *in vitro* the homeobox will interact with a number of other genes. But *in vivo* there are more things to consider than the homeobox and the target region; there are a lot of other competing molecules present and presumably there are subtle differences in the interactions.

Scott: The evidence for promiscuity is quite limited. It is based on a small number of experiments done with proteins made in bacteria that have not been correctly modified. None of the proteins has been completely purified and affinity constants have not been accurately determined. Even those proteins exhibit quite a bit of specificity. I think the proposed lack of specificity has been over-emphasized in these discussions, largely because most of the relevant findings are unpublished. Some proteins can bind to similar or identical sites, but that hardly means there is no binding specificity.

Nüsslein-Volhard: In the case of *bicoid*, the consensus sequence to which *bicoid* binds is quite different from the other consensus sequence for homeobox-containing proteins. We have also made promoter constructs with three binding sites, attached these to a reporter gene and injected those into embryos which are *bicoid* or wild-type. The *bicoid* embryos didn't express the reporter gene, the wild-type embryos did. When we successively removed these binding sites, the activity of the reporter gene fell dramatically. We think this is very close to the *in vivo* situation (Driever & Nüsslein-Volhard 1989).

Jun Ma in collaboration with Wolfgang Driever in our laboratory has shown that if you make a construct with the *hunchback* promoter region linked to a yeast reporter gene and transform it into yeast, and transform *bicoid* into the yeast in addition, then *bicoid* will activate transcription of these constructs. I think the evidence is coming as close as you can wish that this is activating transcription in a sequence-specific manner.

Morata: Has anyone done a simple experiment like linking just the *Ultrabithorax* (*Ubx*) homeobox or the *Antennapedia* homeobox to a heat shock promoter, getting it over-expressed, and seeing whether other homeotic genes are repressed?

Garcia-Bellido: Even doing that with a normal gene, you don't get a clear cut result, because the system has a lot of inertia.

Morata: We have been doing some experiments with the normal *Ubx* gene under heat shock control. In embryonic development there is a clear result, there is very good transformation of the type you would expect when *Ubx* is over expressed. In some cases, at least, experiments with heat shock promoters work well.

Garcia-Bellido: Similar experiments have been done with different constructs of *Ubx* and *Antennapedia* in cell transformation. Apparently, a lot of the normal protein is needed to get self-stimulation of the gene.

Scott: We don't know how much protein is necessary. Mark Krasnow and Gary Winslow have done several of those experiments. No one has accurately measured the amount of protein that's required for those activation events.

Struhl: When you do experiments of that kind, or more *in vitro* types of molecular experiments, it is important to remember that whether or not a protein will activate *in vivo* can be very dependent on the presence of other components, some of which function only transiently or in particular combinations. For example, if the *hunchback* gene or the gene *fushi tarazu*, which has a standard homeobox, is placed under the control of a heat shock promoter and the embryos heat shocked at the critical time when those genes normally function, you see an effect. If you heat shock later, nothing happens. That is, the same protein fails to trigger the response, even though the appropriate target genes are present. So the context critically determines the outcome of the experiment. This means you cannot be sure that the responses you are seeing *in vitro* or in tissue culture are relevant. This relates to Mike McKeown's question. If you want to use these molecular criteria as a key to finding out how these genes work, you are in some danger of studying an artifact.

References

Driever W, Nüsslein-Volhard C 1989 The *bicoid* protein is a positive regulator of *hunchback* transcription in the early *Drosophila* embryo. Nature (Lond) 337:138–143

Fleig R, Sander K 1985 Blastoderm development in honey bee embryogenesis as seen in the scanning electron microscope. Int J Invertebr Reprod Dev 8:279–286

Sander K 1960 Analyse des ooplasmatischen Reaktionssystems von *Euscelis plebejus* Fall (Cicadina) durch Isolieren und Kombinieren von Keimteilen, II. Wilhem Roux' Arch 151:660–707

Sander K 1988 Studies in insect segmentation: from teratology to phenogenetics. Development 104:111–123

Schüpbach T 1987 Germ line and soma cooperate during oogenesis to establish the dorsoventral pattern of egg-shell and embryo in Drosophila melanogaster. Cell 49:699–707

Morphogen gradients and the control of body pattern in insect embryos

Gary Struhl

Howard Hughes Medical Institute, Center for Neurobiology and Behavior, Columbia University College of Physicians and Surgeons, New York, NY 10032, USA

Abstract. The basic body plan of insects is set up in response to determinants initially localized at the anterior and posterior poles of the egg. Early in development, these determinants give rise to a series of morphogen gradients which in turn trigger a cascade of molecular signals determining the body pattern. This signalling process is outlined and recent experiments testing the roles of these gradient systems in determining anterior and posterior pattern are described. The results of these experiments suggest that anterior pattern is controlled by a single instructive gradient, whereas posterior pattern depends on the overlap of several gradients, each providing only one or a few distinct responses.

1989 Cellular basis of morphogenesis. Wiley, Chichester (Ciba Foundation Symposium 144) p 65–91

Since the birth of embryology as an experimental discipline, it has been apparent that the development of cell and body patterns depends on the existence of robust and complex systems of spatial information. Yet, until recently, we had little idea of the physical nature of this information or of the mechanisms by which it is generated and intepreted. Nevertheless, evocative notions have been put forward to account for the response of developing organisms or tissues to various embryological and genetic perturbations (see Huxley & de Beer 1934, Horder et al 1983). Chief among these has been the concept that gradients of form-generating molecules, or *morphogens* (Turing 1952), might control pattern. The beauty of this concept is that it shows how a relatively simple mechanism could suffice to dictate pattern as a function of physical distance. Thus, numerous models have been proposed in which a local source of a morphogen leads to the formation of a stable gradient that dictates pattern by specifying distinct outcomes (e.g. pattern elements) in response to particular concentrations of the morphogen.

The development of both cell and body patterns in insects has provided several examples of gradient phenomena. Indeed, one of the most compelling arguments for the existence of morphogen gradients in any system comes from the classic embryological experiments of Sander and colleagues (Sander 1959, 1960, see also

Sander 1976). Using a variety of techniques to isolate or rearrange the internal components of the early *Euscelis* embryo, Sander was able to show that the information which dictated the segment pattern emanated from the anterior and posterior poles and spread progressively towards the centre of the embryo, leading to the stable determination of the body plan by the cellular blastoderm stage. On the basis of these results, he proposed that the insect body pattern is specified by opposing anterior and posterior morphogen gradients, the relative concentrations of these morphogens determining the boundaries and qualities of different regions of the body.

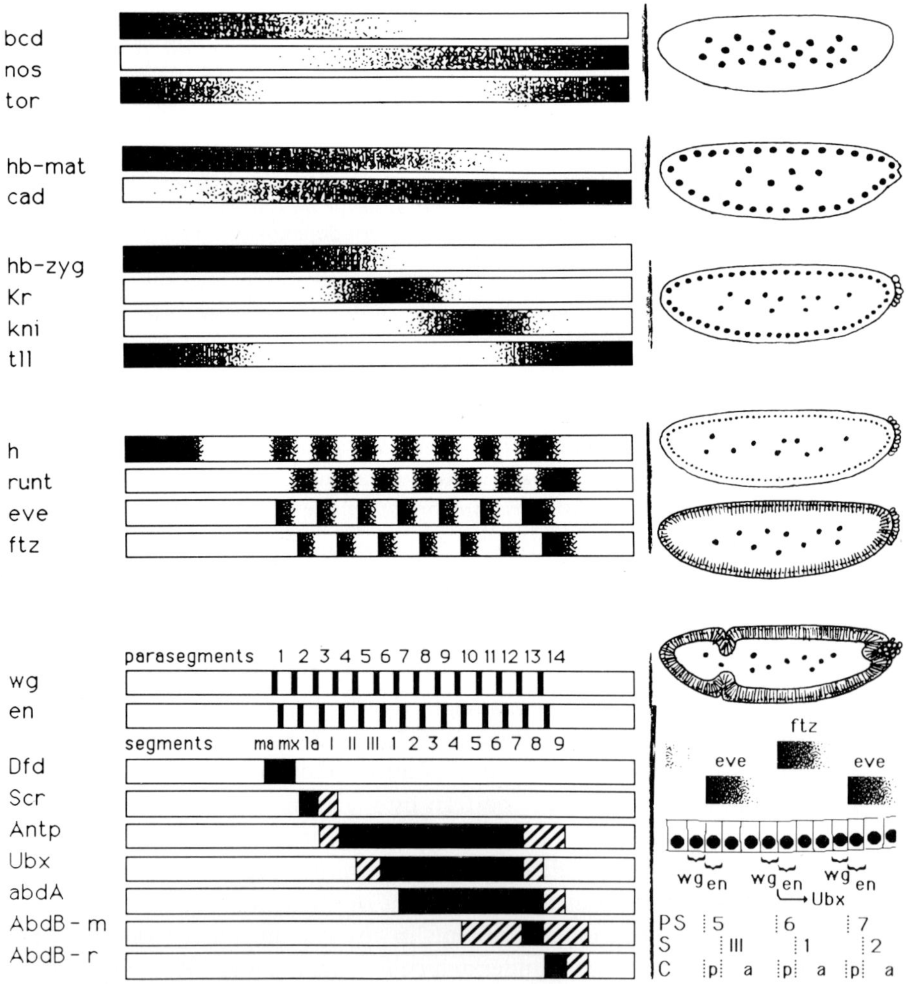

FIG. 1. An outline of the molecular cues controlling the specification of the *Drosophila* body pattern. Genes are listed on the left, the patterns of expression of their active products shown in the middle, and the stages during which they first become active illustrated at right (anterior left, posterior right; please see reviews by Scott & Carroll (1987), Akam (1987), Nüsslein-Volhard et al (1987), and Ingham (1988) as well as the text for details and proper citations. The distributions of active *nos*, *tor*, *kni* and *tll* products have not been determined at the molecular level, but are inferred from genetic and embryological results). The graded distributions of the active forms of the *bcd*, *nos*, and *tor* morphogens are known or presumed to be set up shortly following fertilization, during the period when nuclei are dividing in the yolky middle of the egg (see Nüsslein-Volhard, this volume). Secondary gradients of the cad^{mat} and hb^{mat} proteins arise in response to polarized activity of the *bcd* and *nos* products, respectively, as the proliferating nuclei migrate towards the egg periphery and the pole cells form at the posterior end. The gap genes hb^{zyg}, *Kr*, *kni* and *tll* (as well as *giant* and cad^{zyg}, not shown) are then activated in response to one or more of the primary and secondary cueing systems during the next few rounds of nuclear divisions which occur at the periphery (the *Kr*, hb^{zyg} and cad^{zyg} genes are also activated later during the final nuclear division cycle preceding gastrulation in one or a few relatively narrow domains which are not shown). Over the course of the last nuclear division cycle preceding cellularization and the onset of gastrulation, at least eight pair-rule genes such as *hairy*, *runt*, *even-skipped* and *fushi tarazu* are then activated in periodic zebra patterns. In general, their products are first expressed in poorly resolved patterns, which become progressively more regular and sharply defined as the embryo cellularizes. In some cases, such as *even-skipped*, more complex periodic patterns of 14 stripes arise around the onset of gastrulation (not shown). Note that the patterns of the *even-skipped* and *fushi tarazu* proteins form graded distributions with sharp anterior edges. When cellularization is completed, the embryo begins to gastrulate. At this time, the sharp edges of *even-skipped* and *fushi tarazu* define the boundaries of parasegments 1–14, as well as the activation of adjacent stripes of expression of segment polarity genes, such as *wingless* and *engrailed*, on either side (see Lawrence, this volume). Homeotic genes (e.g. *Deformed*, *Sex combs reduced*, *Antennapedia*, *Ultrabithorax*, *abdominal-A* and *Abdominal-B*; see Busturia et al, this volume) are also activated in particular subsets of these units around the same time. All of the homeotic genes, as well as *engrailed*, are generally expressed in all the cells that descend from the founder groups in which they are initially activated (hatched territories indicate units in which the expression of these genes either changes qualitatively over time or is uncertain). As diagrammed in the lower right corner, cells expressing *engrailed* give rise to posterior compartments (C, p) while the remaining cells generate anterior compartments (C, a); parasegments (PS) and segments (S) are composed of repeating pa or ap units (head segments: ma = mandibular, mx = maxillary, la = labial; I, II and III = pro-, meso- and metathoracic segments; 1–9 = the first nine abdominal segments).

Figure 1 on facing page

During the past several years, the evidence in support of Sander's proposal has become compelling. First, using *Drosophila*, Nüsslein-Volhard and her colleagues (see Nüsslein-Volhard et al 1987, Nüsslein-Volhard, this volume) have identified genes encoding the anterior and posterior morphogens, as well as an additional morphogen controlling terminal pattern. More critically, in the case of the anterior morphogen, *bicoid*, they have established that the active product of the gene is expressed as a gradient during the critical period when the pattern is being specified (Driever & Nüsslein-Volhard 1988a). Thus, it is now evident that such morphogen gradients exist.

The discovery of concrete examples of morphogen gradients provides only the beginning of a satisfactory explanation for how the body pattern arises. To answer these questions, we need to know how the spatial information inherent in these gradients is translated into the final cellular pattern. My main goal in this essay is to describe our current understanding of this process of interpretation. To do so, I first present an outline of the molecules involved in generating and reading the primary gradients. Then, I discuss a set of recent experiments which examine the roles of the primary anterior and posterior gradients in specifying the body plan.

A tentative outline linking primary gradients to the generation of body pattern

The process of interpreting the primary morphogen gradients is fundamentally one of information transfer and refinement, in which a few relatively crude cues—the primary morphogens—activate a cascade of successively finer cues. Many of the key steps are diagrammed in Fig. 1 and described in chronological and causal order below (more detailed descriptions and citations can be found in recent reviews, e.g. Scott & Carroll 1987, Akam 1987, Nüsslein-Volhard et al 1987, Ingham 1988, see also articles by Nüsslein-Volhard, Lawrence and Busturia et al, this volume). It is important to note at the outset that early insect embryos have the unusual property that nuclei proliferate in a common cytoplasm until the end of the blastoderm stage. Hence, virtually all the spatial information controlling the body pattern is created within a single multinucleate cell. Under these conditions, both localized cytoplasmic determinants and, subsequently, nuclei can serve as point sources for cueing molecules which are capable of diffusing from their sites of synthesis and influencing the expression of new cues at a distance.

Polar determinants and the primary morphogen gradients

The first cues responsible for organizing anteroposterior body pattern are generated during oogenesis and the initial phase of nuclear proliferation leading up to the migration of nuclei to the egg periphery. The nature of these cues is described in detail by Nüsslein-Volhard (this volume). There are three primary cues, the products of the *bicoid*, *nanos* and *torso* (*bcd*, *nos* and *tor*) genes, which are responsible, respectively, for anterior, posterior and terminal portions of the body. The *bcd* and *nos* cues appear to be generated from tightly localized sources at the anterior and posterior poles, respectively, of the early embryo and to diffuse or be transported towards the centre of the egg, forming gradients of active product. Less is known about the *tor* cue; the molecule appears to be present throughout the embryo, but differentially active in the vicinity of both poles (Klingler et al 1988). In the case of *bcd*, it has been demonstrated that the initial localized determinant is the *bcd* mRNA, which is synthesized in nurse cells and then transferred to the maturing oocyte where it is trapped upon entry at the anterior pole (Berleth et al 1988). After fertilization, the mRNA is translated and the nascent protein diffuses from its site of synthesis, generating a continuous gradient of the active morphogen, *bcd* protein (Driever & Nüsslein-Volhard 1988a; see Fig. 2A). Considerably less is known about the remaining two determinants, though in the case of *nos*, the active morphogen appears to be synthesized in nurse cells, transported to the oocyte, selectively stabilized in association with cytoplasmic granules at the posterior pole, and then released following fertilization, allowing it to diffuse anteriorly towards the middle of the embyro (Sander & Lehmann 1988).

Formation of secondary morphogen gradients

The first response to the primary morphogen gradients appears to involve the formation of additional gradients. Two such gradients have been observed so far. First, the *bcd* gradient appears to create an opposing gradient of *caudal* (*cad*) protein by differentially repressing the translation of its uniformly distributed mRNA (Macdonald & Struhl 1986 and unpublished data, Mlodzik & Gehring 1987; see Fig. 2B). Similarly, differential activity of the *nos* morphogen in the posterior half of the body generates an opposing anterior gradient of *hunchback* (*hb*) protein by differentially destabilizing maternally derived *hb* mRNAs (Tautz et al 1987, Tautz 1988; see Fig. 2E).

It is important to note that both the *cad* and *hb* gradients arise in opposition to the primary *bcd* and *nos* gradients during the early period of syncytial nuclear divisions when the nuclei are transcriptionally silent. Both of these secondary gradients are generated by mechanisms which repress protein expression from initially ubiquitous maternal transcripts. Subsequently, the *cad* and *hb* genes are also transcriptionally activated in more sharply defined domains within the regions in which their respective gradients are first established. In the case of *hb*, these two responses are controlled by *different* primary morphogens: the *nos* morphogen represses posterior expression of the maternal mRNA;

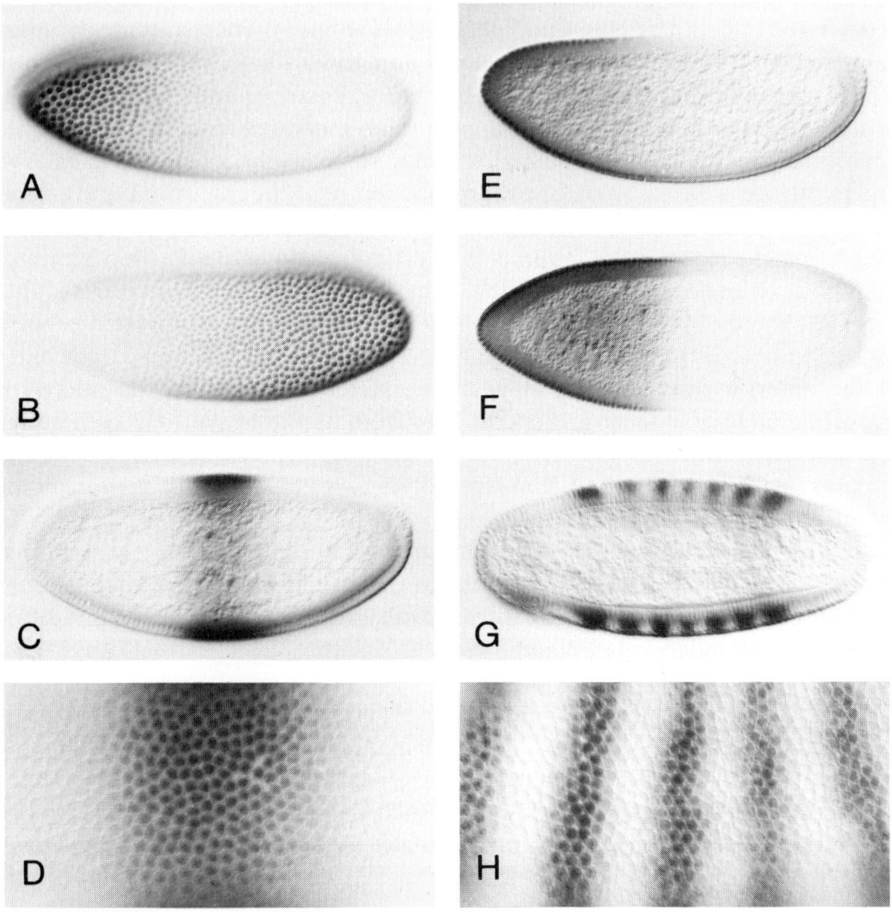

FIG. 2. Some examples of cueing molecules. A) gradient of *bcd* protein; B) responding gradient of cad^{mat} protein; C) approximate step function of *Kr* protein; D) higher magnification of C showing the bell-shaped distribution of *Kr* protein at both edges of the stripe; E) gradient of hb^{mat} protein (shown in a *bcd* mutant embyro which lacks hb^{zyg} protein expression); F) approximate step function formed by $hb^{mat} + hb^{zyg}$ protein; G) periodic expression of *h* protein; H) higher magnification of G showing the bell-shaped distribution of *h* expression in each stripe. All embryos are oriented with their anterior ends on the left and generally with their ventral sides facing the bottom. A, B, D, and H show the surface of the embryo; the remaining photographs are optical sections through the middle showing only the periphery in focus. Note that all of these molecules are found predominantly in nuclei. Antibodies were generously provided by Paul Macdonald and Ken Howard.

subsequently, the *bcd* morphogen activates *hb* transcription in nuclei in the anterior half of the embryo (Tautz 1988). For clarity, I will refer to the maternal and zygotic products of these genes as if they arise separately (e.g. as the products of the hb^{mat} and hb^{zyg} genes), even though the protein products appear to be physically and functionally identical.

Transcriptional activation of gap genes

By the time the proliferating nuclei reach the egg periphery, the *bcd*, *cad* and *hb* gene products are expressed as gradients; similarly, the products of the *nos* and *tor* genes behave as if differentially active along the anteroposterior axis. These molecules collectively provide the cues necessary for the first burst of localized zygotic gene activity. Not surprisingly, at least three of these gradient molecules contain homeobox domains (*bcd*, *cad*) or zinc fingers (*hb*), suggesting that they function directly as transcriptional regulators. Indeed, during the remainder of early embryogenesis, most of the cueing molecules which are generated also have these kinds of structural domains, suggesting that nuclei now communicate by expressing transcriptional regulators which can directly control their neighbours' synthetic activity.

At least four gap genes, hb^{zyg}, *Krüppel* (*Kr*), *knirps* (*kni*) and *tailless* (*tll*) (Nüsslein-Volhard & Wieschaus 1980), are activated in response to these gradients. The hb^{zyg} and *kni* genes respond primarily to the *bcd* and *nos* gradients, respectively, defining broad but separate domains in the anterior and posterior halves of the embryo (Tautz 1988, Nauber & Jackle, unpublished data as cited in Ingham 1988; see Fig. 2F). *Kr* is activated in the centre of the embryo, in a region just overlapping or bounded by the hb^{zyg} and *kni* domains (Gaul & Jackle 1987; Fig. 2C). Finally, the *tll* gene is presumed to be turned on at both ends of the embryo, where it either overlaps or abuts the domains of hb^{zyg} and *kni* expression (Strecker et al 1986, Klingler et al 1988, see Nüsslein-Volhard, this volume). Thus reading across the embryo, the overlapping anterior (*tor* + *bcd* + hb^{mat}) and posterior (*cad* + *nos* + *tor*) cueing systems are translated into crude step functions of terminal (*tll*), anterior (hb^{zyg}), middle (*Kr*), posterior (*kni*) and terminal (*tll*) gap gene activity.

Perhaps the most important attribute of the gap genes is that their patterns of expression are more tightly restricted than those of the gradient molecules which trigger their activation. This probably reflects the fact that the primary gradients are localized solely by mechanisms involving diffusion or transport from highly polarized sources at both ends of the egg, whereas the domains of gap gene products are generated by localized gene activation. Hence, the transition from molecular gradients to the approximate step functions of gap gene activation creates a series of new spatial cues with significantly higher resolution.

It should, however, be emphasized that the domains of gap gene expression do not have sharp boundaries. Instead, their products are graded across several nuclei (Gaul & Jackle 1987, Tautz 1988; see Fig. 2D). This is hardly surprising. First, the boundaries appear to be dictated by molecular gradients; hence it is likely that nuclei close to the critical concentration threshold may respond in an intermediate as opposed to an all-or-none manner to the morphogen. Such a graded response is likely to be both steeper and more locally restricted than the gradient which triggered it. Second, these genes are first expressed 3–4 nuclear cycles (approximately 90 minutes) before cellularization. Hence, diffusion of their mRNA or protein products, if only for a few nuclear diameters, would also contribute to creating a graded distribution.

Defining segmental units: activation of pair-rule and segment polarity genes

The products of the gap genes control the activation of pair-rule genes (Nüsslein-Volhard & Wieschaus 1980) in periodic 'zebra' patterns of seven to eight stripes by a mechanism which is poorly understood at present (see Akam 1987, Scott & Carroll 1987, Ingham 1988; Fig. 2G). This process occurs about 2–3 nuclear division cycles after the initial activation of the gap genes (around the beginning of the 14th and last nuclear division cycle preceding cellularization). Initially, the stripes of pair-rule gene expression do not have sharp boundaries, but form bell-shaped distributions (e.g. Frasch & Levine 1987, Carroll et al 1988; see Fig. 2H). However, during the remainder of the 14th nuclear cycle, these stripes sharpen and in some cases bifurcate, generating a complex system of overlapping periodic signals. This information precisely defines the boundaries of segmental and compartmental units of the embryo (Lawrence, this volume), leading to the tightly localized expression of segment polarity genes, such as *engrailed* (*en*) and *wingless* (*wg*), which sustain these boundaries during subsequent development (Morata & Lawrence 1975, DiNardo et al 1988, Martinez-Arias et al 1988).

Regional specification: activation of homeotic 'selector' genes

In addition to generating periodic cues defining the basic segmental units of the embryo, the distribution of gap gene products also appears to determine the regional character of these units by controlling where homeotic genes, such as those of the bithorax complex (Lewis 1978, see Busturia et al, this volume), are initially activated. In general, these genes are first expressed around the end of the blastoderm stage in broad domains encompassing specific segmental units. These patterns of expression are sustained subsequently, committing the descendant cells within each unit to 'select' a particular developmental path.

Diverse roles for morphogen gradients specifying anterior and posterior body pattern—some recent experiments

It is clear from the outline described above that the primary morphogen gradients dictate the body pattern through their actions on subordinate cueing systems. Moreover, it is equally clear that the subordinate cueing system is itself replete with molecular gradients. Simply knowing that these various cueing molecules are expressed as gradients does not tell us how their graded distributions confer pattern. It is obvious *a priori* that molecular gradients can be interpreted in terms of distinct thresholds, generating discrete responses in space. What we need to know is the 'promorphological' value (Wilson 1925) of this information, i.e. in what sense the gradient prefigures the pattern.

Here I describe a set of experiments which compare and contrast the ways in which the *bcd* and *nos* morphogens control body pattern. The properties of the *bcd* gradient have been extensively characterized (see Nüsslein-Volhard et al 1987, Nüsslein-Volhard, this volume); hence I review only briefly the evidence showing that this gradient plays a dominant, instructive role in organizing most of the anterior body pattern. In contrast, the role of the *nos* gradient has been less well characterized. The experiments described here challenge prior views (e.g. Nüsslein-Volhard et al 1987) that this gradient plays a corresponding instructive role to that of *bcd*; instead, they argue that it provides a relatively simple cue which specifies only where posterior patterning can occur along the body. This conclusion forces us to consider what other factors govern posterior pattern: evidence is presented that these other factors appear to be the shorter-range gradients of the gap genes *Kr*, *kni* and *tll*.

Anterior body patterning

As described by Nüsslein-Volhard and colleagues (Frohnhöfer & Nüsslein-Volhard 1986, 1987, Nüsslein-Volhard et al 1987, this volume), both embryological and genetic experiments indicate that the concentration landscape of *bcd* protein exerts a dominant influence on the body pattern. Two experiments are of particular importance in establishing this point. First, microinjection of *bcd* activity into the middle of an embryo triggers the formation of organized anterior pattern, including head structures at the site of injection and appropriately polarized thoracic segments on either side (Frohnhöfer & Nüsslein-Volhard 1986). Second, mutations which lead to low levels of indiscriminate *bcd* activity throughout the embryo (e.g. *exuperantia oskar* double mutant embryos) develop a pattern of a few enormous gnathal or thoracic segments (the type of segments normally formed in association with low concentrations of *bcd* protein) (Frohnhöfer & Nüsslein-Volhard 1987). Thus, the local concentration of *bcd* protein appears to act 'autonomously' (Driever & Nüsslein-Volhard 1988b) to determine the size, type, number and polarity of the resulting body segments.

In terms of molecular mechanism, the *bcd* gradient appears to control pattern by creating a series of distinct concentration thresholds which dictate specific outcomes at defined relative positions along the body. Indeed, a strong case has been made for the proposal that *bcd* functions as a transcription factor which activates at least two subordinate signalling genes, *hb* and a second gene, x, presumed to be involved in the determination of head segments, in restricted anterior domains (Nüsslein-Volhard, this volume). The *bcd* gradient also generates a third response by controlling the formation of the secondary gradient of *cad* protein from uniformly distributed maternal transcripts (Macdonald & Struhl 1986 and unpublished data, Mlodzik & Gehring 1987). The developmental role of the *cad* gradient is not well understood, though *cad* activity is clearly critical for normal patterning in the posterior half of the embryo. However, either graded expression of the cad^{mat} protein under the control of the *bcd* gradient, or an approximate step function of cad^{zyg} expression resulting from the early transcriptional activation of the gene, is sufficient to ensure normal body patterning (Macdonald & Struhl 1986). Accordingly, the *bcd* gradient may control anterior patterning in part by defining a distinct limit to the anterior spread of *cad* activity.

Thus *bcd* protein appears to function as a classic coordinate-generating morphogen, evoking at least three and possibly more thresholds which are responsible for dictating anterior body pattern.

Posterior body patterning

Differential expression of the posterior morphogen, *nos*, has been proposed to exert a similar organizing influence on body patterning (Nüsslein-Volhard et al 1987). However, the experiments described below suggest that the *nos* gradient acts only to locate where posterior patterning occurs: the pattern is controlled within this domain by more local gradients of the gap gene products *Kr*, *kni* and *tll*.

In terms of its embryological properties, the *nos* morphogen is similar to *bcd* in several respects (Lehmann & Nüsslein-Volhard 1986). Perhaps most importantly, cytoplasmic transplantation experiments indicate that the active morphogen is differentially expressed along the anteroposterior axis, peaking at the posterior pole. In addition, injection of *nos*-containing cytoplasm into nos^- embryos can restore posterior pattern in a concentration-dependent fashion. However, these experiments reveal one significant difference. When ectopic peaks of *bcd* activity are created in the middle of the embryo, they cause an extreme anterior response (e.g. anterior head structures). In contrast, ectopic peaks of *nos* activity generated in the same way are not correlated with an equally extreme posterior response. Instead, the rescued abdominal pattern always forms with the most posterior segments closest to the ends of the embryo, *irrespective of the site of injection*.

Indeed, injection of *nos*-associated cytoplasm into the middle of embryos lacking both *nos* and *bcd* morphogen systems generally leads to the development of 'bicaudal'-type segment patterns composed of two abdomens arranged mirror-symmetrically with their most terminal structures at the poles.

This surprising property of the posterior determinant system can be viewed in either of two ways. First, the putative *nos* gradient could play an instructive role in dictating posterior pattern, much like that of *bcd*. However, one would then have to argue that the *nos* morphogen is 'self-organizing' (Nüsslein-Volhard et al 1987), i.e. capable of moving from the site of injection and recreating its normal graded distribution. Alternatively, localized activity of the *nos* morphogen might not directly control posterior pattern, but rather determine where posterior patterning can occur. In this case, one would have to argue that the particular pattern generated (e.g. mirror-symmetrical bicaudal abdomens) depends on the distribution of other morphogens (see below).

In the absence of clear evidence that the graded distribution of *nos* product corresponds with the final pattern, the embryological experiments establish only that *nos* activity is a necessary prerequisite for posterior patterning to occur. This difference between the *nos* and *bcd* systems becomes more pronounced when the molecular roles of these genes are considered. As described above, the earliest response to posterior *nos* activity in early embryos appears to be the localized destabilization of maternal transcripts of the *hb* gene, and hence the formation of a secondary molecular gradient of hb^{mat} protein (Tautz 1988). Surprisingly, two experiments which circumvent this response and produce uniform *hb* protein expression suggest that formation of this secondary gradient may be the only direct response to the *nos* gradient.

First, I have recently performed a series of experiments in which the *hb* coding sequence is placed under the control of the *hsp70* promoter and the embryos are heat shocked shortly after fertilization. Under these conditions, the resulting embryos express high levels of *hb* protein throughout and develop like nos^- mutant embryos, i.e. they lack abdominal segments (diagrammed in Fig. 3). A similar result has been obtained in a second experiment in which the 3′ untranslated portion of the hb^{mat} mRNA was replaced by that of another gene, an alteration that apparently rescues these transcripts from being degraded by the *nos* product (D. Tautz, C. Schröder, and H. Jackle, personal communication). Thus, inappropriate *hb* activity in the posterior half of the embryo mimics the effect of eliminating the *nos* morphogen, suggesting that the *nos* morphogen normally exerts its influence on the body pattern simply by abolishing expression of the hb^{mat} protein (see Fig. 3).

These results are initially perplexing because they imply that posterior patterning could occur normally even without the *nos* morphogen provided that the hb^{mat} transcript could be inactivated by some other means. This surprising implication may well be correct. Lehmann & Nüsslein-Volhard (1987) have

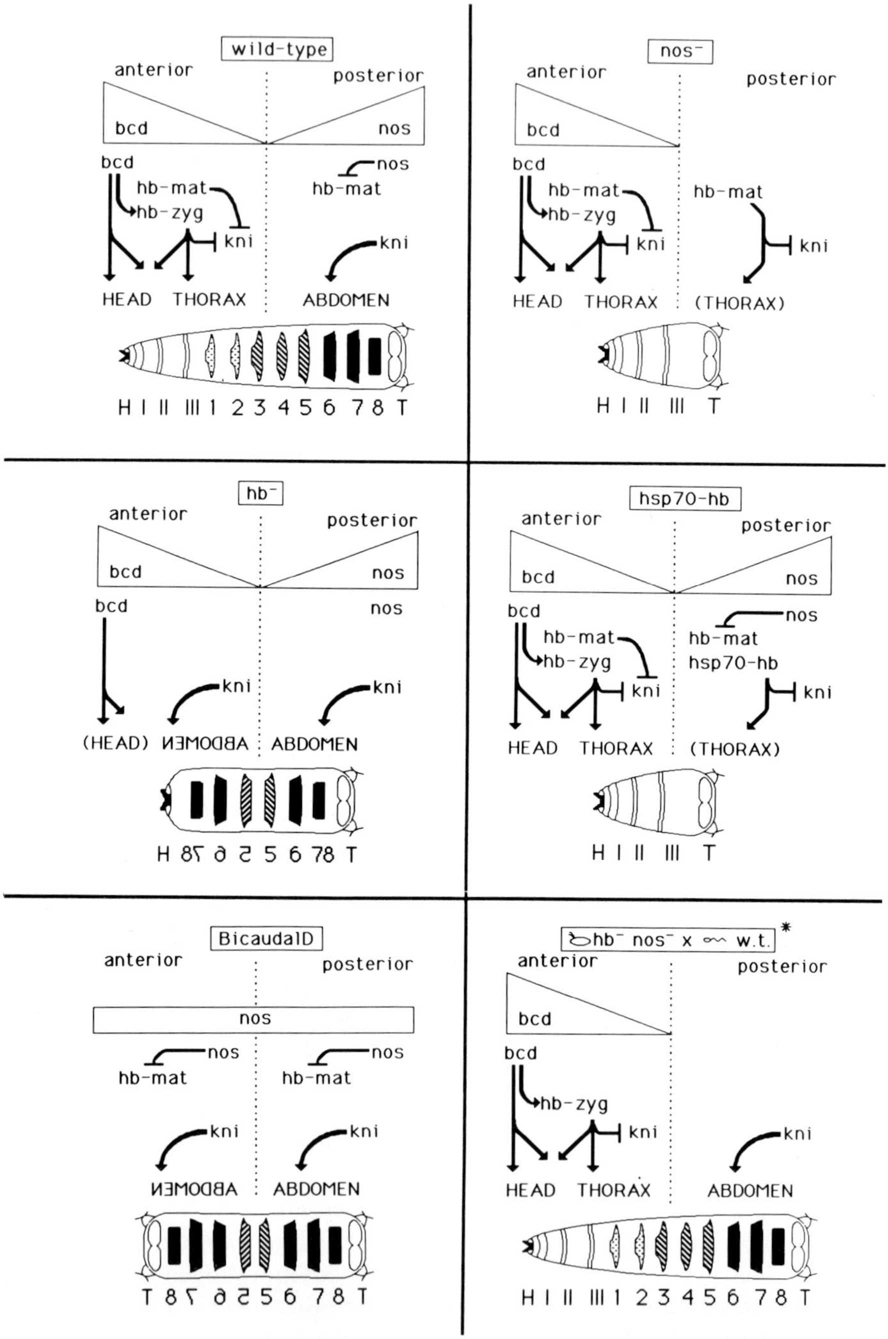
wild-type
anterior
posterior
bcd
nos
bcd
hb-mat
hb-zyg
kni
nos
hb-mat
kni
HEAD THORAX
ABDOMEN
H I II III 1 2 3 4 5 6 7 8 T
nos⁻
anterior
posterior
bcd
bcd
hb-mat
hb-zyg
kni
hb-mat
kni
HEAD THORAX
(THORAX)
H I II III T
hb⁻
anterior
posterior
bcd
nos
bcd
nos
kni
kni
(HEAD)
ABDOMEN
H 8 7 6 5 5 6 7 8 T
hsp70-hb
anterior
posterior
bcd
nos
bcd
hb-mat
hb-zyg
kni
nos
hb-mat
hsp70-hb
kni
HEAD THORAX
(THORAX)
H I II III T
BicaudalD
anterior
posterior
nos
nos
hb-mat
nos
hb-mat
kni
kni
ABDOMEN
T 8 7 6 5 5 6 7 8 T
hb⁻ nos⁻ x w.t. *
anterior
posterior
bcd
bcd
hb-zyg
kni
kni
HEAD THORAX
ABDOMEN
H I II III 1 2 3 4 5 6 7 8 T

FIG. 3. Experiments testing the roles of the *nos* and *hb* molecules in specifying posterior body pattern. wild-type) Anterior (left) and posterior (right) domains of the early embryo are considered separately. In the anterior half, *hb* protein is expressed from two independent sources: (i) hb^{mat} transcripts, present owing to the absence of *nos* activity, and (ii) hb^{zyg} transcripts activated under the control of the *bcd* gradient. Activity of the *hb* protein promotes head (H) and thoracic development (segments I, II and III), and represses expression (or activity) of *kni* protein which promotes abdominal segmentation. In the posterior half, no *hb* protein is synthesized because *nos* activity degrades hb^{mat} transcripts and the concentration of *bcd* protein falls beneath a critical threshold necessary to activate hb^{zyg} transcripts. As a consequence, the *kni* protein is active in generating the posterior pattern (abdominal segments 1–8; T = posterior terminalia). nos^-) Similar to wild-type except that in the absence of the *nos* morphogen, hb^{mat} transcripts are stable throughout the embryo giving rise to inappropriate *hb* protein in the posterior half. As a consequence, *kni* protein is not expressed (or not active), precluding abdominal development. Most of the posterior half of the embryo appears to behave like a large thoracic segment, presumably owing to the activity of *hb* protein in the absence of *bcd* protein. hb^-) No *hb* protein is generated from either hb^{mat} or hb^{zyg} transcripts. In the anterior half, graded activity of *bcd* protein specifies some aspects of head segmentation at the anterior end; however, absence of *hb* protein leads to inappropriate activity of *kni*, specifying abdominal segmentation. Abdominal segments also form in the posterior half owing to the normal activity of *kni* protein (abdominal segments 7 and 8 are fused in both halves, a consequence of the absence of hb^{zyg} expression during the late blastoderm stage). The failure to form full abdominal patterns in either half may be due to high uniform levels of *kni* activity throughout the middle of the embryo (falling levels of *kni* activity may be required to specify the formation of anterior abdominal segments, see Fig. 4). *hsp70:hb*) Similar to nos^- except that in this case the role of the *nos* morphogen in degrading hb^{mat} transcripts is circumvented by the ectopic expression of *hsp70:hb* transcripts induced by heat shock (*hsp70:hb* transcripts are presumably expressed in the anterior half as well, but have no ill effect because *hb* protein is normally active in this portion of the embryo). *Bicaudal D*) This mutation prevents the expression of *bcd* protein and also causes indiscriminate *nos* activity (*nos* activity is diagrammed as uniform throughout the embryo, though it may be more active towards the poles; Lehmann & Nüsslein-Volhard 1986). Under these conditions, neither *hb* nor *bcd* protein is expressed in either half of the embryo, leading to activity of *kni* protein and hence abdominal pattern in both halves. This phenotype differs from that of hb^- in that the absence of *bcd* protein prevents the formation of head structures at the anterior end. hb^- nos^- eggs fertilized by wild-type sperm) Predicted phenotype of removing maternally derived products of both the *hb* and *nos* genes: in the anterior half, expression of hb^{zyg} transcripts under the control of the *bcd* gradient should suffice together with graded *bcd* activity to specify head and thoracic development and to repress *kni* activity; in the posterior half, absence of *nos* activity should have no adverse effects because functional hb^{mat} transcripts are also absent.

Figure 3 on facing page

demonstrated that embryos which receive no functional hb^{mat} products from the mother develop normally, provided that they are fertilized by wild-type sperm and hence capable of activating hb^{zyg} expression anteriorly under the independent control of the *bcd* gradient. If, as inferred above, the *nos* morphogen acts only via its effect on the hb^{mat} mRNA, then absence of functional hb^{mat} mRNA should render the *nos* morphogen superfluous in these embryos. Nevertheless, normal posterior body patterning occurs.

If neither the *nos* nor the hb^{mat} gradients are essential for posterior patterning, what are their roles? Perhaps more importantly, what *are* the factors that dictate the posterior pattern?

The answer to the first question is relatively simple. If one considers the activities of the *bcd* and *nos* morphogens from the perspective of their effects on *hb* gene products, it becomes clear that *both* gradients function to limit early *hb* activity to a restricted anterior domain—*bcd* triggering hb^{zyg} expression anteriorly and *nos* repressing hb^{mat} expression posteriorly. Anteriorly, *hb* protein is involved in controlling head and thoracic segmentation. Conversely, its absence posteriorly is required for abdominal segmentation. Thus, within the context of wild-type development in which hb^{mat} transcripts are deposited throughout the egg, both gradients locate the boundary of *hb* protein expression and hence divide the embryo into distinct anterior (head + thorax) and posterior (abdominal) patterning systems.

This interpretation is supported by two related findings (see Fig. 3). First, if mutations which inactivate the *hb* gene are used to eliminate both hb^{mat} and hb^{zyg} functions from early embryos, the resulting larvae form a pattern of mirror-symmetrical abdominal segments (Lehmann & Nüsslein-Volhard 1987), even though the *nos* morphogen is presumably active in only the posterior half of the body. These embryos do not form completely mirror-symmetrical patterns, as rudimentary head structures controlled by the *bcd* morphogen appear at the anterior end. A similar conclusion is suggested by the phenotype of embryos derived from mothers carrying the *Bicaudal D* mutation. In this case both embryological and genetic experiments suggest that the *nos* morphogen is spread throughout the early embryo (Lehmann & Nüsslein-Volhard 1986) and molecular studies show that the *bcd* gradient does not form (Driever & Nüsslein-Volhard 1988b). Presumably as a consequence, the resulting embryos fail to express detectable hb^{mat} or hb^{zyg} protein early in embryogenesis (P. M. Macdonald & G. Struhl, unpublished data) and form symmetrical double abdomens. Thus, in both cases, failure to generate an anterior domain of *hb* activity leads to the inappropriate development of posterior body pattern.

These results indicate that differential *hb* expression across the anteroposterior axis generates distinct anterior and posterior patterning domains. As described above, the *bcd* gradient actively dictates pattern within the anterior domain both by turning on the hb^{zyg} gene and by triggering other distinct responses at different concentration thresholds. In contrast, the *nos* gradient specifies the posterior patterning domain by preventing hb^{mat} expression but does not appear to play any further role in the patterning process. Thus, we are faced with the second question—what factors do dictate posterior pattern?

A clue to the answer comes from experiments which alter the normal patterns of expression of the gap genes *Kr* and *kni* in otherwise wild-type embryos. If *Kr* activity is abolished by mutation, or generated ectopically in experiments I have recently performed using a *hsp70:Kr* hybrid gene, the pattern of the abdominal segments is altered in a manner suggesting that the particular distribution of *Kr* protein exerts an instructive influence on abdominal pattern (see below and Fig. 4).

The distribution of *Kr* protein cannot be the sole determinant of posterior pattern, since considerable abdominal segmentation is still observed in Kr^- and heat-shocked *hsp70:Kr* embryos. Other determinants are likely to be the neighbouring distributions of the *kni* and *tll* gap gene proteins. The *kni* gene is of particular interest, primarily because kni^- embryos fail to generate most of the abdominal segments (Fig. 4). Indeed, the *nos* morphogen is generally thought to control posterior pattern by activating *kni* in a precisely defined posterior domain (Nüsslein-Volhard, this volume). However, as described above and in Fig. 3, it probably does so indirectly by abolishing inappropriate posterior expression of hb^{mat} protein, absence of *hb* activity leading by default to the activation of *kni*.

Knowing that the relative distributions of *Kr* and *kni* gap gene proteins play an important role in posterior patterning does not in itself explain how they do so. However, if one considers what is known about the distribution of these gap gene products together with the particular segmentation phenotypes caused by inactivating or inappropriately expressing the *Kr* protein, a relatively simple hypothesis emerges.

As diagrammed in Fig. 4, both the *hb* ($hb^{zyg} + hb^{mat}$) and *Kr* proteins are expressed as approximate step functions with soft (graded) edges. Based on both the mutant phenotype caused by *kni* mutations and preliminary molecular evidence of Nauber & Jackle (as cited in Ingham 1988), the *kni* protein is likely to be expressed in a neighbouring posterior domain (see Fig. 4); further, it is likely to form graded distributions at either edge, as is the case for the *hb* and *Kr* patterns of expression. The basic premise of the model, as diagrammed in Fig. 4, is that the local concentrations of the *Kr* and *kni* proteins dictate particular portions of abdominal pattern, presumably by determining where subordinate cueing molecules, such as pair-rule and homeotic genes, are activated. Specifically, the overlap of rising *kni* and falling *Kr* activity creates a series of signals, ultimately specifying abdominal segments 1–4(5); conversely, falling *kni* activity in the absence of *Kr* specifies abdominal segments (5), 6 and 7. Hence in Kr^- embryos, the broad bell-shaped distribution of *kni* activity would specify mirror-image posterior abdominal segments (anterior situated segments having reversed polarity relative to the body as a whole), as is observed. Similarly, in heat-shocked *hsp70:Kr* embryos the bell-shaped distribution of *kni* activity should once again generate mirror-symmetrical abdominal patterns because *Kr* expression is the same on both sides of the bell curve. However, in this case the resulting pattern would be composed of anterior abdominal

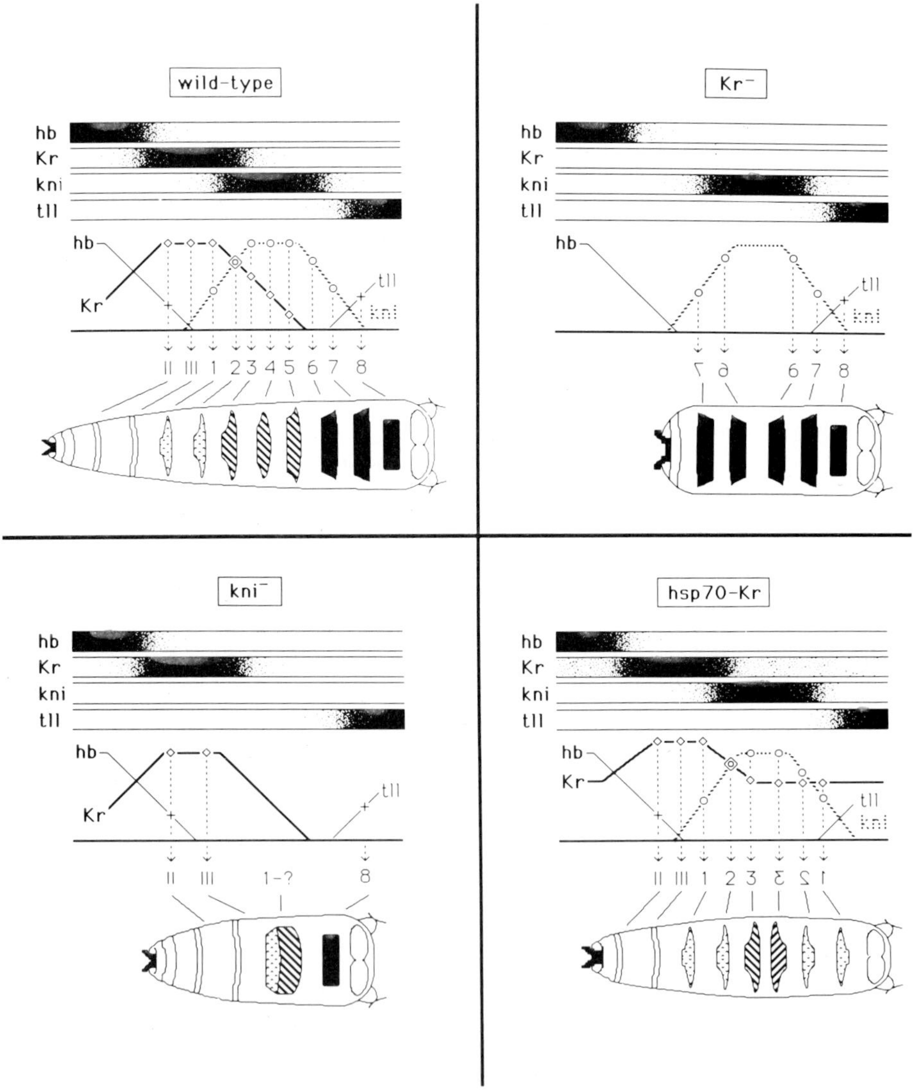

segments (posteriorly situated segments having reversed polarity), as is observed.

If this particular explanation is correct, it would suggest that posterior body patterning is controlled by a cascade of gradients, each specifying only one or a few thresholds, and none acting alone to define a coherent set of coordinates controlling the global pattern (see Fig. 4). Considering for the moment only the patterning of the middle six abdominal segments, a gradient of the *nos* morphogen generates a secondary gradient of hb^{mat} protein by repressing otherwise uniform expression from maternally derived *hb* mRNA. The gradient of hb^{mat} protein in turn sets a threshold defining the anterior limit of *kni* expression, and hence, distinguishes between anterior (*hb*-associated) and posterior (*kni*-associated) patterning systems. Similarly, polarized activity of the primary morphogen *tor* (Fig. 1) is likely to define the limit of terminal *tll* expression (Klingler et al 1988); further, the phenotype of tll^- embryos, in which abdominal patterning spreads to the posterior end of the embryo (Strecker et al 1986), suggests that the boundary of *tll* expression dictates the posterior limit of *kni* activity. Thus, the differential activities of the *nos*, *hb* and *tor* morphogens collectively define the boundaries of *kni* gene expression. Finally,

FIG. 4. Experiments suggesting that graded activities of the *Kr* and *kni* proteins dictate posterior pattern. wild-type) The patterns of active *hb*, *Kr*, *kni* and *tll* proteins (the last two presumed either from genetic or preliminary molecular evidence; see text) are diagrammed as in Fig. 1. Underneath, the concentrations of these proteins are plotted logarithmically as a function of position along the anteroposterior axis. Specific concentrations of the *Kr* and *kni* proteins are proposed to specify the polarity, boundaries and types of abdominal segments by triggering the local activation of pair-rule and homeotic genes. Rising *kni* activity in the presence of falling *Kr* activity dictates abdominal segments 1–4(5); conversely, falling *kni* activity in the absence of *Kr* activity dictates segments (5)6 and 7 (rising *tll* activity seems to specify segment 8). Kr^-) The expected patterns of active *hb*, *Kr*, *kni* and *tll* proteins are shown above the resulting segment pattern. In the absence of *Kr* protein, a bell-shaped distribution of *kni* activity would be expected to specify mirror-symmetrical posterior abdominal segments, as is observed. Although loss or ectopic expression of one of these gap proteins might well affect the local pattern of expression of the other (e.g. Jackle et al 1986), the mutant phenotypes can be accounted for without considering this possibility, as diagrammed. kni^-) In the absence of *kni* activity, most of the abdominal segments cannot be specified, though a domain of abdominal tissue of uncertain type appears posterior to a normal thorax. *hsp70:Kr*) Activation of the *hsp70:Kr* gene by heat shock causes indiscriminate but relatively low levels of ectopic *Kr* protein (G. Struhl, unpublished data). Under these circumstances, the bell-shaped distribution of *kni* activity would form in the presence of low *Kr* activity. The changing local concentrations of *kni* would now be expected to dictate mirror-symmetrical patterns of anterior abdominal segments, as is observed.

the resulting bell-shaped distribution of *kni* protein, in conjunction with short-range gradients of *Kr* and *tll* protein on either side, determines where subordinate pair-rule and homeotic genes are activated. This model can be readily extended to account for the patterning of the adjacent thoracic and terminal segments, if one assumes that the graded distributions of *hb*, *Kr* and *tll* have a similar instructive value in these regions.

Thus the anterior and posterior body patterns are dictated by different strategies, even though both depend on basic gradient phenomena. In essence, the distribution of *bcd* morphogen exerts a dominant, instructive influence on segment size, number, polarity and type, even though the final pattern is mediated by a subordinate cascade of signalling molecules. In contrast, the *nos* morphogen does not; this morphogen serves only to locate the domain within which a separate patterning system operates. Moreover, there is no single coordinating gradient operating within this domain. Instead, the critical morphogens are zygotic functions which arise in response to several different gradients, and govern only subdomains of the pattern.

An evolutionary basis for the different strategies controlling anterior and posterior patterning

If one were designing the systems for controlling the insect body pattern from scratch, it would seem logical to use symmetrical coordinate systems similar to the *bcd* morphogen system in the anterior and posterior halves of the embryo (Sander 1976, Nüsslein-Volhard et al 1987). Yet, this eminently sensible strategy is not used. Moreover, the mechanism which is used seems decidedly perverse. For example, the posterior determinant system involves a relatively complex process to localize a morphogen at the back end of the egg which then diffuses or is transported forward after fertilization just so that it can locally degrade a maternal transcript (hb^{mat}) which appears to serve no positive function in either the egg or early embryo. Indeed, the maternal *hb* transcript is, if anything, detrimental to the early embryo, as failure to clear it from the posterior half of the body disrupts posterior patterning.

A possible explanation for the seemingly arbitrary nature of the signalling strategies used is that they reflect chance mechanisms for creating morphological change during evolution. For example, the initial body pattern generated during the early development of most insects, especially more 'primitive' species, consists of a series of anterior body segments (head and perhaps thoracic segments) followed by a posterior terminus. The remaining thoracic and abdominal segments then arise in succession from a proliferating budding zone positioned posterior to the head/thorax. Clearly, the mechanisms of anterior and posterior body patterning are different in these species. Indeed, one might guess that these animals require only a *bcd*-like anterior gradient system for organizing their initial body plan, the abdomen developing without direct reference to any specific posterior morphogens present prior to gastrulation. Similarly, such embryos might express maternally derived *hb*-like regulatory products which are required for anterior head and thoracic development throughout the blastoderm, differentially degrading or excluding them subsequently from cells constituting the posterior budding zone.

The principal difference between such short germ band insects and dipterans, e.g. *Drosophila*, is that instead of simply laying down the head pattern at the blastoderm stage, the latter dispense with the budding zone and specify the entire pattern at once, a tactic which allows them to develop far more quickly. To do this, they may have to utilize a mechanism that prevents posterior expression of *hb*-like regulatory products which are associated with head and thoracic modes of development. Thus, the posterior determinant system might reflect a relatively recent strategy for differentiating the initial body plan into distinct head + thorax and abdominal domains. However, as in more primitive insects, the strategy for patterning the posterior domain remains distinct.

The specific argument that the evolution of the posterior determinant system is a special adaptation to allow for rapid development suggests a general argument about the use of different signalling strategies. If one considers the molecular mechanisms known to be involved in generating spatial information in early *Drosophila* embryos, it becomes clear that there are almost as many specific mechanisms as there are signals, e.g. mRNA localization, differential translation, differential mRNA degradation and a variety of transcriptional mechanisms (see Nüsslein-Volhard, this volume, and above). It appears as if essentially any mechanism capable of creating a molecular differential can be utilized. Hence, there seems to be no unifying logic to either the mechanisms or the strategies which are used in generating the spatial cues organizing the insect body pattern. Instead, the particular means employed appears to reflect the opportunistic nature of evolution.

Final considerations

Molecular gradients are a fact of life for early *Drosophila* embryos. Many of these gradients are clearly involved in organizing body pattern, hence, their constituent molecules can be regarded as morphogens. However, these gradients can be used in very different ways to control pattern. In essence, we can distinguish two kinds of gradients. The available evidence suggests that gradients like those of the *bcd*, *Kr* and *kni* morphogens exert dominant and instructive influences, dictating pattern in response to distinct concentration thresholds. Alternatively, the *nos*, *hb* and *cad* gradients appear to play permissive roles, defining perhaps only single thresholds that create domains within which other patterning mechanisms can operate. In gradients of this second type, graded expression *per se* is important only in that it sets the critical threshold in the correct place. Yet, altering these gradients, and hence the position of the threshold, can profoundly change the pattern.

In considering the relative significance of these different kinds of gradients, it is important to remember the context in which they are used. One of the most striking aspects of early *Drosophila* embryogenesis is that no single cueing molecule is sufficient to control all aspects of a pattern. Nor does any single gradient appear to confer more than a few distinct responses. Instead, spatial information arises in an aggregate fashion created by the overlaps and cascades of a variety of cueing molecules.

Acknowledgements

I would like to thank all of my colleagues who tolerated and helped me during the course of preparing this manuscript. I also thank Peter Lawrence and Tom Jessell for their firm and much needed criticisms of early versions of the manuscript. I apologize for relying heavily on citing reviews rather than the primary literature, owing to a limit on the number of references. Finally, I thank the Howard Hughes Medical Institute, the McKnight Foundation and the Alfred P Sloan Foundation for financial support.

References

Akam M 1987 The molecular basis for metameric pattern in the Drosophila embryo. Development 101:1–22

Berleth T, Burri M, Thoma G et al 1988 The role of localization of *bicoid* RNA in organizing the anterior pattern of the *Drosophila* embryo. EMBO (Eur Mol Biol Organ) J 7:1749–1756

Busturia A, Casanova J, Sanchez-Herrero E, Morata G 1989 Structure and function of the bithorax complex genes of *Drosophila*. In: Cellular basis of morphogenesis. Wiley, Chichester (Ciba Found Symp 144) p 227–242

Carroll SB, Laughon A, Thalley BS 1988 Expression, function, and regulation of the *hairy* segmentation protein in the *Drosophila* embryo. Genes and Development 2:883–890

DiNardo S, Sher E, Heemskerk-Jongens H, Kassis JA, O'Farrell PH 1988 Two-tiered regulation of spatially patterned *engrailed* gene expression during *Drosophila* embryogenesis. Nature (Lond) 332:604–609

Driever W, Nüsslein-Volhard C 1988a A gradient of *bicoid* protein in *Drosophila* embryos. Cell 54:83–93

Driever W, Nüsslein-Volhard C 1988b The *bicoid* protein determines position in the *Drosophila* embryo in a concentration-dependent manner. Cell 54:95–104

Frasch M, Levine M 1987 Complementary patterns of *even-skipped* and *fushi tarazu* expression involve their differential regulation by a common set of segmentation genes in *Drosophila*. Genes and Development 1:981–995

Frohnhöfer HG, Nüsslein-Volhard C 1986 Organization of anterior pattern in the *Drosophila* embryo by the maternal gene *bicoid*. Nature (Lond) 324:120–125

Frohnhöfer HG, Nüsslein-Volhard C 1987 Maternal genes required for the anterior localization of *bicoid* activity in the embryo of *Drosophila*. Genes and Development 1:880–890

Gaul U, Jackle H 1987 Pole region-dependent repression of the *Drosophila* gap gene *Kruppel* by maternal gene products. Cell 51:549–555

Horder TJ, Witkowski JA, Wylie CC 1983 A history of embryology. Cambridge University Press, London

Huxley JS, de Beer GR 1934 The elements of experimental embryology. Cambridge University Press, London

Ingham PW 1988 The molecular genetics of embryonic pattern formation in *Drosophila*. Nature (Lond) 335:25–34

Jackle H, Tautz D, Schuh R, Seifert E, Lehmann R 1986 Cross regulatory interactions among the gap genes of *Drosophila*. Nature (Lond) 324:668–670

Klingler M, Erdelyi M, Szabad, J, Nüsslein-Volhard C 1988 Function of *torso* in determining the terminal analgen of the *Drosophila* embryo. Nature (Lond) 335:275–277

Lawrence PA 1989 Cell lineage and cell states in the *Drosophila* embryo. In: Cellular basis of morphogenesis. Wiley, Chichester (Ciba Found Symp 144) p 131–149

Lehmann R, Nüsslein-Volhard C 1986 Abdominal segmentation, pole cell formation, and embryonic polarity require the localized activity of *oskar*, a maternal gene in *Drosophila*. Cell 47:141–152

Lehmann R, Nüsslein-Volhard C 1987 *hunchback*, a gene required for segmentation of an anterior and posterior region of the *Drosophila* embryo. Dev Biol 119:402–417

Lewis EB 1978 A gene complex controlling segmentation in *Drosophila*. Nature (Lond) 276:565–570

Macdonald PM, Struhl G 1986 A molecular gradient in early *Drosophila* embryos and its role in specifying the body pattern. Nature (Lond) 324:537–545

Martinez-Arias A, Baker NE, Ingham PW 1988 Role of segment polarity genes in the definition and maintenance of cell states in the *Drosophila* embryo. Development 103:157–170

Mlodzik M, Gehring W 1987 Expression of the *caudal* gene in the germ line of Drosophila: formation of an RNA and protein gradient during early embryogenesis. Cell 48:465–478

Morata G, Lawrence PA 1975 Control of compartment development by the *engrailed* gene in *Drosophila*. Nature (Lond) 255:614–617

Nüsslein-Volhard C, Roth S 1989 Axis determination in insect embryos. In: Cellular basis of morphogenesis. Wiley, Chichester (Ciba Found Symp 144) p 37–64

Nüsslein-Volhard C, Wieschaus E 1980 Mutations affecting segment number and polarity in *Drosophila*. Nature (Lond) 287:795–801

Nüsslein-Volhard C, Frohnhofer HG, Lehmann R 1987 Determination of anteroposterior polarity in *Drosophila*. Science (Wash DC) 238:1675–1681

Sander K 1959 Analyse des ooplasmatischen Reaktionssystems von *Euscelis plebejus* Fall. (Cicadina) durch Isolieren und Kombinieren von Keimteilen. I. Mitt.: Die Differenzierungsleistungen vorderer und hinterer Eiteile. Wilhelm Roux' Arch Entwicklungsmech Org 151:430–497

Sander K 1960 Analyse des ooplasmatischen Reaktionssystems von *Euscelis plebejus* Fall. (Cicadina) durch Isolieren und Kombinieren von Keimteilen. II. Mitt.: Die Differenzierungsleistungen nach Verlagern von Hinterpolmaterial. Wilhelm Roux' Arch Entwicklungsmech Org 151:660–707

Sander K 1976 Specification of the basic body pattern in insect embryogenesis. Adv Insect Physiol 12:125–238

Sander K, Lehmann R 1988 *Drosophila* nurse cells produce a posterior signal required for embryonic segmentation and polarity. Nature (Lond) 335:68–70

Scott MP, Carroll SB 1987 The segmentation and homeotic gene network in early *Drosophila* development. Cell 51:689–698

Strecker TR, Kongsuwan K, Gengyel JA, Merriam JR 1986 The zygotic mutant *tailless* affects the anterior and posterior ectodermal regions of the *Drosophila* embryo. Dev Biol 113:64–76

Tautz D 1988 Regulation of the *Drosophila* segmentation gene *hunchback* by two maternal morphogenetic centers. Nature (Lond) 332:281–284

Tautz D, Lehmann R, Schnurch H et al 1987 Finger protein of novel structure encoded by *hunchback*, a second member of the gap class of *Drosophila* segmentation genes. Nature (Lond) 327:383–389

Turing A 1952 The chemical basis of morphogenesis. Philos Trans R Soc Lond Biol Sci 237:37–72

Wilson EB 1925 The cell in development and inheritance. Macmillan, New York

DISCUSSION

Wieschaus: It was suggested that the maternal *hb* product that is made during oogenesis has no apparent function, but it obviously does have some effect on the embryo, at least in that in *hb* mutants there is a different phenotype depending on whether or not those maternal transcripts are present.

Struhl: If you have just maternal *hb* product, the phenotype is a weak version of that which results from a complete absence of both maternal and zygotic *hb* product (Lehmann & Nüsslein-Volhard 1987).

When the embryo completely lacks *hb* product you would expect *kni* to be active everywhere, leading to the entire animal developing in an abdominal fashion. In fact that is not quite what you see, because *bcd* is still active at the anterior pole, activating some head-determining genes in a way that doesn't depend on *hb*. So there are head structures in the anterior end. If maternal but not zygotic *hb* product is present, there is essentially less abdominal development in the anterior half, as if you are partially suppressing the activity of *kni* there. Nevertheless, some inappropriate *kni* activity remains, leading to an expanded domain of abdominal pattern at the expense of thoracic pattern.

Wieschaus: If the only role of *nos* is to remove the posterior activity of the maternal *hb* product, then you would predict that *nos* would have no effect on the phenotype of embryos derived from a *hb* germline clone.

Struhl: That's the critical test. The clear prediction is that *hb* product is the substrate—if you remove the substrate, then it doesn't matter what you do with *nos* product itself.

Meinhardt: I thought that *Kr* was some sort of ground state, that *nos* activates *kni*, and *bcd* activates *hb*. If *bcd* and *nos* are removed, you get an extension of *Kr* expression (as shown by Gaul & Jäckle 1987), so why do you think that *kni* is the default state and not *Kr*?

Struhl: I think that *kni* is the default state when *hb* is missing because of my observations and the results of Ruth Lehmann and others who have looked at the *hb* phenotype and the phenotypes of these various mutant combinations (e.g. Fig. 3). Namely, whenever *hb* is present you get head and thorax and whenever you take *hb* away you get abdomen. In terms of what is happening with the *Kr* product, it may be that the *nos* gradient also affects *Kr* expression and that is a second response to the posterior determinant system. That's the reason I said the view that the posterior system only controls *hb* is rather extreme. This is typical of the current state of the field. We have discovered an enormous amount but many missing pieces remain: there is a lot of ground for future work.

Melton: You said *bcd* controls *hb* by transcriptional activation and it controls *cad* by turning off translation. This implies that *bcd* protein has two quite different molecular activities. Why do you think that the protein has those two activities, as opposed to one or the other being mediated by an indirect effect, for example failure of *bcd* to turn on *hb*, leading to the translational control of *cad*?

Struhl: This work was done by Paul Macdonald and myself (Macdonald & Struhl 1986, see also Mlodzik & Gehring 1987). The *cad* protein gradient arises at a time when other genes are not being transcribed; it is formed even in unfertilized eggs. When a fly lays an egg which isn't fertilized the nucleus remains stagnant and no genes are transcribed, but there is formation of a *cad* gradient. So the *cad* gradient cannot be a transcriptional response.

So how do we know that *hb* gene activation is not an indirect response to the *cad* gradient? Janni (Nüsslein-Volhard) has presented evidence that *bcd* acts as a transcription factor with the capability to bind *hb* DNA. Paul Macdonald, Kevin Struhl and I have been doing experiments where we have taken parts of the *hb* promoter region, put them in front of naïve TATA boxes, put them back into *Drosophila* and shown that they can activate in response to *bcd*. We have even taken the same target sequences, put them into yeast, and shown that a small region of about 125 bp, including one of the binding sites that Janni has identified, mediates transcriptional activation of an otherwise naïve yeast promoter in response to *bcd* protein. Although that's not compelling evidence that *bcd* acts as a transcriptional activator in early *Drosophila* embryos, we believe it provides a very strong argument that the molecule has the ability to be a transcriptional activator in addition to being a translational regulator.

Melton: To extend that further—if you accept that the homeobox contains the transcriptional activator function, would you predict that you could deletion map part of the *bcd* protein and find a structural domain that can repress translation of *cad* mRNA?

Struhl: That is possible in principle. Janni was kind enough to send us a number of the *bcd* mutants; three of them turned out to be either mis-sense mutants or small in-frame deletions within the homeobox. All three of those mutants knock out both the translational regulation and the transcriptional activation. So there is a possibility that both the translational regulation and the transcriptional activation are mediated by the same structural domain, which is the homeobox.

Nüsslein-Volhard: We have determined a consensus sequence for *bcd* protein and the *hb* promoter. The consensus sequence occurs in the *cad* RNA but we don't know whether the *bcd* protein really binds to *cad* RNA.

Struhl: Marik Mlodzik and Walter Gehring (1987) showed that *cad* RNA is uniformly distributed at the time the protein gradient forms. So selective degradation of the *cad* RNA is not the mechanism for formation of the gradient. That has to be distinguished from the apparent effect of *nos* on *hb*, which does seem to be achieved via degradation.

Kornberg: The question of universality has been raised a number of times already. You asked what lessons people working in vertebrate development would draw from this kind of hierarchy. I would turn the question around and ask you what lessons can we infer from these patterns in *Drosophila* embryos to later stages during *Drosophila* development?

Struhl: Both systems, vertebrate development and later *Drosophila* development (e.g. imaginal discs, the eye), are cellular systems. The way in which information is generated and elaborated in early *Drosophila* embryos is unique, because there are no cell boundaries. That creates a much easier system to deal with.

For me, the ramification of what we have learned about early *Drosophila* embryogenesis is that perhaps the same principles which apply to interpreting morphogenetic gradients in this simple system will also apply to gradient phenomena in more complicated cellular systems. For example, gradients might dictate complex patterns in a direct instructive fashion by providing many different thresholds triggering distinct responses, as is the case for the control of anterior body pattern by the *bcd* gradient. Or they may provide only one or a few thresholds which are read by a reductive cascade of subordinate signalling steps, as is the case for posterior body patterning under the control of the *nos*, *tor* and *hb* gradients. For example, a few initial gradients could create one or a few thresholds which trigger distinct responses as a function of position. These distinct responses could lead to the creation of new, shorter-range gradients which in turn trigger more local responses. This process could be repeated until sufficient spatial information is generated to organize pattern within the system. Taken to its most extreme limit, such an interpretative cascade might lead to specification of the final cellular pattern by interactions involving direct contacts and all-or-none signalling between adjacent cells, as occurs in the developing *Drosophila* retina.

I believe that gradients may play the same fundamental role in all of these systems, namely to provide a system of spatial information in which the changing concentration of one or more signalling molecules triggers discrete responses as a function of position. However, this information may be used in diverse ways. Hence, the particular number of thresholds provided by a given gradient system is not of general significance—what is generally significant is the phenomenon of generating a discrete response to a continuously changing signal. Everything can be built on that, you don't need many different thresholds to make a morphogen gradient.

Nüsslein-Volhard: That is a matter of definition. Gradient mechanisms have been defined by various people. You can't just say that one is enough.

Struhl: Earlier in the meeting (Introduction) a gradient was defined as a distribution of a substance which has (in the words of E B Wilson) a pro-morphological consequence, that is, its distribution is related to the pattern.

To me, the important point about gradients is that they generate discrete responses in a concentration-dependent fashion. I find it a minor point that one system could produce several different responses, because even a single threshold can indirectly produce several responses. So it's only a question of whether the gradient is read directly or indirectly.

Nüsslein-Volhard: It may not be important for you, but it is important for the mechanism. We have to be precise about what terms we use. We agree that there are differences in the systems: the posterior system probably just defines one particular thing and the anterior system probably defines more than one. There is a consensus that gradient mechanisms have to have at least two different qualities.

Struhl: I would challenge that definition.

Nüsslein-Volhard: Polarity has always been explained by gradients. In your case of one value per threshold, you cannot explain polarity within the pattern defined by your gradient.

Struhl: If you have more than one gradient, as in *Drosophila* embryos, then their overlapping signals can create the initial cues responsible for generating polarity. Further, as I showed in the *Kr* experiments (Fig. 4), the interpretative system of gap gene activities may be what generates polarity in response to the initial graded cues.

Nüsslein-Volhard: But depending on the interaction with a neighbouring pattern, you can get opposite polarities with the same gradient in the same orientation.

Cooke: A simple terminological confusion seems to be arising. There was an era when we thought that the most important feature was the threshold, which is a dividing line between one sort of response and another. One threshold led to two of what Janni (Nüsslein-Volhard) calls qualities, or a quality is bounded by two different thresholds. But people like Hans Meinhardt now propose that positional gradients control not thresholds between gene activities but the optimum for activation of different genes, which then compete.

Nüsslein-Volhard: A threshold alone doesn't give polarity.

Struhl: But if you have more than one gradient, which is clearly happening in the *Drosophila* embryo where you have several overlapping gradients, then even single thresholds can give polarity.

Nüsslein-Volhard: Let's take the concrete example of the posterior system. You say you have such a gradient and depending on the situation in the embryo, whether you have *Kr* or *kni* or *hb* or *tll*, you know that the gradient is from one end of the embryo to the other, and the resulting pattern can be in either orientation—so in this case the gradient doesn't determine the polarity.

Struhl: Because you have removed the interpretative response.

Nüsslein-Volhard: But in the posterior system you do not have a direct concentration-dependent realization, because the polarity is . . .

Struhl: It may depend on the concentration of *Kr* or *kni*.

Nüsslein-Volhard: It depends on the surrounding conditions, it doesn't depend on the concentration of your gradient.

Meinhardt: Can we consider this another way? There are several possible ways of subdividing a field into regions, let us say, A, B and C. One is the classical interpretation of positional information. Another is that the gradient is only used to distinguish the C region and the ground state is the expression of A. This leads to an A/C confrontation which causes the intercalation of B. In this case B would be not under direct control of the gradient.

Let me make another proposal for the special case of *Drosophila*. We are accustomed to the concept that there is a gradient which is interpreted at different levels. We should consider another mechanism of gradient interpretation. Imagine that there are several gradients with different steepnesses produced by a source at a particular position. For instance, a very steep gradient activates gene C and a shallower gradient activates gene B. In addition, a position-independent activation of gene A may occur. The activities of the genes A, B and C are locally exclusive and the dominant activation determines which gene is active at which position. The steep gradient dominates only in the region close to the source. The shallow gradient has some dominating influence a bit further away from the source, where the influence of the steep gradient wanes. At points distant from the source, all graded concentrations are low and position-independent signals predominate. I think that's what happens in *Drosophila*. The *Kr* gene does not respond to the gradients, its expression is some sort of default state. There is a presumably shallower gradient, *nos*, which activates the *kni* gene (and switches off *Kr*). A steep gradient, presumably *tor* or *torsolike* activates the most terminal genes. So we can use different gradients with different steepnesses to activate genes in particular regions. I think we should not focus exclusively on the idea that a single gradient activates particular genes.

Struhl: The classical definition of gradients was based on the phenomena that were being observed at the time. Namely, that there seem to be many different outcomes and they seem to obey a single graded property. Now that we have a better molecular description of the interpretative events that are going on, it is possible that the classical description is misleading. The fundamental phenomenon could be discrete responses to different concentration thresholds, even in the most limiting extreme case.

Wolpert: You can't divide something into less than two!

References

Gaul U, Jäckle H 1987 Pole region dependent repression of the Drosophila gene Kruppel by maternal gene products. Cell 51:549–555

Lehmann R, Nüsslein-Volhard C 1987 hunchback, a gene required for segmentation of an anterior and posterior region of the *Drosophila* embryo. Dev Biol 119:402–417

Macdonald P, Struhl G 1986 A molecular gradient in early *Drosophila* embryos and its role in specifying the body pattern. Nature (Lond) 324:537–545

Mlodzik M, Gehring W 1987 Expression of the *caudal* gene in the germ line of Drosophila: formation of an RNA and protein gradient during early embryogenesis. Cell 48:465–478

General discussion I

Lawrence: The genes which come on early, gap genes and even earlier ones than those, don't have sharp boundaries to their zones of expression. From that we argue that none of them defines and allocates particular cells to a developmental pathway. The first genes to do this are probably *fushi tarazu* (*ftz*) and *even-skipped* (*eve*). Gary (Struhl) also made the point that they couldn't have sharp boundaries because there is no way of confining the distribution of those gene products to individual cells before cells are formed. Therefore those genes cannot simply allocate individual cells to developmental pathways but must in some way be defining zones that are thereafter used to allocate cells.

Wolpert: There is a distinction between can't and doesn't. What is the argument for can't?

Struhl: The argument is very simple. If there are no cell boundaries, the potential for a mRNA or a protein to diffuse from its site of synthesis seems to be quite high. This isn't a compelling argument, because some transcripts or proteins might be selectively trapped as soon as they are synthesized and hence unable to move.

Garcia-Bellido: But the problem is to know the responding mechanisms. Although you are looking at gradients of gene products, you still cannot tell where the concentration is effective. The issue is, when the gradient is operative. Because it may operate, not as a threshold but to reinforce the interactions between other genes downstream. This means it works by ordering patterns of expression of different widths in register. So the border for *hunchback* expression may have to be in register with that for *ftz* and eventually with that for the segment polarity genes. If you don't see limits at the level of individual cells, it may be that the operative limits come later by interactions between cells.

Nüsslein-Volhard: I wanted to comment on the difficulty of translating a smoothly graded signal to a more or less sharply defined pattern. Even to achieve the sharpness we see in gene expression, a stripe three cells wide or something, requires the assumption of cooperativity in making this boundary. The degree to which this is cooperative may make the transition from the graded signal to the sharp boundary more or less steep. Gary, you said that the distribution of *Krüppel* is bell shaped, but there was a plateau in between, wasn't there?

Struhl: My main point was not that it was a perfect bell-shaped distribution, the gradation covers eight or more cells.

Nüsslein-Volhard: You can account for that without assuming that the protein diffuses, just by saying that transcription is not turned on or off in an all-or-none fashion.

Cooke: If you have competition, there is a possibility of cooperativity in the action of one gene product, for instance, because it is an allosteric protein with multiple binding sites. But if there is competition between different things which have fairly steep bell-shaped distributions, even at that stage the boundary for subsequent gene activation could be at the level of a single cell.

Lawrence: We are talking about the products of the genes identified by Janni Nüsslein-Volhard and Eric Wieschaus (1980). We are looking at particular examples, so I suggest that if cells are allocated by a particular gene, then the pattern of expression of that gene will show a sharp boundary.

Garcia-Bellido: But you will only notice that gene during the time that it is expressed. My point is that the bell-shaped distributions or gradients are just devices to generate diversity that is only read or interpreted as a threshold by interactions between other kinds of genes, which are downstream in the developmental process.

Lawrence: But haven't those downstream genes been found? Aren't they the gap genes?

I used to suspect that the border of *Krüppel* expression would coincide exactly with the border of *ftz* or *eve* expression. Ulrike Gaul, Paul Johnston and I observed that *Krüppel* overlies several *ftz* stripes, particularly parasegment 6, but does not have sharp boundaries (Lawrence 1988). It is difficult to see how *Krüppel* itself could indicate which cells occupy particular states. There may be other genes downstream of *Krüppel* which do.

Garcia-Bellido: But that would be the same as saying that from the *bicoid* pattern you can never tell where the signal is going to be interpreted by *hunchback*.

Gardner: It seems to me that so long as you are assaying the distribution in concentration of these proteins by antibody staining rather than by assaying their biological activity, it is impossible to decide which features of their spatial patterns are significant and which are not.

Struhl: Relevant to that is when you do the heat shock *Krüppel* experiments and stain with the anti-*Krüppel* antibody, you can barely detect the indiscriminate expression. So at a level which is on the borderline of what you can detect by immunohistochemistry, there are major effects on the polarity of the system.

Lawrence: You are saying the boundary is even more graded than it appears.

Wolpert: Because it is acting at an even lower level.

Weisblat: It could be the opposite, that everywhere you can see the protein it is exerting its maximal effects so the biological effect is not graded at all.

Garcia-Bellido: The question is whether this kind of gradient is operative and whether all the differentials of the gradient are effective thresholds. We cannot answer it but it is an important question.

Wolpert: There is a case to be made for vertebrates. If you take feather follicles in birds, there is good reason to believe that the spacing of those follicles involves a different mechanism than the specification of the character of the feathers. To what extent do the people working on *Drosophila* think that the mechanism for specifying the boundaries between *Krüppel* and the cardinal genes is of the same class as that which is specifying those segmental stripes?

Struhl: It's basically an open question.

Wolpert: That's what I wanted to know. Concerning polarity, what do you actually mean by polarity?

Struhl: I think there is both polarity of the cells, as exhibited, for example, by the anteroposterior orientation of the hairs they secrete, and polarity, in terms of the sequence of segments. Both reflect some inherent polarity of the system which can be influenced by various changes in the distributions of these products.

Wolpert: When do you think that individual cells acquire polarity? They clearly have it when they make cuticle. Do you think they acquire it at the beginning from the *bicoid* gradient or gradually?

Struhl: Gradually.

Nüsslein-Volhard: I agree.

Martinez-Arias: In *Krüppel*$^-$ embryos that I have looked at, I have never seen a complete reversal of polarity except in that particular segment where there is a duplication. I have never looked at heat shock-*Krüppel* embryos, so I cannot say anything about them.

Wieschaus: The largest polarity reversal that I have seen in an animal homozygous for a deletion of *Krüppel* is bigger than one segment—the maximum is probably about two segments.

Martinez-Arias: Can you tell us more about the terminal zones and the global polarity of the egg?

Nüsslein-Volhard: The idea is as follows (Nüsslein-Volhard et al 1987). We start with a *bicoid*$^-$ *oskar*$^-$ or *bicoid*$^-$ *nanos*$^-$ mutant embryo which has the terminal regions and transplant *nanos* activity into the middle. This embryo at the ends would induce the expression of *tailless*, which is dependent on the *torso* system, and in the middle it would express *knirps*, as a response to the *nanos* activity. So this embryo can only make the ends. When you transfer the posterior activity to the middle, you get the classical mirror-image type bicaudal embryo. The posterior substance was placed in the middle of the embryo but the most posterior regions form at the ends.

You could postulate a complicated distribution mechanism but you can also say it may have something to do with the expression of this terminal system. In order to test that, Ruth Lehmann and Hans Georg Frohnhöfer repeated the experiment in a triple mutant embryo where they didn't have the terminal system. They transplanted *nanos* activity into an embryo which was *bicoid*$^-$ *nanos*$^-$ and *torso*$^-$. This embryo has no information for pattern and polarity whatsoever. The outcome of the experiment is something where, unfortunately, we cannot interpret the identity of the segments. But the symmetrical abdominal pattern that forms has its most posterior part in the middle of the embryo, and the polarity is opposite to that of a bicaudal embryo. So in the first case, where *torso* was present, we get an embryo that has, in terms of gradient, the highest values for 'posteriorness' at the two poles. In the second case, when *torso* was absent, the maximum is in the middle of the embryo. I think that in the first case, *nanos* is expressed in the middle and *tailless* at both ends. The *nanos* activity induces *knirps*, the *torso* activity induces *tailless*. The juxtaposition of *tailless* and *knirps* gives posterior abdomen.

The neutral embryo doesn't have any of these systems present, it only expresses *Krüppel*, which is not dependent on either of these two systems. So when you transplant *nanos* into this embryo, you displace *Krüppel* in the middle because it is repressed by *nanos*, and then you have *knirps* in the middle and *Krüppel* at the ends. This gives the reverse polarity. *Krüppel* and *knirps* as neighbours produce anterior abdomen, whereas *tailless* and *knirps* in juxtaposition give posterior abdomen. This occurs with the same concentration and distribution of the *nanos* product.

Wolpert: I can see the polarity in the one without *torso*, but I can't see the polarity in the first one—where you said it's exactly the same shape but you have the ends.

Nüsslein-Volhard: Let's assume that we could identify the other segments, then we would see that if A7 was in the middle, it would go to A6.

Wolpert: Sure, but the other one where you have exactly the same distribution of morphogen in the centre, and yet the polarity is the opposite.

Nüsslein-Volhard: If A7 was in the middle, it would go to A8, because of expression of *tailless*.

Meinhardt: The gradient doesn't play a role.

Nüsslein-Volhard: The argument is that it's not the concentration of *nanos* which says 'this is abdominal segment N^{0}N'. With the same concentration profile of *nanos* activity we get two different qualities in this experiment. There are other data which indicate that it's not the direct concentration of *nanos* which determines whether it is anterior abdomen or posterior abdomen.

Struhl: What those experiments also tell you, is that it is the relative position or relationship of the regions of expression of gap genes, like *Krüppel* and *tailless* and perhaps *knirps*, that carries the information of polarity at that stage. But normally those gap genes are instructed to be active—where they are active depends on the initial determinants. So the primary gradients set up the patterns of expression of the gap gene products, and the overlapping distributions of these products confer polarity.

Nüsslein-Volhard: I don't see the point you are making.

Struhl: My point is that polarity is something provided by a gradient, but in an indirect fashion.

Nüsslein-Volhard: The gradient mechanism does it in a *direct* fashion. When you make B by juxtaposition of different qualities A and C, you call that induction, you don't call that a gradient mechanism—that's my notion.

Struhl: I guess that's not my notion.

Sander: Could you describe the pattern of overexpression of *Krüppel* and explain how the mirroring of polarity arises?

Struhl: I don't want to do that in detail, because I basically don't understand it well enough. Briefly, I think the particular concentrations of *Krüppel* and *knirps* proteins dictate the activation of subordinate pair-rule and homeotic genes. Hence, changing their distributions could well change polarity (e.g. Fig. 4, Struhl, this volume).

Nüsslein-Volhard: We don't understand why the juxtaposition produces this polarity.

Struhl: There is another point, which is that if you just have the *nanos* product as a gradient alone, as in *bicoid*$^-$ *oskar*$^-$ *torso*$^-$ embryos injected with posterior pole plasm, you do see a polarized response, so the *nanos* gradient is capable of conferring polarity.

Nüsslein-Volhard: But *Krüppel* is still there; the product is coming in from the posterior end, so *Krüppel* is displaced to the anterior of the egg.

Struhl: My point is that the gradient of *nanos*, under those conditions where it is the sole determinant, provides a polarized response of the appropriate type. To me that is evidence that it can function, even in your classical view, as a gradient. However, it is only when you tamper with the interpretative system that this ability becomes apparent.

Meinhardt: We have now discussed at length that at a particular stage the information about polarity resides in the borders produced by juxtaposition between *hunchback* and *Krüppel, etcetera*. How is the polarity transmitted from this type of border to the next level? After the gap genes, the next hierarchical level is that of the pair-rule genes. I have proposed that one can describe the pattern of the pair-rule genes essentially as two binary patterns, the *hairy/runt/hairy/runt/* . . . pattern on one hand and *eve/ftz/eve/ftz/*. . . on the other, and that a phase shift exists between these patterns. Most of us would agree that it is the phase shift which determines the polarity of the outcoming segmental pattern. In the normal embryo we know that *eve* is expressed slightly posterior to the *hairy* gene. If it were on the anterior side, the segmental pattern would have a reversed polarity. We have to ask how the borders between *hunchback* and *Krüppel,* or between *Krüppel* and *knirps*, fix the phase shift between the *eve/ftz* pattern on the one hand and the *hairy/runt* pattern on the other.

Wieschaus: The smallest unit of polarity in an epithelial sheet is probably the single cell. The mechanisms that control its polarity presumably must also work in larger units within the sheet, but I don't think it will be units of random size. It seems to me that there are certain units of polarity reversal that can be established, depending on the kinds of influences you put on the two ends.

Martinez-Arias: I am not sure that you can equate the acquisition of co-ordinates of polarity in the egg to the kind of polarity changes observed in *Krüppel* mutants.

Davidson: I have a naive molecular biologist's reaction. One of the genes that we are working on is an ordinary worker gene at the downstream end of the kind of system we have been talking about today. It codes for a cytoskeletal protein and has at least ten sites of DNA–protein interaction, for which there is *in vivo* evidence that they are important. This is not exceptional, for example, the chick β-globin gene has at least eight well characterized sites. When the gene functions, the mechanism in some way sums the states of interaction at all the different sites where proteins are bound. I would be surprised if these 'upper level' hierarchy genes are significantly simpler in this than other genes. Consequently, when one does experiments such as change the concentrations of individual participants in these molecular interaction summing systems, it is difficult for me to see how one-to-one mechanistic interpretations can be derived concerning the significance of a given gene product that interacts with the regulatory region of another regulatory gene. Imagine, for example, three sites where the DNA has to fold-up and form a knot with three proteins interacting. So I think that quantitative arguments based on levels of specific products that affect the activity of the gene which in turn regulates a wholly unknown downstream set of genes are on the fringe of rationality.

Lawrence: We have to try!

Garcia-Bellido: I agree with Eric (Davidson), that if we are going to look for the molecular mechanisms of, for example, alterations in the configuration of DNA at sites of transcription, we are going to have major problems. To return to polarity, there should be some rules for change of polarity. I believe that the cells are polarized somehow and their polarity is related to differences in the amounts of products on opposite surfaces of the cell. (There are some indications of polarity, although superficial, like the cell autonomous orientation of the elements of the cuticle, that show up after cell dissociation and reaggregation, or in morphogenetic mosaics of polarity mutants.) If there is a change of polarity in a *Krüppel* mutant, this is alright, because the background is wild type and the segment polarity genes are still present. But polarity does not depend on just one system, it can be changed by mutation in the gap genes, the pair-rule genes or in the segment polarity genes. What is polarity? There is a major disagreement whether it is a confrontation between some kind of specific signalling or a consequence of graded differences between neighbouring cells.

Wolpert: There are systems where one really has a chance to tackle this problem successfully, for example in chemotaxis in slime moulds. There is a quite small concentration gradient and yet the cell becomes highly polarized.

Struhl: When you talk about the sheet of cells being polarized and when are cells polarized in the blastoderm, it is important to remember that there aren't cells until the onset of gastrulation. You could say that the whole syncitial cell is polarized and we would all agree to that. But at the time that *Krüppel* is first activated, when the pair-rule genes are responding, there are no cells. So in what sense do you mean the cells are polarized?

Lawrence: They have membranes, the polarity could reside in the membrane; or in the position of some organelles.

Scott: One of the ways I think about it from a different angle is how astonishing it is that the *ftz* stripes correspond at all to the pattern that you see in the phenotype later on. It is still not clear exactly how *ftz* expression corresponds, or leads, to the phenotype of *ftz* mutant embryos. That's even more astonishing for a gene such as *hairy* that is higher in the hierarchy. That you get such a simple outcome—deletion of alternate segments—from a whole series of gene interactions is surprising. We know that *hairy* influences *runt*, *eve ftz*, *engrailed* and others, yet the final pattern you end up with is a very simple change from the wild-type pattern. This gives me some optimism in response to Eric's point. We might well be missing some things because we don't know all the relevant components of the system, but if we were missing a large number of relevant genes, I think there would have been less success than there has been in finding regulatory interactions among the genes we know about. However, it is certainly true that in the majority of these interactions, it is not known how direct the interaction is.

Davidson: It was the extension of quantitative arguments that I found most difficult. I think that black and white perturbation has obviously been a great success—but I was talking about some of the discussion that was here this afternoon.

Wolpert: In my introduction I drew divisions between the left and the right—I now begin to feel that the distinctions are between the optimists and the pessimists!

References

Lawrence PA 1988 The present status of the parasegment. Br Soc Dev Biol Dev Suppl 104:61–65

Nüsslein-Volhard C, Frohnhöfer HG, Lehmann R 1987 Determination of anteroposterior polarity in *Drosophila*. Science (Wash DC) 238:1675–1681

Struhl G 1989 Morphogen gradients and the control of body pattern in insect embryos. In: Cellular basis of morphogenesis. Wiley, Chichester (Ciba Found Symp 144) p 65–91

The establishment of regional identity in the *Xenopus* blastula

J. Heasman, A. Snape*, A. Turner and C. C. Wylie

*Department of Zoology, University of Cambridge, Downing Street, Cambridge CB2 3EG, UK and *Laboratory of Molecular Genetics, National Institute of Child Health and Human Development, Bethesda, Maryland 20892, USA*

Abstract. The molecules involved in the commitment of *Xenopus* cells to particular germ layers are unknown. The question has been investigated for the cells of the blastula in *in vivo* cell transplantation assays and *in vitro* aggregation assays. Using the former technique, we have shown that vegetal cells become committed before gastrulation, even when placed in inappropriate sites. We could find no evidence of regional determination within the endoderm germ layer at the early gastrula stage. In aggregation assays, animal and vegetal cells sorted incompletely at the mid-blastula stage and more efficiently at the gastrula stage. Their behaviour may be mediated by differential expression of adhesion molecules, which so far remain unidentified.

1989 Cellular basis of morphogenesis. Wiley, Chichester (Ciba Foundation Symposium 144) p 99–112

One fundamental objective in developmental biology is to understand the factors which control the earliest steps of embryonic development in vertebrates. For invertebrates, developmentally important genes have been identified and analysed using the powerful combination of classical genetic and modern molecular techniques. This approach is not feasible for vertebrate embryos. Instead, vertebrate, and specifically amphibian, embryos offer the important advantage of ease of manipulation at the single cell level. So experiments on *Xenopus* may reveal the phenotypic changes in embryonic cells whereby morphogenesis can take place.

The cellular process of morphogenesis for *Xenopus* is apparently simple until the time of gastrulation. The fertilized egg forms a hollow ball of cells consisting of an animal cap, equatorial zone and vegetal mass. After about 10 hours the complex tissue movements of gastrulation reshuffle these areas into ectoderm, mesoderm and endoderm germ layers, respectively. We know from fate mapping experiments the normal fate of the different cells of the 16 and 32-cell stage embryo (Moody 1987, Dale & Slack 1987). The question we would like to answer is what is the mechanism by which cells in particular regions become restricted in potency so that their progeny are committed to particular germ layers? We have shown previously that in the animal cap and vegetal mass this restriction has occurred by the time of gastrulation. Here we describe some characteristics of the behaviour of blastula cells *in vivo* and *in vitro*, which suggest cell surface differences are involved in commitment.

The state of commitment of blastula and gastrula cells

We studied this by transferring single fluorescently labelled cells into the blastocoel cavities of unlabelled host embryos. If descendants of the original cell were found in a wide variety of tissues derived from more than one germ layer, then the donor cell was considered to be pluripotent at the time of transplantation. But if progeny were found in derivatives of only one germ layer, the original cell was considered to be committed to that germ layer at the time of transplantation into the blastocoel. The results of these experiments showed that cells of the animal and vegetal poles are pluripotent at the early blastula stage and gradually become committed to ectoderm and endoderm germ layers, respectively, a process which is completed by the gastrula stage (Heasman et al 1984, Snape et al 1987).

Committed vegetal cells sort to the endoderm region wherever they are placed in the blastocoel

In the first series of experiments, donor cells were injected onto the blastocoel floor of the late blastula hosts. The normal fate of cells from this region is to form endoderm, so it was conceivable that the results described above were obtained simply because the smaller size of later stage vegetal cells lessens the probability of their reaching other germ layers from the blastocoel floor. We tested this by comparing the effect of placing donor early gastrula vegetal cells on the blastocoel roof (i.e. in an environment of cells fated to become ectoderm), with that of placing them on the floor (where the cells are destined to be endoderm). To put the cells onto the roof of the blastocoel, we simply inverted the host embryos after injecting the labelled cells and kept them in the inverted position for 30 minutes. We found that when we sectioned embryos which had been fixed after this period of time the injected cells were on the blastocoel roof in 70% of cases. The results show that the state of commitment of vegetal progeny is independent of the initial location of the donor cells (Table 1), and that gastrula cells from the vegetal pole can enter the host's endoderm even when transplanted to an inappropriate starting position.

TABLE 1 Results showing the effect of host inversion and donor cell fixation on the fates of early gastrula (stage 10) subcortical vegetal pole cells transferred into stage 9–10 hosts

Expt	*Donor cells*	*Host orientation*	*Total number of hosts*	*Hosts with labelled cells*	*Percentage of hosts containing labelled cells in the following combination of germ layers*						
					end + *mes* + ect	*ect*	*ect* + *mes*	*mes*	*end* + *ect*	*end* + *mes*	*end*
1a	Unfixed	Non-inverted	26	25	12%	—	—	4%	—	8%	76%
		Inverted	27	26	—	—	—	11.5%	—	11.5%	77%
1b	Fixed[a]	Non-inverted	19	19	5%	—	5%	26%	—	42%	21%
		Inverted	19	15	20%	—	—	7%	—	40%	33%
2	Unfixed	Non-inverted	23	21	—	—	—	—	—	—	100%
		Inverted	27	26	—	—	—	—	—	4%	96%

One stage 10 vegetal cell was injected into each host. Each blastomere was labelled with tetramethyl rhodamine isothiocyanate. Hosts were sectioned at stage 38 and serial sections were examined for fluorescent cells. [a]Cells were fixed with 4% formaldehyde and washed three times with 2% lysine before injection. end, endoderm; mes, mesoderm; ect, ectoderm.

There is no detectable regional commitment within the endoderm germ layer during the blastula and early gastrula stages

Experiments using vital stains have shown that there is some regionalization of the fate map of the prospective endoderm of the gastrula (Keller 1975). For example, cells of the vegetal pole generally form posterior endoderm in the floor of the archenteron, while those of the suprablastoporal endoderm form the dorsal endoderm, above the archenteron. We wondered whether the donor cells which we isolated at the early gastrula stage were committed not only to endoderm but also to a particular region therein. To test this, we isolated cells from different regions of the vegetal hemisphere (Fig. 1a) and tested their state of commitment by transferring them to host blastocoels, one per blastocoel (Table 2). We found that whatever the source of the cells, the majority of labelled progeny were in the ventral gut region, generally close to the site where the blastocoel obliterated. We found no evidence from blastomere transfer experiments for regional commitment of single cells from different areas of the prospective endoderm. This suggests that regionalization within the endoderm germ layer occurs after the early gastrula stage, or is a reversible process up to this time.

Cells from different regions of the blastula sort *in vitro*

Our transplantation experiments show that early gastrula vegetal cells which are initially inserted so that they adhere to animal cells do nevertheless populate

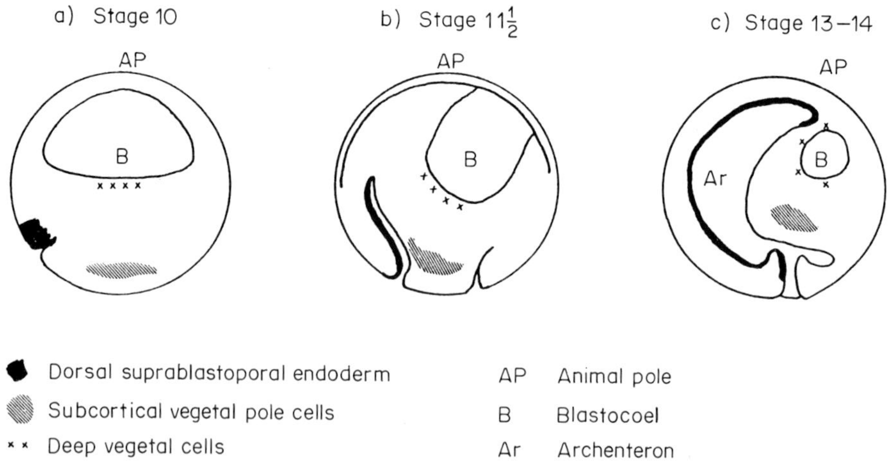

FIG. 1. Diagram showing the expected fates of different regions of the vegetal mass after gastrulation.

TABLE 2 The location of endoderm-committed cells deriving from various regions of the vegetal hemisphere after transfer into stage 9–10 hosts

Experiment number	*Source of donor cells*	*Total number of hosts with labelled cells in endoderm only*	*Number of hosts having labelled cells in endoderm in the following location*			
			dorsal	*ventral*	*lateral*	*head*
1	Stage 8 embryos subcortical vegetal pole	10	1	9	—	—
	Stage 7 embryos blastocoel floor	8	—	7	—	1
2	Stage 8 embryos blastocoel floor	7	1	6	—	—
	Stage 9 embryos blastocoel floor	23	1	22	—	—
	Stage 10 embryos subcortical vegetal pole	19	—	17	2	—
3	Stage 10 embryos superficial dorsal blastopore lip	11	1	7	3	—

the endoderm. They could do this either by migrating away from the injection site, or by remaining in place but becoming attached preferentially to vegetal cells when the blastocoel obliterates. The fact that they lose attachment to animal cells in preference to vegetal by either mechanism suggests that cell surface differences have developed between animal and vegetal cells, which enable them to find the correct germ layer by the gastrula stage.

One prediction of the hypothesis that commitment involves the development of cell surface differences between cell populations in the blastula is that randomly mixed, committed animal and vegetal cells should be able to sort from each other *in vitro*, while uncommitted cells should not. We tested this in dissociation/reaggregation experiments similar to those of the classical studies of Holtfreter (Townes & Holtfreter 1955). Whereas Holtfreter concentrated on long-term aggregates of 10–15 hours and used gastrula and neurula cells, we have examined the characteristics of blastula cells, using the same time-course over which we know changes occur *in vivo*.

Methodology

We dissected animal cap, marginal zones and vegetal masses from embryos of different stages in normal Barth's saline (NBX) (Gurdon 1977) and dissociated the cells in Ca^{2+}- and Mg^{2+}-free NBX. We labelled populations of cells by immersion in tetramethyl rhodamine isothiocyanate (10 μg/ml for 2–4 minutes) and then placed approximately equal volumes of labelled and unlabelled cells in random aggregates. Aggregation was started by the addition of 10% fetal calf serum to the saline. After 1–4 hours in stationary culture on agar, the aggregates were fixed in 2% trichloroacetic acid, dehydrated, embedded in polyethylene glycol distearate and sectioned at 15 μm.

A set of four-hour aggregates was selected to provide standards to define degree of sorting (Fig. 2). This period was used because it is the time a mid-blastula embryo takes to reach gastrulation in our conditions, and the time over which commitment occurs, according to our transplantation studies. Unsorted aggregates scored zero and completely sorted ones three on this scale. All experiments were scored blind by two people using these standards for comparison. Some of the results of these experiments are shown in Table 3. The main findings were as follows:

1) After four hours the average sorting score of animal/vegetal mid-blastula aggregates was 1.7 (Fig. 3a) compared to that of 0.58 for control aggregates consisting of labelled and unlabelled mid-blastula vegetal cells and a score of 0.21 for animal control aggregates. Where any pattern could be discerned, the animal cells were in the centre of the aggregates.

2) This sorting phenomenon is detectable before gastrulation. We checked that segregating behaviour is a property of blastula cells and not a later development

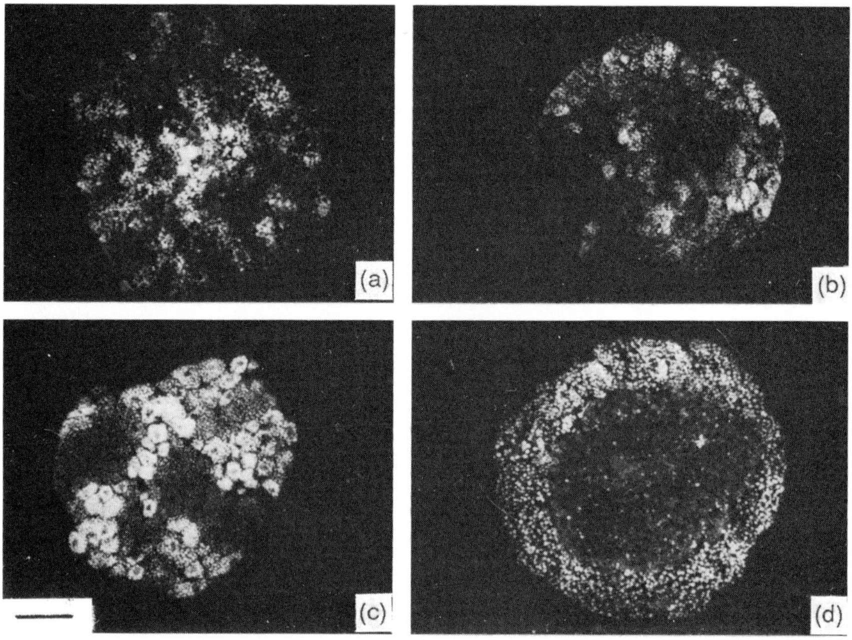

FIG. 2. Standard aggregates for scoring of sorting. Sectioned four-hour aggregates of labelled and unlabelled cells were selected to represent a) the unsorted state (score 0); b) limited sorting (score 1); c) substantial sorting (score 2); and d) complete sorting (score 3). All subsequent scoring used this series as a standard. Bar = 100 μm.

by fixing aggregates after only one hour of aggregation, well before sibling embryos gastrulated. Again sorting was distinct, although by no means complete.

3) Early gastrula cells sort out more completely than blastula cells, yielding sorting scores of 2.63 compared to 1.7 (Fig. 3b).

4) We tested whether homotypic combinations of mid-blastula and early gastrula cells would sort; we found substantial sorting with respect to animal cells (2.16) and incomplete segregation for vegetal cells (1.34).

These findings are in accord with the predictions of our *in vivo* transplantation studies, where cells taken from mid-blastula donors are only partially committed and might, therefore, be expected not to sort completely, while gastrula cells are completely committed and segregate from each other very efficiently. It is particularly interesting that homotypic combinations of different stages sort out from each other over a four hour period, animal cell combinations more so than vegetal. Interpreted according to the surface changes hypothesis, it would suggest that changes to the animal cell surface are more dramatic over this time than those to the vegetal surface.

TABLE 3 Sorting out in aggregates after four hours

Aggregate type	*No. of aggregates*	*Average score*	*Standard error*
St 8A/St 8V	35	1.70	0.09
ST 8A/St 8A	7	0.21	0.15
ST 8V/St 8V	10	0.58	0.22
St 10A/St 10V	18	2.60	0.13
St 10V/St 10V	11	0.07	0.06
St 8A/St 10A	20	2.16	0.16
St 8V/St 10V	20	1.34	0.23
St 9M/St 9A	14	1.39	0.09
St 9M/St 9V	17	2.12	0.05

Aggregates were scored on a scale of 0 (unsorted)–3 (sorted) by comparison with standard aggregates (Fig. 2). Each aggregate was scored blind by two people and the scores averaged.

The mechanism of sorting *in vitro*

The simplest explanation for cell sorting *in vitro* is that distribution is dictated purely physically, by differences in cell size. In all the experiments described above, the cell sizes of the two populations are different and, where it is possible to distinguish a pattern, the smaller cells are on the inside of the aggregates. If this explanation were correct, then we would expect that the larger the ratio of size difference between a pair of cell types, the higher would be the sorting score. We looked at this by plotting a graph comparing the degree of sorting and the ratio of surface areas (Fig. 4). The results suggest that while cell size may be a factor, it alone cannot explain the sorting results. For example, the ratio of surface areas for

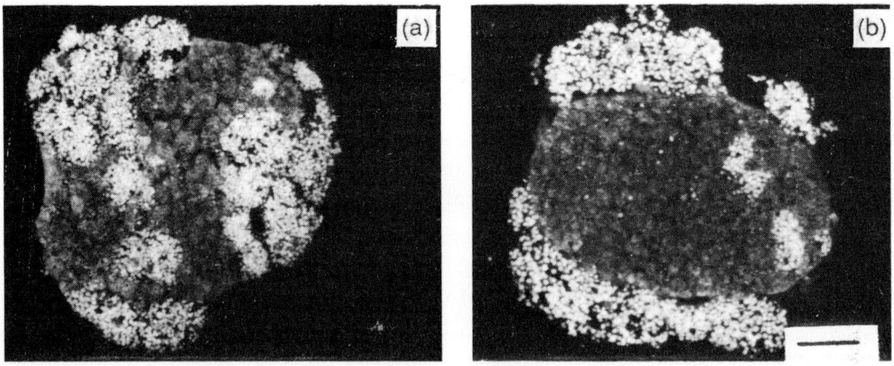

FIG. 3. Sorting out of a) mid-blastula (stage 8) animal and vegetal cells, and b) early gastrula animal and vegetal cells. Animal cells were labelled with tetramethyl rhodamine isothiocyanate in both cases. Aggregates were fixed and sectioned after four hours. Bar = 100 μm.

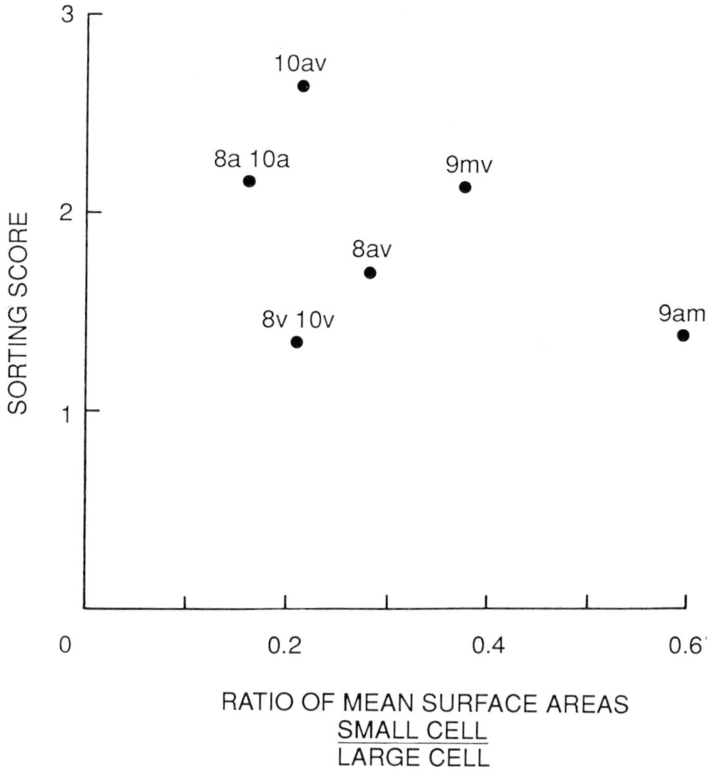

FIG. 4. The relationship between sorting and the ratio of cell sizes. 100 each of dissociated animal, marginal and vegetal cells were taken from mid-blastula and early gastrula stages and their diameters measured. The average surface area of cells from all three zones was calculated for the mid-blastula and early gastrula stages, and expressed as a ratio of smaller cell type over larger cell type. If size were the only factor involved in sorting, then the points on the graph should be simply related. a, animal cells; m, marginal cells; v, vegetal cells.

stage 10 animal/stage 10 vegetal is the same as that for stage 8 vegetal/stage 10 vegetal, but the sorting scores are quite different.

It seems likely that one of the mechanisms at work here is one where cell surface differences appearing before gastrulation distinguish the animal and marginal populations from the vegetal. These may be quantitative differences in the concentration of one type of adhesion molecule common to both types of cell, or qualitatively different animal- or vegetal-specific molecules. Evidence from other systems points to the importance of specific cell adhesion molecules in tissue organization in morphogenesis (Gallin et al 1986, Hatta & Takeichi 1986). Recently, Takeichi et al provided the first direct evidence that cadherins

(Ca^{2+}-dependent adhesion molecules) may be involved in cell sorting during morphogenesis (Nose et al 1988, Takeichi et al, this volume). While a *Xenopus* cadherin has not so far been identified, we and others have noted that adhesion of early *Xenopus* embryos is mediated by a Ca^{2+}-dependent and not a Ca^{2+}-independent mechanism. We have found that sections of late blastula *Xenopus* embryos stain positively with an antibody which recognizes a highly conserved part of the C-terminal region of chicken A-CAM, (data not shown). This suggests that one or more cadherin-type molecules are present in early *Xenopus* embryos.

Shiokawa et al have shown that Ca^{2+}-dependent reaggregation of *Xenopus* blastomeres involves the translation of maternal rather than newly synthesized mRNA (Shiokawa et al 1983). We have recently prepared a panel of monoclonal antibodies which prevent the Ca^{2+}-dependent adhesion of dissociated animal and vegetal cells. These stain the cell surfaces of blastulae (not in a vegetal- or animal-specific manner), and they also have specific staining patterns in oocytes. They are, therefore, strong candidates for antibodies which recognize the Ca^{2+}-dependent adhesion system(s). Direct evidence for the importance of such molecules in *Xenopus* morphogenesis awaits messenger RNA inhibitor studies, in which we can test the effect of particular messages in embryonic development by blocking their activity in the oocyte in a specific manner.

It seems quite possible from the timing of commitment shown in our experiments that germ layer segregation is the result of the expression of new message, transcribed from the embryonic genome, in the animal or vegetal hemisphere, superimposed on or replacing the Ca^{2+}-dependent adhesion system already in place. It is equally possible that commitment is the result of modification of the existing surface molecules. The combination of immunological and nucleotide microinjection methods using the *in vivo* and *in vitro* assays described here should provide a useful approach to identifying the molecules involved.

Acknowledgement

This work was supported by the Wellcome Trust.

References

Dale L, Slack JMW 1987 Fate map for the 32-cell stage *Xenopus* embryo. Development 99:527–551

Gallin WJ, Chuong CM, Finkel LH, Edelman GM 1986 Antibodies to liver cell adhesion molecule perturb inductive interactions and alter feather pattern and structure. Proc Natl Acad Sci USA 83:8235–8239

Gurdon JB 1977 Methods for nuclear transplantation in amphibia. Methods Cell Biol 16:125–139

Hatta K, Takeichi M 1986 Expression of N-cadherin adhesion molecule associated with early morphogenetic events in chick development. Nature (Lond) 320:447–449

Heasman J, Wylie CC, Hausen P, Smith JC 1984 Fates and states of determination of single vegetal pole blastomeres of *Xenopus laevis*. Cell 37:185–194

Keller RE 1975 Vital dye mapping of the gastrula and neurula of *Xenopus laevis*. I. Prospective areas and morphogenetic movements of the superficial layer. Dev Biol 42:222–241

Moody SA 1987 Fates of the blastomeres of the 16-cell stage *Xenopus* embryo. Dev Biol 119:560–578

Nose A, Nagafuchi A, Takeichi M 1988 Expressed recombinant cadherins mediate cell sorting in model systems. Cell 54:993–1001

Shiokawa K, Tashiro K, Oka T, Yamana K 1983 Contribution of maternal mRNA for maintenance of Ca^{2+}-dependent reaggregating activity in dissociated cells of *Xenopus laevis* embryos. Cell Differ 13:247–255

Snape AM, Wylie CC, Smith JC, Heasman J 1987 Changes in the states of commitment of single animal pole blastomeres in *Xenopus laevis*. Dev Biol 119:503–510

Takeichi M, Hatta K, Nose A, Nagafuchi A, Matsunaga M 1989 Cadherin-mediated specific cell adhesion and animal morphogenesis. In: Cellular basis of morphogenesis. Wiley Chichester (Ciba Found Symp 144) p 243–254

Townes PL, Holtfreter J 1955 Directed movements and selective adhesion of embryonic amphibian cells. J Exp Zool 128:53–120

DISCUSSION

Melton: I recall from the experiments of Lee & Kirschner that fibronectin is expressed in all cells in the blastula and that it localizes to the roof of the blastocoel. Your evidence suggests that integrin is also localized there. Do you know if integrin is expressed everywhere and then somehow sequestered in the roof of the blastocoel or is it expressed only by those cells?

Heasman: I think the latter is the case, although Doug de Simone hasn't looked at the temporal expression of the integrin mRNA yet, so one can't say for certain whether the integrin is expressed from the oocyte through to the blastula stages. The antibody staining of fibronectin, in our hands, in the early blastula is quite symmetrical, it is not just at the roof of the blastocoel. That concentration happens towards the end of the blastula stages or the early gastrula stage.

Weisblat: If two groups of cells had different mobilities, they might separate from each other and yet I wouldn't call that sorting. Can you exclude that possibility?

Heasman: In our assays all the cells move. We can't exclude that migration is involved in sorting but there isn't one population that's stationary and one that's moving.

Lawrence: You described the acquisition of apparent surface properties in the endoderm. Have you done similar experiments with cells from other germ layers?

Heasman: We haven't done experiments in as much detail as those in the vegetal hemisphere. We have compared the sorting of dorsal and ventral marginal zone cells. We wondered whether there was some regional specification already in the marginal zone in blastula stages. However, there is no sorting between cells of the dorsal and ventral parts of the marginal zone at the mid-blastula stage. We haven't done that experiment at the later blastula stage. There doesn't appear to be a surface difference between dorsal and ventral marginal cells or, if there is one, it is not big enough to detect in our assay.

Lawrence: Can you can give us an overall picture of how different germ layers or cell types are determined stage by stage in the embryo?

Heasman: I think until we know more about the molecules involved, it's not possible to extrapolate in that way.

Wolpert: If you take mesoderm, does it behave in the same way? What about the ectoderm?

Heasman: The marginal zone cells, which are fated to become mesoderm, will sort from vegetal cells during the blastula stages but do not sort significantly from animal cells.

Gardner: Could you say anything about the temporal relationship between the mid-blastula transition and the onset of commitment? Is it important for desynchronization of cell cycles to occur before or around the time that cells become committed?

Heasman: I don't know whether the commitment process depends on desynchronization in the mid-blastula stage. The desynchronization is, at a cellular level, a slowing down in the vegetal hemisphere compared to the animal cap. The commitment process that we see may be controlled by maternal molecules and not depend on the expression of the embryonic genome. We have transferred cytoplasm from the animal cap into the vegetal pole of the fertilized egg. In preliminary studies, we are able to change the commitment of the cells from the vegetal hemisphere, such that they now will sort with animal cells. This suggests that the commitment process is not dependent on embryonic transcription but is dependent on the maternal expression.

Jessell: Once you have dissociated the cells, can you distinguish, without knowing anything about the molecules that might be involved, whether adhesion is mediated by calcium-dependent or calcium-independent molecules?

Heasman: The predominant adhesion mechanism in *Xenopus* embryos at these stages is calcium dependent. All of the antibodies that we have that block adhesion, block calcium-dependent adhesion. We can't exclude completely that there is a calcium-independent mechanism. One possible scenario is that initially blastomeres adhere together by a calcium-dependent mechanism during the blastula stages; then superimposed or replacing that in one particular group of cells there is a calcium-independent or just a different mechanism.

Takeichi: Do the antibodies which inhibit aggregation of animal pole cells also recognize the endoderm cells?

Heasman: Yes, unfortunately.

Takeichi: What happens later in development, for example in the formation of the neural tube?

Heasman: By the time the neural tube is forming, calcium-independent mechanisms are operating, for example N-CAM is present. *In vitro*, our cells continue to follow a normal course of commitment in that the temporal expression of some antigens is the same as that *in vivo*. It is quite possible that when we see cells sorting very completely from each other—for instance homotypic combinations of early gastrula and mid-blastula animal cells—what's probably happening is that commitment is continuing, differentiation of surface is continuing with molecules such as N-CAM appearing, and the early cells just cannot catch up with the later cells in the aggregate.

With the vegetal pole we know that morphologically differentiating cells don't appear until the swimming tadpole stage, at which point there is a columnar epithelium. So it is possible that vegetal cells from early blastula mixed with those from early gastrula are able to catch up to the same state of commitment and therefore do not sort as well in the assay. The usefulness of the assay is simply that we can test antibodies *in vitro* as to whether they stain surfaces in a polarized fashion. It is difficult when you look at a whole embryo to know whether an antibody is staining differentially or not. We can also see if the antibodies block adhesion or sorting in an aggregate.

Cooke: Focusing on slightly later stages in development, some of us are interested in the possible distinction between cell commitment at the level of the individual cell by some irreversible genetic-type switch, and ensemble properties, whereby cells in their normal tissue organization sense that they are going to belong to one part of the body pattern, but that sense doesn't survive disaggregation of cells. Have you done any experiments on this general paradigm? For example, take cells from some part of the mesoderm which, by its mechanical behaviour, indicates that in normal development it would contribute to somite, and test whether a single such cell, when put back into an earlier embryo, will give rise to progeny that populate more diverse tissues.

Heasman: We have done some experiments similar to that in the vegetal hemisphere. At the early gastrula stage, the cells seem to be determined not only to form endoderm but a particular region of the endoderm (Keller 1975). The suprablastoporal endoderm is going to end up in the roof of the archenteron. The cells of the vegetal pole are going to end up in the ventral endoderm posteriorly. We asked whether there is any regional commitment within the endoderm at the early gastrula stage. The answer from these sorts of experiments is no, a cell is not committed to a particular region of endoderm. In general, all of the cells that we have put in the blastocoel produce progeny that are found in the ventral endoderm. If we use suprablastoporal fated cells, they still end up in ventral endoderm, they are not found in the roof of the archenteron.

Wolpert: Is there any other assay that one could use to detect these differences earlier? Or is this the earliest operational criterion for recognizing an endoderm cell?

Heasman: I think it is the earliest.

Struhl: I thought endoderm cells were associated with more yolk. If you put them on a gradient, they have a different density, is that right?

Heasman: You could split up the whole embryo in that way, but there would be a graded distribution of the cells. You would not be able to tell where it changed from marginal zone to endoderm cells.

Wolpert: What is the distribution of these cells? If you start with cells from the most vegetal pole and then move up the embryo, is there a sharp boundary where you find cells that don't show this phenomenon?

Heasman: We don't know.

Wolpert: So you are only really looking at the most vegetal?

Heasman: We have compared the vegetal with the endodermal mass in the floor of the blastocoel, and with the suprablastoporal endoderm at the early gastrula stage. They all show the same sorting phenomenon when placed in the blastocoel of a host embryo.

Cooke: You haven't yet got an independent endodermal marker?

Heasman: We analyse the cells at the tailbud stage, for which we do have markers. We can take a pluripotent animal pole cell and it will produce progeny in the endoderm. We can stain it with a differentiation marker for endoderm, such as VC1 (Heasman et al 1985) which stains only endoderm cells, and those cells that were fluorescently tagged when they were transplanted also stain with VC1 markers. So they differentiate according to the germ layer they are in, they are not just trapped in the wrong place and unable to differentiate.

Garcia-Bellido: Is this time dependent? If you block cell division, would larger cells from this region behave in the same way?

Heasman: We haven't done those experiments.

Garcia-Bellido: This relates to whether this information is purely maternal or whether it has to be processed zygotically. If there is zygotic processing, one would imagine that the signal would increase with time.

References

Heasman J, Snape A, Smith JC, Holwill S, Wylie CC 1985 Cell lineage and commitment in early amphibian development. Phil Trans R Soc Lond B 312:145–152

Keller RE 1975 Vital dye mapping of the gastrula and neurala of *Xenopus laevis*. I. Prospective areas and morphogenetic movements of the superficial layer. Dev Biol 42:222–241

Factors specifying cell lineages in the leech

David A. Weisblat and Stephanie H. Astrow*

*Department of Zoology and *Graduate Group in Neurobiology, University of California, Berkeley, California 94720, USA*

Abstract. As in arthropods, several major organ systems in leeches, including body musculature, nervous system and nephridia, are organized into a fixed number of longitudinally iterated units called segments. Many cells, especially neurons, can be uniquely identified from segment to segment. Leech embryos comprise identified cells, which facilitates developmental analysis. So far as it is known, cell lineages in leech are largely determinate. Prior to first cleavage, cytoplasmic reorganization generates domains of yolk-deficient cytoplasm called teloplasm. *In situ* hybridization experiments suggest that teloplasm is enriched for polyadenylated RNAs. During the first three, unequal cell divisions, teloplasm is segregated to macromere D′; normally, this cell alone cleaves further to generate five bilateral pairs of embryonic stem cells, M, N, O/P, O/P and Q teloblasts. Centrifugation experiments have shown a causal link between inheritance of teloplasm and the cleavage pattern that generates teloblasts. Teloblasts undergo highly unequal divisions, generating a longitudinal array of segmental founder cells called m, n, o, p and q blast cells, from which the definitive segmental tissues arise via further stereotyped cell divisions. Microinjecting new-born teloblasts or their precursors with polyadenylic acid induces the formation of supernumerary teloblasts. This discovery permits further analyses of factors specifying the five cell lines generating segmental tissues of the leech.

1989 Cellular basis of morphogenesis. Wiley, Chichester (Ciba Foundation Symposium 144) p 113–130

In this essay, we give a brief overview of development in the glossiphoniid leech, *Helobdella triserialis*, concentrating on the contributions of distinct cell lineages, and then describe our studies of the factors that cause one lineage to become different from others in the early embryo.

Leeches constitute the class Hirudinea in the phylum Annelida, segmented worms. As such, their major organ systems, including body musculature, nervous system and nephridia, are organized into a fixed number of longitudinally iterated units called segments. Many cells, especially neurons, can be uniquely identified from segment to segment; this greatly simplifies the morphological analysis of organisms that are otherwise too complex to study at the level of individual cells. Three factors make glossiphoniid leech embryos advantageous for studies of early development: first, the egg contains all the nourishment needed to complete development, which makes the cells relatively large and also means that the embryos can be cultured in simple media; second, the egg undergoes a stereotyped series of complete cleavages, giving rise to identifiable blastomeres in the early embryo; and third, the cell lineages leading from these early blastomeres to their definitive progeny are also largely determinate.

It is important to correct here a common misconception, that 'identified blastomeres' and 'determinate development' are two different ways of saying the same thing. This is not true. In development of the zebrafish, *Brachydanio rerio*, for example, stereotyped cleavages give rise to identifiable blastomeres in the early embryos, and yet the specific fates of these cells are not reproducible from embryo to embryo in normal development (Kimmel & Law 1985a,b, Kimmel & Warga 1987). Conversely, one can imagine an early embryo the early blastomeres of which are *not* identifiable, and yet, by statistical analyses of cell lineages from many embryos at the same stage, could be shown to have highly predictable cell fates.

It is also important to define explicitly the term 'developmental determinacy'. As used here, this term refers only to the predictability of cell fates in normal development, without implying anything about the mechanisms by which cell fates are acquired, or the extent to which the normal cell fates can be modified (Stent 1985, Weisblat 1988). Questions such as these, which involve the elucidation of causal relationships in development, can be addressed only by observing the fates of cells in embryos that have been subjected to experimental interventions of various sorts.

The overall scheme of development in *Helobdella* is essentially as described by Whitman in the late nineteenth century for embryos of the glossiphoniid species, *Clepsine complanata* (Whitman 1878). The salient points are summarized here and in Fig. 1, using a revised version of Whitman's original terminology (Fernandez 1980, Stent et al 1982, Bissen & Weisblat 1989). Fertilization occurs internally and the zygotes, which are about 400 μm in diameter, arrest in early meiosis until they are laid. Prior to and immediately after laying, the zygotic cytoplasm is homogeneous with respect to the distribution of yolk platelets. Soon after the second polar body is extruded from the animal pole of the zygote, latitudinal rings of yolk-deficient cytoplasm, called teloplasm, begin to form beneath the cell cortex in both the animal and vegetal hemispheres. Similar and presumably homologous phenomena occur in the zygotes of various annelids and have been examined in an oligochaete, *Tubifex hattai* (Shimizu 1982, 1984, 1986), as well as in another glossiphoniid leech species, *Theromyzon rude* (Fernandez et al 1987). These yolk-deficient domains

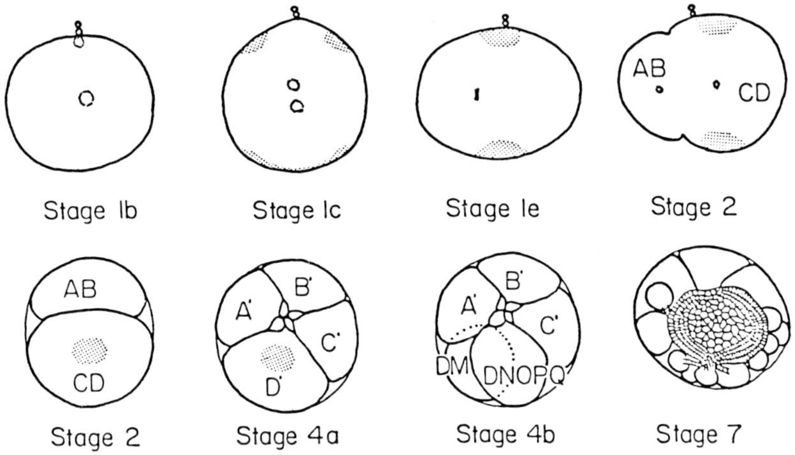

FIG. 1. (Upper) Teloplasm formation in *Helobdella*. Meridional sections through embryos during the first cell cycle. Animal pole is at the top in each panel; polar bodies are represented by small circles at the top of each embryo. Small zones of clear cytoplasm (dashed circles) contain the pronuclei. The onset of teloplasm accumulation is accompanied by a slight latitudinal contraction, about two-thirds of the way from the equator to the animal pole, giving the animal hemisphere a slightly domed shape (Stage 1c). Teloplasm is evident as two ringed domains of clear cytoplasm (stippling), only parts of which are visible in these meridional profiles. Compact domains of teloplasm are formed as the rings of teloplasm move poleward. The embryo then elongates along the future dorsoventral axis; by metaphase, the chromatin (black bar) is located eccentrically with respect to the animal–vegetal axis (Stage 1e). The first cleavage furrow is unequal (Stage 2), segregating most of both pools of teloplasm into the larger daughter cell, CD, from which the teloblasts, and hence the definitive segmental tissues, eventually arise. (Lower) Views of chosen early stages looking down on the animal pole.

within the zygote seem to arise from a rearrangement of cytoplasmic constituents and not from the metabolic conversion of yolky into non-yolky cytoplasm. Evidence for this is twofold: first, mitochondria labelled with the vital dye rhodamine 123 and observed by fluorescence microscopy in the living embryo can be seen to move across the surface of the embryo and accumulate in the rings (Shimizu 1986, S. H. Astrow & D. A. Weisblat, in preparation); second, time-lapse video recordings reveal clear movements of yolk platelets beneath the surface of the embryo during teloplasm formation (our unpublished observations). During the latter part of the first cell cycle, the animal and vegetal rings of teloplasm coalesce at their respective poles. As the rings of teloplasm form, the embryo is cylindrically symmetrical about the animal–vegetal axis, with respect to both the position of the male and female pronuclei and the distribution of teloplasm. By late metaphase in the first cell cycle, the chromatin has become eccentrically located with respect to the animal–vegetal axis (K. Liu & D. A. Weisblat, unpublished observations). As a result, the first cleavage, although basically meridional, is slightly unequal, and the larger cell, CD, inherits a much greater proportion of the teloplasm than does its smaller sibling cell, AB. The second cleavage is perpendicular to the first and is also slightly unequal in cell CD, so that teloplasm is inherited almost exclusively by cell D. The third cleavage is highly unequal and occurs with a dextrorotatory spiral orientation characteristic of annelids and molluscs, producing four micromeres (a-d, at the animal end of the embryo) and four macromeres (A′-D′); macromere D′ retains the bulk of the teloplasm. At the fourth cleavage, macromeres A′-C′ again divide close to the animal pole, producing a secondary trio of micromeres, but the cleavage plane of macromere D′ is obliquely equatorial, generating a vegetal blastomere, mesodermal precursor DM, and an animal blastomere, ectodermal precursor DNOPQ, of about equal size (stage 4b; Fig. 1).

Through a further series of stereotyped cleavages, blastomeres DM and DNOPQ generate five bilateral pairs of relatively large embryonic stem cells called the M, N, O/P, O/P and Q teloblasts, as well as additional micromeres. Teloblasts each undergo several dozen highly unequal divisions, generating longitudinal arrays (bandlets) of segmental founder cells (m, n, o, p and q blast cells) (stage 7; Fig. 1). Definitive segmental tissues arise from blast cells via a further series of stereotyped cell divisions (Zackson 1984, Shankland 1987a,b, Bissen & Weisblat 1989). Using tracer substances microinjected into individual teloblasts to identify the definitive progeny of the injected cell, it has been shown that iterated, individually identified neurons and other cells invariably arise as part of the M, N, O, P or Q kinship group of cells descended from the corresponding cell line (Kramer & Weisblat 1985, Weisblat & Shankland 1985). Macromeres A′–C′ produce additional micromeres and are then eventually enveloped and digested by the developing gut.

Surrounding the eccentrically situated nucleus in each teloblast is a region called the fenestra, comprising yolk-deficient cytoplasm that marks these cells as descendants of macromere D′. Since Whitman's time, the obvious correlation between the unique fate of the D cell line and its unique cytoplasm has formed the basis for the assumption that cytoplasmic determinants contained in teloplasm and differentially segregated to the various teloblasts during cleavage are responsible for differences between the five kinship groups that make up the segmental tissues of the leech. For example, it seemed likely that teloplasm somehow was responsible for the differences in timing and orientation of the cell divisions by which macromere D′ generates the teloblasts. Moreover, it was believed that the animal and vegetal domains of teloplasm contained distinct ectodermal and mesodermal determinants, respectively, that were separated from each other at the approximately equatorial fourth cleavage by which cells DM and DNOPQ arise. As pointed out above, these issues can be addressed only through observing the effects of specific experimental manipulations.

Teloplasm contains factors specifying the D cell line

To test the hypothesis that teloplasm in the D′ macromere influences the special fate of that cell, 2-cell embryos were subjected to moderate centrifugation (350 ***g*** for five minutes) shortly before the cleavage of cell CD (Astrow et al 1987). This treatment partially stratified the cytoplasm of the embryo. Yolk-deficient cytoplasm accumulated in a layer at the centripetal pole of the embryo (Fig. 2) and since the animal–vegetal axis was roughly parallel to the gravitational vector, the meridional cleavage furrow sometimes bisected that layer such that daughter cells C and D inherited equal amounts of teloplasm. Most centrifuged embryos made normal micromeres; but in centrifuged embryos in which C and D inherited equal amounts of teloplasm, macromeres C′ and D′ both made teloblasts. Variations in the orientation and positioning of the cleavage furrow allowed us to conclude that it was indeed the inheritance of teloplasm and not size that was correlated with the change in fate of macromere C′. Moreover, since centrifuged embryos frequently made supernumerary teloblasts and bandlets of blast cells, it seems that teloplasm acts more qualitatively than

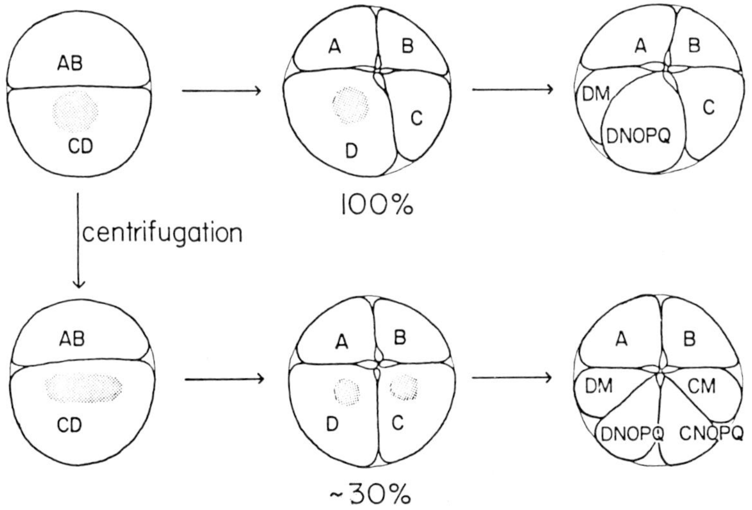

FIG. 2. Centrifugation redistributes teloplasm and induces abnormal cleavages in *Helobdella*. All embryos are represented as viewed from the animal pole. During centrifugation, the embryos orient so that the gravitational vector is roughly parallel to the animal–vegetal axis and directed into the plane of the figure. Stippling indicates the distribution of teloplasm in normal and centrifuged 2-cell and 8-cell embryos.

quantitatively to influence macromere fate; if the quantity of teloplasm available was the factor limiting the total number of teloblasts produced in the embryo, then centrifuged embryos should have made only ten teloblasts, even if some came from macromere C′ and others from macromere D′.

The question of whether there are distinct mesodermal and ectodermal determinants in the vegetal and animal teloplasms is problematical because, while it is not easy to demonstrate the absence of something in a logically compelling fashion, all available evidence indicates that this is *not* the case. First, we observed that, even though the animal and vegetal domains of teloplasm had presumably mixed as a result of centrifugation and centrifuged embryos failed to develop definitive progeny, progeny of macromeres C′ and D′ could still be distinguished as mesodermal and ectodermal precursors by various criteria (Astrow et al 1987). Moreover, contrary to the accepted idea that animal and vegetal teloplasms normally remain distinct until they are segregated into the DNOPQ and DM cells, respectively, we have observed a second cytoplasmic rearrangement in *Helobdella* (Holton et al 1989). In this hitherto undescribed rearrangement, vegetal teloplasm moves to the animal pole between the second and fourth rounds of cell division. Subsequently, teloplasm is inherited by both cells DM and DNOPQ through the oblique orientation of the fourth cleavage furrow. As judged by distribution of fluorescent mitochondria in embryos labelled with rhodamine 123, the animal and vegetal teloplasms mix and cells DM and DNOPQ both inherit some of each. A similar cytoplasmic rearrangement has been observed in *Tubifex* (T. Shimizu, personal communication). Further evidence on this point comes from experiments by Devries (1973), working with the oligochaete, *Eisenia*. By compressing zygotes to reorient the first cleavage plane, he was able to separate animal from vegetal teloplasm in the 2-cell embryo. Both cells, AB and CD, formed teloblasts and each generated mesodermal and ectodermal structures.

On the basis of these observations, we favour a two-step model for the determination of mesodermal and ectodermal precursors in *Helobdella*. According to this model, cells that inherit teloplasm gain the capacity to cleave further to make pro-teloblasts, while those that do not can make more micromeres. But whether or not a given pro-teloblast is an ectodermal precursor or a mesodermal precursor seems to be determined by other factors. One possibility is that an animal–vegetal gradient of a determinative substance could exist in the cortex of the zygote, where it would be resistant to the effects of centrifugation. Another possibility is that the differences between ectodermal and mesodermal precursors are induced by interactions, or the lack thereof, with micromeres at the animal pole. These possibilities are currently under investigation.

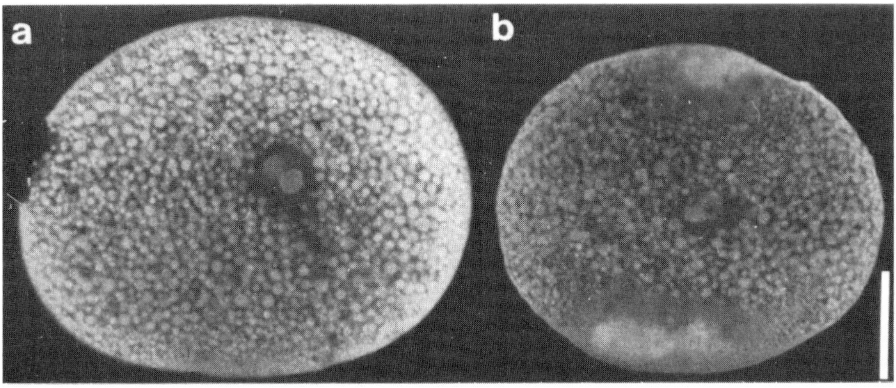

FIG. 3. Effects of a microtubule inhibitor, *cis*-tubulazole, on teloplasm formation. Photomicrographs of meridional sections along the animal–vegetal axis of embryos, viewed by fluorescence microscopy. a)Embryo bathed in *cis*-tubulazole contains yolk-deficient cytoplasm about the nucleus, but lacks teloplasm. b)Sibling embryo, bathed in *trans*-tubulazole, contains both perinuclear plasm and teloplasm. Scale bar, 100 μm.

The mechanism of teloplasm formation

Having established that teloplasm is a factor which specifies certain aspects of cell lineage in *Helobdella*, two further questions of interest arise: (1) how does teloplasm form? and (2) what in it affects cell fate? The cytoskeletal basis of the cytoplasmic rearrangement by which teloplasm arises in the zygote has been studied in *Tubifex* by Shimizu (1984, 1985, 1986). By observing the binding of rhodamine-labelled phalloidin and by electron microscopy, he demonstrated the presence of an extensive network of actin microfilaments in the cortex of the egg that seemed to contract across the surface of the embryos toward the poles during teloplasm formation. In support of this hypothesis, he found that cytochalasin D, an inhibitor of actin polymerization, in the bath medium blocks teloplasm formation in *Tubifex*, and that colchicine, an inhibitor of microtubule polymerization, had no effect. In a series of experiments initially expected to be a routine replication of Shimizu's observations, we were surprised to find that in *Helobdella*, cytochalasin fails to block teloplasm formation, even when microinjected into the zygote at ten times the concentration required to block cleavage (S. H. Astrow & D. A. Weisblat, in preparation). On the other hand, two inhibitors of microtubule polymerization, nocodazole and *cis*-tubulazole, block teloplasm formation; colchicine is less effective and *trans*-tubulazole (which is not supposed to affect microtubule function and therefore serves as a control for non-specific effects of the active *cis*-isomer) had no effect (Fig. 3). Cortical microtubules could be readily observed immunohistochemically in

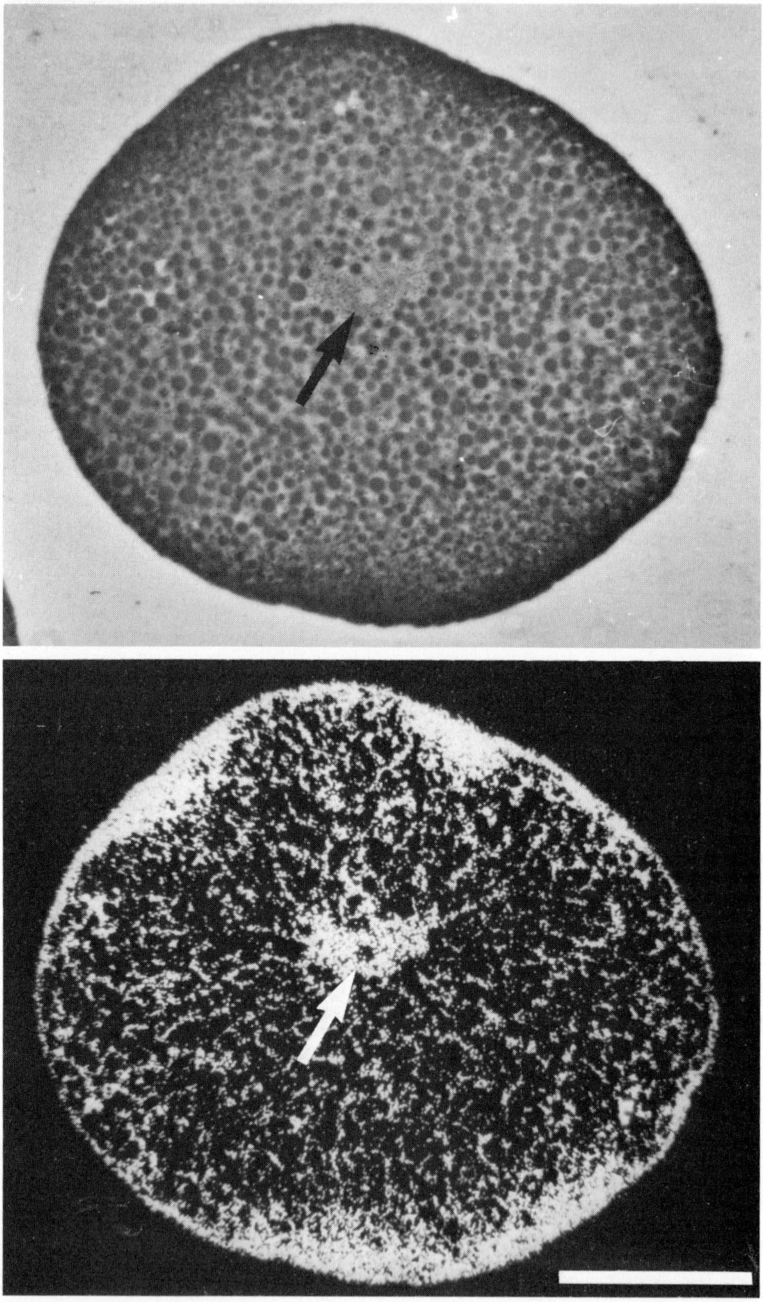

the cortices of normal zygotes and zygotes treated with *trans*-tubulazole, but were much reduced in the cortices of zygotes treated with *cis*-tubulazole. Neither nocodazole nor tubulazole blocked teloplasm formation in *Tubifex* embryos. We are forced to conclude that the process of teloplasm formation in *Helobdella* differs to some extent from that in *Tubifex*, despite the fact that these are presumably analogous processes. These results remind us that care must be taken when extrapolating from one organism to another and thus support the notion that, to get a true understanding of development, it is important to apply as many techniques as possible to each species of interest.

The identification of active factors in teloplasm

What substance(s) in teloplasm actually specify the D cell line in *Helobdella*? There is no answer to this question at present. Fernandez & Stent (1980) showed that teloplasm in *Theromyzon* contains all the components of normal cytoplasm. We also showed that yolk-deficient cytoplasm in *Helobdella* is enriched for a diverse set of proteins relative to yolky cytoplasm (Astrow et al 1987). It is important to consider the possibility that any or all of these components may contribute to the effects of teloplasm on specifying the fate of the D cell line. Nonetheless, we are concentrating on the roles of RNAs in this process. Our rationale for this bias is not much better than that of the drunk looking for his lost wallet under the streetlamp because that is where the light is best, but two lines of evidence at least partially validate this approach.

The first is that, as shown by acridine orange staining, teloplasm is enriched for RNAs and RNA is redistributed with the teloplasm by centrifugation (Astrow et al 1987). In addition, *in situ* hybridization experiments with tritiated polyuridylic acid indicate that polyA$^+$RNAs accumulate in teloplasm as it forms and co-segregate with it throughout the cell divisions leading to teloblast formation (Fig. 4). Preliminary experiments indicate that little transcription goes on in the embryo until after blast cell production has begun (S. T. Bissen, personal communication). Thus, the signal obtained by *in situ* hybridization with tritiated polyuridylic acid must represent either maternal poly A$^+$RNAs or the polyadenylation of maternal RNAs.

FIG. 4. Localization of polyadenylated RNAs by *in situ* hybridization with tritiated polyuridylic acid. Brightfield (top) and darkfield (bottom) photomicrographs of sections made from an embryo fixed midway through teloplasm formation. Fixed embryos were permeabilized and subjected to the hybridization protocol, then sectioned and processed for autoradiography. During teloplasm formation, both animal and vegetal domains of yolk-deficient cytoplasm label extensively, as do perinuclear cytoplasm and radially oriented strands of cytoplasm. The pronucleus (arrows) is unlabelled, however. Scale bar, 100 μm.

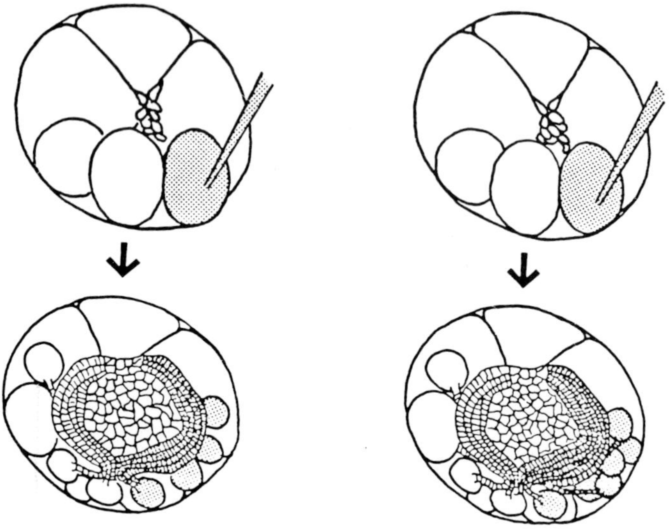

FIG. 5. Polyadenylic acid (polyA) induces the formation of supernumerary teloblasts and corresponding bandlets of blast cells. Diagrammatic representation of control (top left) and experimental (top right) embryos, in which the right NOPQ pro-teloblast (stage 5) was injected with normal lineage tracer or lineage tracer plus polyA, respectively. By stage 7 (below), the control embryo has generated the full normal complement of ten teloblasts and bandlets, including four righthand ectoteloblasts and bandlets from the injected cell (stippling). But polyA-injected pro-teloblasts typically make supernumerary teloblasts and bandlets (here, a total of six). The supernumerary cells can correspond to any of the normal ectodermal cell lines, i.e. N, O, P or Q.

The second line of evidence stems from a strange and fascinating observation, namely that microinjecting new-born teloblasts or their precursor blastomeres with polyadenylic acid (polyA) induces the formation of supernumerary teloblasts (Fig. 5; Ho & Weisblat 1987, R. K. Ho, unpublished, B. H. Nelson & D. Lans, personal communication). The effect is confined to the progeny of the injected blastomere and cannot be obtained by injecting monomeric adenylic acid. Injecting polyG, polyC or other polyanions has no such effect on cell fates. The effect of polyA injection is temporary; eventually the teloblasts arising from a blastomere injected with polyA begin making bandlets of blast cells in the normal manner. By contrast, cell lines derived from blastomeres injected with polydeoxyadenylic acid, which should be resistant to hydrolysis, continue making equal divisions indefinitely. Thus, it seems that the effect of polyA injection depends on the polymer, not on any low molecular weight breakdown product, and that the effect may be terminated by hydrolysis of the polyA.

The supernumerary teloblasts and their progeny blast cells assume one of the five normal identities, according to the identity of the injected blastomere. Thus, N or M teloblasts make only more M teloblasts, NOPQ blastomeres can make supernumerary N, O/P or Q blastomeres. Finally, by this technique, it is possible to generate embryos containing supernumerary definitive progeny of the duplicated cell line.

On the basis of these observations, we are hopeful that the effect of polyA injection is not an artifact, but rather reflects one or more specific effects on the production, localization, translation or survival of endogenous RNAs. Further experiments should enable us to distinguish these possibilities and thus make progress towards identifying factors within teloplasm that specify cell fate in *Helobdella* and also the mechanisms by which these factors act.

Acknowledgements

The research summarized here has been supported by grants DCB-8409785 and DCB-8711262 from the US National Science Foundation to DAW. SHA was supported from NIH training grant GM 07048.

References

Astrow SH, Holton B, Weisblat DA 1987 Centrifugation redistributes factors determining cleavage patterns in leech embryos. Dev Biol 120:270–283

Bissen ST, Weisblat DA 1989 The durations and compositions of cell cycles in embryos of the leech *Helobdella triserialis*. Development 106:105–118

Devries J 1973 Aspect du determinisme embryonnaire au cours de premiers stades de la segmentation chez le lombricien *Eisenia foetida*. Ann Embryol Morphog 6:95–108

Fernandez J 1980 Embryonic development of the glossiphoniid leech *Theromyzon rude*: characterization of developmental stages. Dev Biol 76:245–262

Fernandez J, Olea N, Matte C 1987 Structure and development of the egg of the glossiphoniid leech *Theromyzon rude*: characterization of developmental stages and structure of the early uncleaved egg. Development 100:211–225

Fernandez J, Stent GS 1980 Embryonic development of the glossiphoniid leech *Theromyzon rude*: structure and development of the germinal bands. Dev Biol 78:407–434

Ho RK, Weisblat DA 1987 Replication of cell lineages by intracellular injection of polyadenylic acid (Poly A) into blastomeres of leech embryos. In: O'Connor JD (ed) Molecular Biology of Invertebrate Development. Alan R Liss, New York, p 117–131

Holton B, Astrow SH, Weisblat DA 1989 Animal and vegetal teloplasms mix in the early embryo of the leech, *Helobdella triserialis*. Dev Biol 131:182–188

Kimmel CB, Law RD 1985a Cell lineage of zebrafish blastomeres II. Formation of the yolk syncytial layer. Dev Biol 108:86–93

Kimmel CB, Law RD 1985b Cell lineage of zebrafish blastomeres III. Clonal analysis of the blastula and gastrula stages. Dev Biol 108:94–101

Kimmel CB, Warga RM 1987 Indeterminate cell lineage of the zebrafish embryo. Dev Biol 124:269–280

Kramer AP, Weisblat DA 1985 Developmental neural kinship groups in the leech. J Neurosci 5:388–407

Shankland M 1987a Differentiation of the O and P cell lines in the embryo of the leech I. Sequential commitment of blast cell sublineages. Dev Biol 123:85–96
Shankland M 1987b Differentiation of the O and P cell lines in the embryo of the leech II. Genealogical relationship of descendant pattern elements in alternative developmental pathways. Dev Biol 123:97–107
Shimizu T 1982 Ooplasmic segregation in the *Tubifex* egg: mode of pole plasm accumulation and possible involvement of microfilaments. Roux's Arch Dev Biol 191:246–256
Shimizu T 1984 Dynamics of the actin microfilament system in *Tubifex* during ooplasmic segregation. Dev Biol 106:414–426
Shimizu T 1985 Movements of mitochondria associated with isolated egg cortex. Dev Growth & Differ 27:149–154
Shimizu T 1986 Bipolar segregation of mitochondria, actin network and surface in the *Tubifex* embryo: role of cortical polarity. Dev Biol 116:241–251
Stent GS, Weisblat DA, Blair SS, Zackson SL 1982 Cell lineage in the development of the leech nervous system. In: Spitzer N (ed) Neuronal Development. Plenum, New York, p 1–44
Stent GS 1985 The role of cell lineage in development. Philos Trans R Soc Lond B Biol Sci 312:1–38
Weisblat DA 1988 Equivalence groups and regulative development. In: Easter SS, Barald KF, Carlson BM (eds) From message to mind. Sinauer, Sunderlund MA, p 209–224
Weisblat DA, Shankland SM 1985 Cell lineage and segmentation in the leech. Philos Trans R Soc Lond B Biol Sci 312:39–56
Whitman CO 1878 The embryology of *Clepsine*. Q J Microsc Sci 18:215–315
Zackson SL 1984 Cell lineage, cell-cell interaction, and segment formation in the ectoderm of a glossiphoniid leech embryo. Dev Biol 104:143–160

DISCUSSION

Tomlinson: Have you any done experiments on transplanting teloplasm?

Weisblat: Stephanie Astrow tried in conjunction with Kathryn Anderson but the teloplasm is too viscous. It was frustration with those experiments that led Robert Ho to try the polyA injection. Having seen the results of the centrifugation, which show that there is a causal relationship between teloplasm inheritance and cleavage pattern, I would be more interested in fractionating the teloplasm than just trying to transplant it.

Heasman: Do we know anything more about teloplasm apart from the fact that there is RNA there? Are there antibodies against it?

Weisblat: In all respects it appears to be normal cytoplasm, except that it doesn't have yolk platelets. We are interested in how it is formed. The teloplasm is uniformly distributed among yolk platelets in the beginning, then there is movement after second polar body formation. Whether it is active movement of yolk platelets away from that region or active movement of granular cytoplasm into that region, we don't know.

Heasman: Drawing a parallel to Doug Melton's talk, it seems that the cytoskeleton is important in organizing asymmetries in the cytoplasm.

Weisblat: I am still perplexed at the apparent difference between *Tubifex* and leech. We expected them to be exactly the same.

Cabrera: People working with *C. elegans* for a long time used similar arguments about cytoplasmic localization to explain the segregation of cell fates. However, when they did a stringent test, by simply rotating two non-equivalent cells, they found that cell–cell interactions play a major role in the assignment of cell fates (Priess & Thomson 1987). Can you do the same in this embryo, switch the position of two cells and see what happens?

Weisblat: We have tried this sort of experiment in leech, using the differences in fate between the mesodermal and ectodermal pro-teloblasts, which inherit vegetal and animal teloplasm, respectively. We thought that if the two teloplasms were the same, may be the differences between mesodermal and ectodermal fates were just position-dependent differences. So we have tried to switch the positions of these two cells as they arise from cell D′, but without success. We have also tried eliminating micromeres to see whether there are things in the micromeres that induce differences between ectodermal and mesodermal tissue, again without success.

Garcia-Bellido: One surprising feature of the leech is that the dorsoventral distribution of the progeny of teloblasts is very similar to that in *Drosophila*. If you take the different NOPQ teloblast progeny kinship groups and distribute them over the embryo, the pattern of dorsal, medial and ventral histiotypes is like the corresponding one in *Drosophila*.

Weisblat: I agree. For example, the neuroectoderm at the ventral midline in *Drosophila* is somewhat analogous to the N bandlets in the leech which generate the bulk of the nervous system and make very small contributions to the epidermis. But the Q bandlets, which occupy future dorsal territory, contribute little to the central and peripheral nervous systems and mainly to dorsal and lateral epidermis.

Garcia-Bellido: In the case of experimentally produced duplicated teloblasts, are the kinship groups retained or do they appear to be scrambled?

Weisblat: If you inject NOPQ, you can generate from 1–4 or even more supernumerary teloblasts. Brad Nelson is looking at what happens to these extra cells. He finds that extra cells can be one of the ectodermal types, N, O/P or Q, and that the extra O/P teloblasts go on to make specific O or P bandlets. Thus you can get supernumerary bandlets of any of the four ectodermal types, surprisingly.

Garcia-Bellido: But that means that there is a multiplication of discrete subsets, not reorganization of a continuous field. So when blastomeres divide, they sample discontinuous rather than continuous determinants in the egg.

Weisblat: It would appear so—although it is surprising, because the previous experiments that I did with Seth Blair (1984) and Marty Shankland (1984) suggested that these O/P teloblasts are equivalent and that in normal development the fates of their progeny are determined by spatial interaction. Whichever one was closer to the animal pole assumed the P fate, which thus seemed to be the primary fate, and it appeared that any O/P-derived cell line not in this favoured position would be constrained to assume the O fate. We therefore expected that any supernumerary bandlets would also take on the secondary fate, the O fate. But that is not the case—you can get extra bandlets of either type.

Garcia-Bellido: Again it means that once they are singled out, they do not check with their neighbours, because you are duplicating part of the kinship group.

Weisblat: But this cell normally does communicate with its neighbours—that's the amazing thing.

Wieschaus: If you get a supernumerary N cell, does it come off in the lineage immediately after the division producing the N cell?

Weisblat: Brad Nelson is doing those experiments. The normal division of NOPQ pro-teloblasts is somewhat asymmetric, it gives a larger cell (OPQ) and a smaller cell, the N teloblast. In polyA-injected embryos, this division seems to occur normally but the N cell then sometimes divides. We expect that this is how supernumerary N cells arise, but we don't know yet.

Wieschaus: Can you distinguish an N cell only from its lineage or does it also sit in a specified position relative to other teloblasts?

Weisblat: In normal development there is a specific position that we associate with the N teloblast.

Garcia-Bellido: Is the spatial cleavage of the D′ blastomere fixed?

Weisblat: It is fairly stereotyped; it is obliquely equatorial, a modified spiral cleavage. From the 4-cell leech embryo the micromeres are extruded at the animal pole and are ectodermal precursors. So the nucleus is at the animal pole in all four of the macromeres; after the third cleavage the D′ nucleus drops back down into the centre of the embryo so that the fourth cleavage is obliquely equatorial, forming two large cells, the ectodermal (DNOPQ) and mesodermal (DM) precursors. The other three nuclei stay at the animal pole and make a few more micromeres in the A′, B′ and C′ cell lines.

Garcia-Bellido: Have you tried perturbing that cleavage, to see whether the resulting blastomeres or teloblasts would have the same kinship groups?

Weisblat: If you do the centrifugation experiment, you mix the teloplasms and some goes into this cell, so that it also divides. These animal cells still generate what we think are ectodermal teloblasts and these vegetal cells still generate mesoteloblasts.

Davidson: E. B. Wilson did that experiment with another annelid, *Nereis*, in 1892 by putting it under pressure.

Garcia-Bellido: The idea is to perturb that cleavage, then label the cells by injection and follow the progeny, and see whether the kinship groups are the same as in normal embryos.

Weisblat: These centrifuged embryos don't develop far enough in the leech to generate definitive kinship groups. With the polyA-injected embryos, we do get normal kinship groups.

Garcia-Bellido: Even in the supernumerary teloblasts?

Weisblat: Apparently yes; for example, you can duplicate N teloblasts by injection of polyA and later show that the segmental ganglia contain an extra complement of serotonin-containing neurons, cells that are normally produced by the N cell line (Ho & Weisblat 1987).

Garcia-Bellido: Have you traced that? Have you labelled these cells and looked at the offspring, or is it simply that the end pattern is duplicated?

Weisblat: We haven't traced the complete lineages, but the initial mitoses are characteristic of N,O,P or Q. When you duplicate the N line, you can duplicate the definitive progeny. But we haven't gone all the way, even in normal development.

Lawrence: Are there any changes in the other neighbouring teloblast lines when you duplicate the N line?

Weisblat: No, as far as we can tell, the effect is always confined to the cell line that was duplicated.

Lawrence: How precise is the lineage? Do you duplicate all the normal progeny of N when you have two N teloblasts?

Weisblat: That we don't know, there could easily be regulation. For example, if you ablate one N teloblast, so that there is a single N line, the progeny of the surviving line become distributed on both sides of the embryo, across the midline (Blair & Weisblat 1982). If you do the N teloblast duplication, so that there are three N lines, it appears that there is also a bilateral distribution (Ho & Weisblat 1987).

Lawrence: So you think that each adventitious N teloblast produces a normal set of progeny, although they are distributed bilaterally.

Wolpert: Is there a unique correspondence between the teloplasm and this phenomenon? How much teloplasm must a cell have in order to make teloblasts?

Weisblat: If you inject polyA into an A, B or C macromere, they never make teloblasts that make blast cells, presumably because these cells lack sufficient teloplasm. When you do the centrifugation experiment and look at the 8-cell stage, you can select fairly reliably those embryos in which both cells C and D will continue cleaving to form teloblasts, on the basis of which embryos have teloplasm in both cells C and D. There are also variations in the relative sizes of cell D and cell C, but that is not a predictor. It is not cell size but inheritance of teloplasm.

Gardner: In the centrifugation experiment where the teloplasm is re-distributed to two blastomeres, as I understand it, none of these embryos go through and regulate, they all fail.

Weisblat: The embryos where both C and D cleave fail. The mirror-image embryos, in which C cleaves but D does not, are very rare but those will develop normally.

Gardner: Where you get C and D cleaving, what is the basis of the failure? Is there any regulation or have you effectively deleted the normal lineage of C?

Weisblat: I don't know. We assume that the normal C lineage is deleted, but we cannot say that this is why the embryo fails. Another possibility for failure is that you do not get a total of exactly 10 teloblasts from C and D after centrifugation. If you redistribute the teloplasm by centrifugation, you get supernumerary teloblasts. I think the reason these embryos fail is that the extra bandlets screw up the normal morphogenic movements of the embryo. Similarly, if you duplicate the M (mesodermal) teloblast selectively, you can't get the definitive progeny because the embryo again fails to undergo its normal morphogenic movements. Duplicating an M teloblast on one side, if you look late enough in development, is equivalent to having ablated *this* side, rather than having duplicated it, which was a disappointment.

Melton: Have you tried other ways of initiating segregation of the teloplasm? I am thinking about Bill Jeffery's experiments on ascidians, where he can induce the myoplasm to segregate by using a calcium ionophore.

Weisblat: We tried some calcium ionophores and they had no effect. An important difference between the two systems may be that leeches normally develop in very dilute medium and have evolved very impermeable membranes and an independence of external ions.

Davidson: How much polyA do you inject compared with the amount of RNA in the egg?

Weisblat: We inject a lot, approximately 10 ng per NOPQ cell, for example. This is probably of the same order of magnitude as the RNA content of the cell.

Davidson: What do you think the polyA is doing, besides soaking up histones?

Weisblat: It seems unlikely that it's acting as a polyanion blotter, because other polyanions don't have this effect.

Davidson: They have similar effects, they stop cleavage.

Weisblat: They either stop cleavage or have no effect whatsoever. We could be soaking up enzymes that are responsible for degradation of maternal mRNA.

Davidson: The other polyanions that you injected, are they size controlled?

Weisblat: No, this is all commercial material.

Davidson: PolydA is often much longer than commercial polyA, so I don't know if that's a good control. It could be a completely non-specific effect, the polyA could be saturating charge sites used for mobilization of the cytoskeletal apparatus that initiates cleavage.

Struhl: Do you think that normal polyA is being used as a signal?

Weisblat: My argument is that the injected polyA in some way affects either the localization, the metabolism or the stability of endogenous messenger RNA.

It might be delaying transcription or delaying degradation of maternal messenger RNA. Affecting any of these processes could require stoichiometric quantities of injected RNA, because they are relatively non-specific.

Kimble: How are you going to find out what the effect of this polyA is?

Weisblat: First we would like to look at the metabolism of the material we inject. We are trying to get a uniform-sized polyA to see whether the termination of the polyA effect is due to its degradation. We would also like to examine the effect of injected polyA on the stability of messenger RNA, starting by injecting labelled exogenous mRNA and examining its stability in embryos that have or have not been injected with polyA.

Meinhardt: I would like to make a more general remark concerning lineage-dependent systems, i.e. systems in which cell division plays an important role in the segregation of two pathways. We have to keep in mind that cell division itself is an extremely important pattern-forming event. Imagine a more or less homogeneous cell; for division, it must develop two poles—always two poles, never one and never three. From the viewpoint of pattern formation, to count precisely to two is not a trivial problem. I propose that this problem has a simple solution: the two poles are different, there is always a plus pole and a minus pole, even if there is an equal cell division. Then the cell does not have to count to two, because the formation of each pole is an all-or-nothing event.

If both poles are different, then it is easy to use this to explain lineage systems: if genetic material is pulled to the plus pole, it receives a different signal to the genetic material at the minus pole. Thus, the daughter cells become different from each other. I would expect that in every cell division the signal to separate those two poles is always the same. Since cell division is evolutionarily a very old process, I believe that if we can understand pattern formation that leads to cell division, we will have a much better basis for understanding pattern formation in general. Do you see any way in which the leech system can be used to find out what distinguishes the plus and the minus poles? I don't think that it is something like microtubule or spindle formation, because these are presumably subsequent processes, not the initiating one.

Weisblat: There are many possibilities which will be very difficult to sort out. It is not even certain that there will be only one.

Wolpert: I think Hans (Meinhardt) is wrong. I don't think there is any evidence that would suggest that the two poles of the spindle are labelled plus or minus.

References

Blair SS, Weisblat DA 1982 Ectodermal interactions during neurogenesis in the glossiphoniid leech *Helobdella triserialis*. Dev Biol 91:64–72

Ho RK, Weisblat DA 1987 Replication of cell lineages by intracellular injection of polyadenylic acid (polyA) into blastomeres of leech embryos. In: O'Connor JD (ed) Molecular biology of invertebrate development. Alan R Liss, New York, p 117–131

Priess JR, Thomson JN 1987 Cellular interactions in early c-elegans embryos. Cell 48:241–250

Shankland M, Weisblat DA 1984 Stepwise commitment of blast cell fates during the positional specification of the O and P cell lines in the leech embryo. Dev Biol 106:326–342

Weisblat DA, Blair SS 1984 Developmental indeterminacy in embryos of the leech *Helobdella triserialis*. Dev Biol 101:326–335

Cell lineage and cell states in the *Drosophila* embryo

Peter A. Lawrence

Medical Research Council, Laboratory of Molecular Biology, Hills Road, Cambridge CB2 2QH, UK

Abstract. Much of this essay is a cautionary tale about the inherent difficulty of building theories to explain pattern formation in embryos. In particular, I discuss the relationship between cell state and cell lineage. The idea that embryonic cells can occupy discrete genetic states is an old one but, because of the mass of new information from *Drosophila* developmental genetics, it has been given new life. I propose an operational definition of a cell state and argue that a true cell state should be propagated faithfully through cell lineage for at least some divisions. The genes involved should have a controlling role and the borders between populations of cells in two different states should be so sharp that cells fall clearly into one state or another.

1989 Cellular basis of morphogenesis. Wiley, Chichester (Ciba Foundation Symposium 144) p 131–149

"Argumentation cannot suffice for the discovery of new work, since the subtlety of Nature is greater many times than the subtlety of argument"

Francis Bacon

"Natural selection is not a clean designer"

Francis Crick

The mechanisms responsible for the formation of patterns in developing animals are still elusive. Over the hundred years or so that the problem has been clearly recognized there has been an uneasy truce between expectation and observation. Expectations have stemmed from our empirical—and presumptuous—ideas of how *we* would design and construct an animal; as some ideas have more intuitive appeal than others, these have been preferred for no other reason. The long-standing controversy about mosaic and regulative eggs is an example: the hypothesis that the mother lays down a complex mosaic of specific determinants in the egg, to prefigure the body plan of the embryo, is an old and strangely attractive one—so attractive it has been replaced only reluctantly under the growing weight of contrary evidence. An important extension of the mosaic theory is the thesis that, once specified, cells normally propagate their type faithfully through cell division but any further diversifications occur through switches of cell state at specialized cell divisions. These ideas have given rise to theories of pattern formation that feature sharply demarcated cell states and branching trees of cell lineage.

Early studies of cell lineage were hampered by inadequate methods, so that wishful thinking and poor microscopy became indispensable parts of the process of discovery. One example of this concerns the compound eye of insects where the prejudice that each ommatidium must be a clone led to a description of the cell divisions of a mother cell (Bernard 1937). Better methods are available now—indeed some were available then!—and we believe that ancestry and lineage relationships play no part in the final allocation of eye cells (Ready et al 1976, Lawrence & Green 1979).

The discovery of the '*Minute* technique' in *Drosophila* meant that a cell could be genetically marked and also made more vigorous than the other cells of the embryo (Morata & Ripoll 1975). This led to the knowledge that the insect embryo is constituted, at least partly, of groups of equivalent cells, the cells in each group together constructing a precisely defined part—a compartment—of the adult body. Within the developing set of cells there are no lineage restrictions, any embryonic cell can give rise to any part of the compartment in the adult. However, between compartments there are sharply delineated boundaries that growing clones respect; these lineage boundaries are always identically placed in different individuals (reviewed in Garcia-Bellido et al 1979).

The scientific community found some compartment boundaries more acceptable than others: the segment had traditionally been thought of as the basic building block of insects and so the discovery that the segmental boundary could be objectively defined by cell lineage experiments (Lawrence 1973) came as no surprise. However, the anteroposterior compartment boundary within the wing (Garcia-Bellido et al 1973) met much more scepticism because it had been assumed that the wing would be made in one piece. Ironically, we now believe that the fundamental metamere of insects is not the segment but the parasegment (Martinez-Arias & Lawrence 1985) and that the parasegment boundary (which ultimately becomes the boundary *within* the wing) is established earlier than the segment boundary and, because of this and patterns of gene expression, is a more fundamental unit of design. The unexpectedness of the parasegment illustrates as well as any how unwise it is to allow prejudice to influence interpretation—at least in the field of pattern formation.

The establishment of pattern in the *Drosophila* embryo

Thanks largely to the work of Nüsslein-Volhard and Wieschaus and their pupils we now can begin to tackle pattern formation by genetic and molecular methods.

They have shown that the anteroposterior and dorsoventral patterns are under independent genetic control. Both depend on gradient systems set up by the mother and require the function of a small number of zygotic genes for implementation (see Nüsslein-Volhard et al 1987, Nüsslein-Volhard, this volume, Struhl, this volume). I will briefly consider the anteroposterior pattern, limiting the discussion to questions of cell lineage and cell state: What is the significance of cell lineage in early *Drosophila* development, and how many cell states are there?

The ground pattern: a description

By the beginning of gastrulation, all cells in the main part of the body are allocated to parasegments, apparently with no ambiguity. We can see the allocation by studying the stripes of expression of the *fushi tarazu* (*ftz*) and *even-skipped* (*eve*) genes, whose sharp and stable anterior boundaries demarcate the parasegment boundaries. These boundaries are lineage boundaries, such that cells either side of them have different fates in the larva and the adult; they also coincide cell-by-cell with the anterior boundaries of expression of the *engrailed* gene (Lawrence et al 1987). At this stage, each parasegment averages about 3.5 cells across (Sullivan 1987), but as the parasegment boundaries are wiggly, the width of the parasegment varies from 2–5 cells. Those cells which express *engrailed* are in the anterior part of the parasegment and consist of a stripe about one cell across, although it can vary from 0–2 cells in width.

At about this time the transcription of selector genes begins, for example *Ultrabithorax* (*Ubx*) is transcribed early in parasegment 6; however, the pattern of transcription rapidly becomes more complex and differs in the epidermis, central nervous system, somatic mesoderm and visceral mesoderm—four 'germ layers' that have been defined as different domains in the dorsoventral axis. In the ectoderm, the patterns of expression of the homeotic or selector genes are complicated, but remain simple parasegmental blocks in the mesoderm. For example, *Ubx* is expressed in parasegments 5–13 of the epidermis and central nervous system, 6–12 of the somatic mesoderm (Akam & Martinez-Arias 1985) and parasegment 7 in the visceral mesoderm (Bienz et al 1988).

The ground pattern: interpretations

The immediate task of the anteroposterior patterning system is to polarize the embryo and divide it into the correct number of parasegments. Two observations illustrate that this is not done by counting cells but by some mechanism sensitive to scale and proportion. Thus, if the mother has extra doses of *bicoid*$^+$, the cephalic furrow and the anterior margin of parasegment 2 are moved posteriorly by as much as 20% of the egg length. This means the number of cells available to form the parasegments 2–14 is reduced by that amount. The stripes of *ftz* are more crowded than usual and each parasegment has an average width of less than three cells—nevertheless, development continues and the hatched larva is viable and normal (Driever & Nüsslein-Volhard 1988). Likewise, in embryos mosaic for patches of cells of different sizes, the *ftz* stripes look as if they have been painted with a brush, they are of even width and take no account of the cell size (Sullivan 1987) (Fig. 1).

The zygotic gap genes, as well as *runt*$^{+}$ and *hairy*$^{+}$, are required for the proper striping of *ftz*$^{+}$ and *eve*$^{+}$ (reviewed in Akam 1987). The main function of *eve*$^{+}$ and *ftz*$^{+}$ is probably to delineate the parasegment borders (Lawrence et al 1987); once they have finished their work, most of the embryo has been transformed from a supracellular field into defined groups of cells, each cell being unequivocally allocated to a particular parasegment.

Some 20 'segmentation' genes are divided by Nüsslein-Volhard & Wieschaus (1980) into two classes. They are required for proper metamerization—if any one of these genes is mutant, the larvae develop without parts of every metamere (segment polarity genes) or without parts of alternate metameres (pair-rule genes). The patterns of expression of these genes echo this classification, some segment polarity genes are expressed as stripes in every metamere, some pair-rule genes in every alternate metamere. However, all is not so orderly as it might

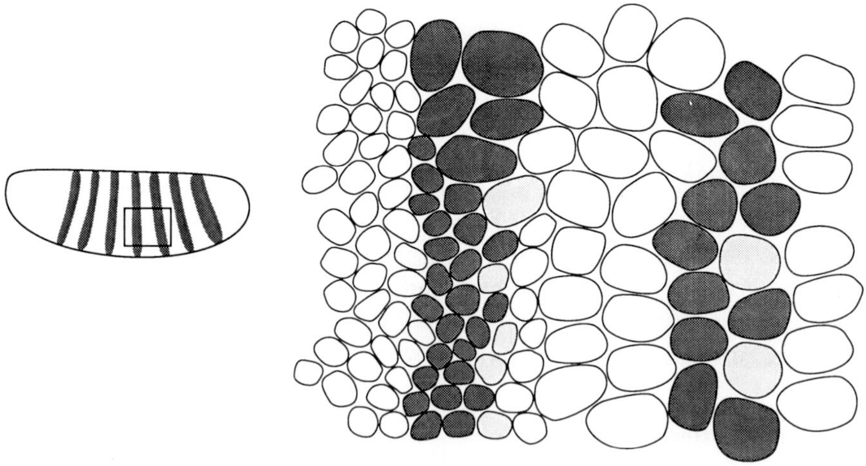

FIG. 1. Diagrammatic summary of the results of Sullivan (1987). At low resolution (left) the *ftz* stripes look normal, but in the enlarged panel, (right) there are cells of very different sizes; the stripes are of normal width but appear to be independent of cell size. I have indicated the presumed asymmetry of the stripes (not reported by Sullivan) by showing more weakly labelled nuclei on the posterior edge of the stripes. Anterior to left, posterior to right.

seem: the *engrailed* gene, which was originally classified as a homeotic selector gene, later as a pair-rule gene, is expressed in every metamere and is therefore now referred to by some as a polarity gene. This instability in the status of *engrailed* is due more to the influence of fashion in biological thought than to any real increase in understanding.

Cell states

Several attempts to build a hypothesis of metamerization have been made; most assume that the main task of the segmentation genes is to allocate the cells of the blastoderm to specific states. For example, it was proposed much earlier that the ectoderm of the embryo consists of alternating stripes of anterior (A) cells and posterior (P) cells. Both A and P cells propagate their state to their offspring so that ultimately, in the adult, A cells form the anterior compartments and P cells form the posterior compartments. All P cells express and require the *engrailed* gene and we called the P cell state a 'genetic address' (Garcia-Bellido et al 1979). Several experiments with genetic mosaics suggest that A cells outnumber P cells by two or three to one, and this is confirmed by the expression of *engrailed* (Kornberg et al 1985). It is therefore well established that there are at least two cell states in the embryo. These cell states meet one important criterion; there is a cell lineage restriction in that A and P cells generate exclusively A and P daughters, respectively.

The idea of a cell state can be rather vague and a working definition is necessary before we can ask if there are more cell states in the embryo. In my opinion, a cell state should not be defined so loosely that if two cells express different genes they can be allocated to two states. The cell state should signify something more permanent that is propagated through cell lineage, if not forever, at least for some divisions. Also, if the cell state is to be marked by, or depend on, the expression of a particular gene, then that gene should be clearly expressed in some cells and not in others. The boundary between expressing and non-expressing cells should be sharp and unequivocal; for, surely, the idea of a cell state has little meaning if cells can occupy it only partially. In practice, this requirement means that, with the resolution available now, *in situ* hybridization is an inadequate method for mapping cell states. A method with the resolution to assess each cell is required to show if boundaries are sharp. This, for the moment, means using an antibody specific for the gene product of interest. Finally, a developmental cell state should be indicated by, and depend on, the expression of a gene with a controlling function. The expression of an ordinary gene (such as *yellow*$^+$) in stripes in the larva should not be taken to mean that the cells expressing *yellow*$^+$ must occupy a different state than those that do not.

Some of these points were recognized by Meinhardt (1984), who proposed a three-cell-state model, in which he suggested that the metameric repeat consists of three compartments, S, A and P. A and P are the anterior and posterior compartments as before. The need for an additional S compartment depended on an *a priori* argument: under the two-compartment model, a segment boundary forms where P cells meet A cells (a P/A junction). The model would also predict a boundary to form, with opposite polarity, at an A/P junction, but this does not happen. Meinhardt therefore posited the S compartment, to give a SAPSAP sequence with segment boundaries forming only at the P/S interface. However, cells are polar structures and this provides a natural alternative to the S compartment: we know that the insect epidermis is made of polarized cells (planar cell polarity, see Nübler-Jung 1987 for references) and this polarity may well be present in the cells of the blastoderm (e.g. Sander 1975). In this case A/P is not the same as P/A—just as GC in a DNA sequence is not the same as CG. Of course, this is not an argument against the three-cell-state model; it is only an argument against the argument for such a model.

More recently, the rapid increase in knowledge of when and where segmentation genes are expressed has led to numerous attempts to explain the results. These have mostly been digital descriptions, in that cells are described as if they either express or do not express the gene at a particular time. This would mean that the boundaries of the stripes of gene expression should be sharp at the level of individual cells, but with the exception of *ftz*, *eve* and *engrailed* this has not been looked at carefully. For *ftz* and *eve* it is not the case; while the anterior boundaries become sharp and stable, the posterior boundaries do not, presumably because the stripes narrow continuously (Lawrence et al 1987). By the criteria I have suggested, it follows that all the cells expressing, say, *ftz* do not occupy a specific cell state which distinguishes them from all the cells not expressing *ftz*.

Nevertheless, because the number of cells per metamere is about 3.5, and because four is perhaps a more attractive number than three, there has been a tendency to develop models with four cell-states. These are used not only as a way of describing the wild-type pattern of gene expression, but also as an aid to explaining the changed pattern of expression of one gene in an embryo mutant for another gene. The most detailed of these models has recently been advanced (Martinez-Arias et al 1988); it eschews the theory of a segmental gradient where the scalar value determines differentiation and the local vector determines polarity (see Lawrence 1973 for review). Four cell states are proposed, one E state coinciding with the posterior compartment proposed earlier (marked by the expression of *engrailed*) and three others, W, N and P, each to coincide with the expression of, or requirement for, a gene (*wingless*, *naked* and *patched*, respectively). The order of the states runs WENPWENP. The acquisition of these four cell states and their maintenance is thought to depend on interactions between neighbouring cells, e.g. in the absence of *patched, wingless* (which is normally expressed just anterior to the parasegment boundary, Baker 1987) is expressed in a broader band. This is modelled as NPWEN becoming NWWEN, causing the unnatural juxtaposition of N and W. This could induce the restoration of proper neighbours by 'intercalation' and a new cell with state E. This yields the palindrome, NEWWEN, consistent with the observation that an adventitious *engrailed* stripe develops. Thus, in the model, 'pattern arises from the states of the cells and from the interactions between different states, without the need for an overall morphogen whose concentration is interpreted by the cells'. Polarity also arises from the order and arrangement of cell states through interactions between neighbours and not as a vector in the concentration landscape of morphogen.

These postulated states are not of the type I defined earlier, because they are not stable through cell lineage. Martinez-Arias and colleagues even take the view that the cell state expressing *engrailed* may be unstable because, although the origin of anterior and posterior compartments in the adult is well established, their existence in the embryo is not completely proven. This example shows both how far one can go with cell-state models in which cells of discrete type interact with each other, and also some of the difficulties (particularly the need for *ad hoc* rules to explain initial conditions in a mutant, intercalation and polarity) when one tries to do without traditional gradient theories. Nevertheless, models of this type cannot be ruled out; indeed, the allocation of cells in the compound eye may well depend in part on 'digital' interactions between neighbouring cells (Tomlinson, this volume).

It may be premature to build such detailed formal models and instead we first need to seek the wild-type function of the genes, to go beyond their epistatic relations with other genes and find with which processes they are involved. For example, *wingless* is expressed in the posterior region of each parasegment from soon after gastrulation. As the cells in the embryo divide, the *wingless* stripe remains narrow, and this implies that some of the daughters of cells expressing *wingless* stop expressing the gene (Martinez-Arias et al 1988). By the criteria suggested above, I do not think *wingless* expression itself marks a cell state. Moreover, the non-autonomy of the mutant phenotype (Morata & Lawrence 1977) and the sequence of the *wingless* gene indicate a secreted protein that may well affect neighbouring or nearby cells (Rijsewijk et al 1987). Although the absence of the *wingless* product leads to a mutant phenotype similar to that when other segment polarity genes are lacking, we have little idea how this comes about—it only shows that the *wingless* product is one of several functions crucial for establishing and/or maintaining metameres in the embryo.

Cell states in the mesoderm?

Theories of metamerization tend to ignore the mesoderm, yet lineage studies suggest that the somatic mesoderm also consists of parasegments that are defined early in development. However, the parasegments are not divided into A and P compartments, and the *engrailed* gene is expressed only transiently there and apparently not required at all (for references see Lawrence & Johnston 1984). The block-like expression of homeotic genes such as *Ubx* and *Sex combs reduced* (*Scr*) suggests that each parasegment in the mesoderm may consist of cells that all occupy a unique cell state. For example, expression of *Ubx* in the somatic mesoderm of parasegment 6 is clearly defined at the cellular level and has a sharp anterior boundary. This boundary in the mesoderm underlies the equivalent boundary in the ectoderm almost exactly. It coincides cell-by-cell with the anterior boundary of parasegment 6, as defined by the expression of a *ftz*-β-galactosidase gene construct, presumably because parasegment borders exist before expression of selector genes and confine expression of these genes to particular parasegments (Lawrence et al 1987). Probably this boundary is the lineage boundary deduced from cell marking experiments and, if so, defines the border between cells of one state (expressing *Ubx* in parasegment 6) and cells in another (not expressing *Ubx* in parasegment 5). This border is inherited by the muscles in the larvae, where *Ubx* is expressed in the muscles of the first abdominal segment (belonging to parasegment 6, Martinez-Arias & Lawrence 1985) but not in the muscles of the third thoracic segment (parasegment 5).

This argument can be extended to the visceral mesoderm, where *Ubx*, *Scr* and other selector genes are also expressed in blocks, each the size of a parasegment and each placed in the position of a single parasegment (Bienz et al 1988). Little is known of cell lineage in the visceral mesoderm, but it seems likely that the visceral mesoderm will prove to be segregated into parasegmental compartments by lineage, these boundaries coinciding cell-by-cell with the boundaries of homeotic gene expression.

Conclusion

In this essay, I have argued that experimental results are a better guide than intuition in understanding how patterns are formed in developing animals. In attempting to define a cell state, I have suggested the concept should be restricted to examples where expression of a controlling gene is faithfully propagated through at least some cell divisions. The boundaries between cells in one state and those in another should be sharp. In the *Drosophila* embryo, the first lineage boundaries in the anteroposterior axis are probably the parasegmental borders. At about the same time as these are established, the embryo is divided dorsoventrally into ectoderm, somatic mesoderm and visceral mesoderm. The ectodermal cells of each parasegment, but not the mesoderm, are allocated to two cell states corresponding to the anterior and posterior compartments. So far experimental evidence calls for no more than these two cell states.

References

Akam ME, Martinez-Arias A 1985 The distribution of *Ultrabithorax* transcripts in *Drosophila* embryos. EMBO (Eur Mol Biol Organ) J 4:1689–1700

Akam M 1987 The molecular basis for metameric pattern in the *Drosophila* embryo. Development 101:1–22

Anderson KV 1987 Dorsal–ventral embryonic pattern genes of *Drosophila*. Trends Genet 3:91–97

Baker NE 1987 Molecular cloning of sequences from *wingless*, a segment polarity gene in *Drosophila*: the spatial distribution of a transcript in embryos. EMBO (Eur Mol Biol Organ) J 6:1765–1773

Bernard F 1937 Recherches sur la morphogénèse des yeux composés d'arthropodes. Bull Biol Fr Belg (Suppl) 23:1–162

Bienz M, Saari G, Tremml G, Züst B, Lawrence PA 1988 Differential regulation of *Ultrabithorax* in two germ layers of *Drosophila*. Cell 53:567–576

Driever W, Nüsslein-Volhard C 1988 The *bicoid* protein determines position in the *Drosophila* embryo in a concentration-dependent manner. Cell 54:95–104

Garcia-Bellido A, Ripoll P, Morata G 1973 Developmental compartmentalization of the wing disc of *Drosophila*. Nat New Biol 245:251–253

Garcia-Bellido A, Lawrence PA, Morata G 1979 Compartments in animal development. Sci Am 241:102–110

Kornberg T, Siden I, O'Farrell P, Simon P 1985 The *engrailed* locus of *Drosophila*: In situ localization of transcripts reveals compartment-specific expression. Cell 40:45–53

Lawrence PA 1973 A clonal analysis of segment development in *Oncopeltus* (Hemiptera). J Embryol Exp Morphol 30:681–699

Lawrence PA, Green SM 1979 Cell lineage in the developing retina of *Drosophila*. Dev Biol 71:142–152

Lawrence PA, Johnston P 1984 On the role of the *engrailed* gene in the internal organs of *Drosophila*. EMBO (Eur Mol Biol Organ) J 3:2839–2844

Lawrence PA, Johnston P, Macdonald P, Struhl G 1987 Borders of parasegments in *Drosophila* embryos are delimited by the *fushi tarazu* and *even-skipped* genes. Nature (Lond) 328:440–442

Martinez-Arias A, Lawrence PA 1985 Parasegments and compartments in the *Drosophila* embryo. Nature (Lond) 313:639–642

Martinez-Arias A, Baker NE, Ingham PW 1988 Role of segment polarity genes in the definition and maintenance of cell states in the *Drosophila* embryo. Development 103:157–170

Meinhardt H 1984 Models for positional signalling, the threefold subdivision of segments and the pigmentation pattern of molluscs. J Embryol Exp Morphol 83 Suppl:289–311

Morata G, Lawrence PA 1977 The early development of mesothoracic compartments in *Drosophila*. Dev Biol 56:40–51

Morata G, Ripoll P 1975 *Minutes*: Mutants of *Drosophila* autonomously affecting cell division rate. Dev Biol 42:211–221

Nübler-Jung K 1987 Tissue polarity in an insect segment: denticle patterns resemble spontaneously forming fibroblast patterns. Development 100:171–177

Nüsslein-Volhard C 1989 Establishment of body axes in insects. In: Cellular basis of morphogenesis. Wiley, Chichester (Ciba Found Symp 144) p 37–64

Nüsslein-Volhard C, Wieschaus E 1980 Mutations affecting segment number and polarity in *Drosophila*. Nature (Lond) 287:795–801

Nüsslein-Volhard C, Frohnhöfer HG, Lehmann R 1987 Determination of anteroposterior polarity in *Drosophila*. Science (Wash DC) 238:1675–1681

Ready DF, Hanson TE, Benzer S 1976 Development of the Drosophila retina, a neurocrystalline lattice. Dev Biol 53:217–240

Rijsewijk F, Schuermann M, Wagenaar E, Parren P, Weigel D, Nusse R 1987 The *Drosophila* homolog of the mouse mammary oncogene *int-1* is identical to the segment polarity gene *wingless*. Cell 50:649–657

Sander K 1975 Pattern specification in the insect embryo. In: Cell patterning. Elsevier Science Publishers, Amsterdam (Ciba Foundation Symp 29) p 241–263

Struhl G Establishment of body pattern in the zygote. In: Cellular basis of morphogenesis. Wiley, Chichester (Ciba Found Symp 144) p 65–91

Sullivan W 1987 Independence of *fushi tarazu* expression with respect to cellular density in *Drosophila* embryos. Nature (Lond) 327:164–167

Tomlinson A 1989 Cellular communication in the developing *Drosophila* eye. In: Cellular basis of morphogenesis. Wiley, Chichester (Ciba Found Symp 144) p 281–289

DISCUSSION

Davidson: You are warning that one should beware of *ad hoc* intuition, I think that you should beware of the application of Occam's razor!

Lawrence: I accept that, but we do need something to keep our ideas in check.

Melton: I have a simple question which concerns the experimental evidence for the range over which these cell interactions occur in stripe formation. You showed an antibody staining for *wingless*, but it wasn't clear if it was staining protein in a cell which is expressing *wingless*, or if that protein is secreted and touches on adjacent cells. Have clones been made in *wingless* to show the range over which the gene product acts? Or can the antibody staining experiments answer that question?

Lawrence: Gines Morata and I made clones a long time ago which suggested strongly that *wingless* is non-autonomous in its effects. The mutant was non-autonomous in patches of cells, how the wild-type protein operates we don't know.

Melton: How many adjacent cells are affected?

Lawrence: The phenotype of the disc in the case of *wingless* is either all transformed or not transformed. *Wingless* lethal clones usually behave quite like wild-type ones, in most cases the disc develops normally. One cannot make clones that fill the whole disc.

Concerning the antibody, M. van den Heuvel, R. Nusse and I are trying hard to find out exactly where the product is, but we don't have the answer yet.

Melton: Do you know if the protein is secreted?

Lawrence: From the sequence, the molecular biologists can't decide for sure: the general dogma is that it is secreted.

Martinez-Arias: On the issue of cell states, you said you don't know why one particular cell in the absence of *patched* should express *wingless*, that such a statement was *ad hoc*; it is not *ad hoc*, it is what happens and in Fig. 7 of our paper (Martinez-Arias et al 1988), we illustrate the observations, highlighting some aspects of the pattern generation processes in some of the segment polarity mutants. We also point out that those observations cannot be explained by gradient models and suggest that more attention should be paid to interactions between nearest neighbours in the generation of pattern.

Lawrence: You have not shown at the level of individual cells that it is precisely those cells which used to express *patched* that now express *wingless*.

Martinez-Arias: It seems to me that your definition simply takes the concept of compartments and applies it to single cells. This does not help to explain pattern formation. Within a compartment cells are different, they behave differently and give rise to different structures. Now we are learning that they express different gene products and that absence of these gene products leads to important and reproducible effects on pattern formation. The concept of cell states as discussed in our paper and by others (Gergen et al 1986), reflects this fact and emphasizes, at the cellular level, that within a compartment or a metameric unit the cells have different combinations of important genes (not selector genes) which are probably responsible for pattern formation. Compartments and selector genes go some way toward explaining metamerization and morphological diversification of metameric units, but they tell us little about proper pattern formation. I think it is too early to be rigid about definitions. All the lineage experiments from which you draw your definition have been performed in the adult. We know very little about the embryo and it might be worthwhile to do some lineage studies in the embryo. Carlos (Cabrera) mentioned the experiments in *C. elegans* where people thought that segregation of determinants was an important mechanism, and we now know that cell interactions play a very important role in the same process.

Wieschaus: It seems that the fundamental issue is what we mean by cell states. One way of defining a cell state is purely descriptive. You could look at the individual cells in a blastoderm stage embryo and say what gene products are being expressed, what the levels of those gene products are and how one blastoderm cell differs from its immediate neighbour. As more data are accumulated, we probably come to the conclusion that all individual blastoderm cells are different. By this definition, there are a lot more cell states than just those defined by anterior and posterior compartments.

The real question, the one that is hard to answer, is which of those molecular differences is relevant for development. In one way it is attractive to reduce the number of important differences and say there are really only two, anterior and posterior, that are important. That may be true at later stages, but it's not clear to me that at the blastoderm stage or in early gastrula, the only relevant cell states are anterior or posterior compartments. Or, that the only relevant cell state at the blastoderm stage is one which is maintained in the same way with the same molecular markers at later points in development. I also don't see why cell states must be defined by sharp boundaries, because I don't see why quantitative differences in a relevant gene product might not also be significant for defining cell states.

Morata: Concerning the existence of an early lineage restriction of posterior compartment cells, Alfonso (Martinez-Arias) said that this has only been shown for structures in the adult. This is true, but the lineage analyses that Peter Lawrence and I (Lawrence & Morata 1977) and also Steiner (1976) did were performed at or near the blastoderm stage. Therefore we were testing the lineage of blastoderm cells, although the outcome was observed only in the adult structures. The experiments clearly suggested a fixed lineage segregation of posterior cells around the blastoderm stage—unless one makes the additional argument that presumptive imaginal cells are already segregated from the rest at that early stage, which is unlikely.

Cabrera: We might be polarizing certain aspects of this discussion unnecessarily. Nobody is denying that there are cell interactions in the blastoderm. But to say that every time you see a gene being expressed in a cell implies the acquisition of a different cell state is a bit extreme.

The cell states were proposed on the basis of the embryonic phenotypes that result from the lack of function of the segment polarity genes. If you look at the phenotype of *wingless*, for example, you see a band of cells is missing in each segment. Therefore you say that *wingless* is needed to provide a developmental address for those cells. But why not also consider the head, which is completely missing in *wingless* embryos? By that argument you would have to say that *wingless* determines the whole head of the embryo.

Cooke: I just wanted to take up this issue of the formal possibility of defining polarity in a repeating pattern without the need for three cell states by saying that the cells have polarity to start with. At the stage that we have been talking about, the numbers of rows of cells constituting each repeating unit represent a tiny proportion of the whole field of the blastoderm. May be it is asking a bit much to expect that cells at that stage could be reading the polarity of a single, blastodermal gradient to give themselves the polarity. It seems more plausible that there are three cell states after all, and the polarity of the segmental gradient comes afterwards as the embryo divides and produces more cells and can produce local gradients.

Lawrence: You find it more empirically satisfying—the question is, is it right?

Meinhardt: From all the experiments we have, there are good reasons to believe that the sequence of cell states determines the polarity of the cells and not the other way round. Let me ask a specific question—what is the signal to make a segment border in the two-cell-state model?

Lawrence: What do you mean by a segment border?

Meinhardt: An indentation which separates one segment from the next.

Lawrence: The whole pattern of the segment depends on a factor, probably a graded factor from the transplantation experiments of Locke (1959) and others (reviewed in Lawrence 1973). Locke drew it conventionally as a gradient that reiterates in every segment with a vertical slope at the segment border. Equally well, with a small posterior compartment, it could be a downward slope, say, in the anterior compartment and a steep upward slope in the posterior. Indeed, Katharina Nübler-Jung and I have observed that the posterior compartment of *Dysdercus* is small—only 5–10% of the segment's width. We did this using the anti-*engrailed* antibody (courtesy of N. Patel, K. Coleman, T. Kornberg and C. Goodman).

We don't know what the factor is, any more than Klaus Sander knew what it was when he drew his embryonic gradients. Later, the segmental gradient landscape might be set up and read with respect to the boundaries established by *ftz* and *eve*. It may be that *ftz* and *eve* are the first genes to allocate cells in the system, all genes that are expressed before those defining zones but not allocating cells.

Meinhardt: I think there is a very good argument that this is wrong. If *ftz* is off or *eve* is off, every second segment border is still formed in a nice way. Your model would predict that all borders are gone.

Lawrence: No, because *ftz* is responsible for the anterior borders of the even-numbered parasegments, and *eve* is responsible for the anterior borders of odd-numbered parasegments. So in *ftz*$^-$ embryos, the anterior borders of the odd-numbered parasegments will persist.

Wolpert: If you have neither *ftz* nor *eve*, do you get no borders?

Lawrence: Yes. *eve* is more difficult, because mutants in *eve* lack the sharp borders of the *ftz* stripes, as well as totally lacking the *eve* stripes (Lawrence & Johnston 1989).

Nüsslein-Volhard: A word of caution when using different levels of gene expression to define cell states. *ftz* and *eve* are two very clear examples of pair-rule patterns that are so clear that one always uses them when describing these things: we forget there are lots of other pair-rule genes which are never used in these models. This is even more true for the segment polarity genes. The only two genes where we have antibodies available are *wingless* and *engrailed*; this is why they occur so frequently in these models. There are six other genes which could be equally or more important.

Peter, you would like to argue that we have two cell states and in the anterior compartment position is determined by a gradient. In the case of *ftz* and *engrailed* you would like to argue that the boundaries are the parasegments. The notion of gradients in segment determination was derived from these transplantation experiments where the segmental boundary was the high point and the point of discontinuity. There has to be a point of discontinuity otherwise you cannot talk of segmental gradients. But you have now shifted this point to the parasegment boundary.

Lawrence: Possibly.

Nüsslein-Volhard: Why does it have to have a polarity?

Lawrence: Because all the experiments are concerned primarily with polarity.

Nüsslein-Volhard: You said that the experiments were done only in the anterior compartment and that you don't know anything about the posterior compartment.

Lawrence: There are some data (Wright & Lawrence 1981) which suggest that the posterior part of the segment of *Oncopeltus* has an opposite polarity. As you know, the denticles in the first line of *Drosophila* have an opposite polarity to the remaining part of the segment.

Garcia-Bellido: I think that both positions are compatible. In segmentation, it is crucial to delimit borders, because they become references. But the fields between borders do not, in principle, have a fine internal structure. This structure requires further specification, which may (1) directly delimit, simultaneously or in a cascade reaction, subregions with mutually exclusive cell states, or (2) define periodicities with wave functions that can partially overlap in individual cells and become exclusive to cells only after cell proliferation.

I think the problem is the operational definition of a cell state. Are these exclusive, specific cell states or do they correspond to combinations of functions within the same cells? Are cell states clonally transmitted, specifying subsets of compartments, or does clonal specification refer only to selector genes of compartments (like *engrailed*) and segment polarity genes (like *wingless*)—neither has fixed borders of expression nor are they clonally active? In the wing, the latter situation seems to be the case.

Morata: The four-cell-state model suggests that each of the genes concerned, *naked*, *patched*, *wingless* and *engrailed*, will have a restricted domain of expression. In the case of *engrailed*, this is clearly approximately one cell wide. Are *patched* and *naked* also expressed in narrow stripes?

Martinez-Arias: We do not yet know the patterns of expression of *patched* and *naked*; the genes are still being cloned. However, from our work there is no clear indication that their expression is restricted to particular regions of metameric units. We can say that the first effects of removing either of these genes are very local, but then other effects follow in other groups of cells and those might also reflect the absence of, and probably a requirement for, those gene products. In our paper (Martinez-Arias et al 1988), nowhere do we imply that the expression of *patched* and *naked* is restricted. As for a four-cell-state model, we make very clear that what matters is that neither two nor three cell states can satisfy certain conditions of pattern-forming systems. A minimum of four is required, but the situation is bound to be more complicated.

Struhl: I strongly disagree with Antonio's statement that the notions of gradients and cell states can be compromised; they could not possibly be more different. In one case you are arguing that the generation of pattern requires qualitative distinctions between cells (that's the whole notion of a cell state), in the other case you are arguing for quantitative distinctions between cells and that to my mind is not cell state. Even if four cells arranged in a line are each expressing different combinations of genes, they may not be in different cell states. We need to know if the presence of absence of the products of these genes controls the developmental behaviour of the cells, or simply reflects that their behaviour is being controlled. Peter's question of how to define a cell state is critical, because you have to ask how do we know one cell is in a different state from the next cell. It cannot be done simply by measuring the quantity or presence or absence of different gene products. This is a particularly important question in terms of more mature segments, where there are many more cells with many more differences, both qualitative and quantitative, between them. Do we have to invoke the progressive elaboration of many qualitatively distinct cell states to generate complex patterns, as suggested by Alfonso and his colleagues (Martinez-Arias et al 1988), or can quantitative differences in the levels of a few signalling molecules suffice to control the pattern of populations of cells, such as compartments, which have a common cell state?

Wolpert: A further distinction seems to be that in their case, future development is dependent on the cell's neighbours and local interactions rather than on the gradient of a morphogen.

Struhl: No one denies that there are local interactions. Andrew Tomlinson's work on the *Drosophila* retina deals with cells where there is a lot of local signalling.

McKeown: In contrast to the situation with segmentation, somatic sex determination in *Drosophila* offers a system in which we have a clear idea of the meaning of cell state. There, we know the primary inducing signal for cell state, the gene which responds to and remembers this signal, the mechanism by which the signal is remembered, the downstream genes which respond to the signal and the mechanism by which they are regulated. The primary signal for determining the sex of individual cells is the ratio of the number of X chromosomes to the number of sets of autosomes. If this ratio is 2:2, as in females, an embryonic activity of *Sex lethal* (*Sxl*) gene is induced. This requires the function of the maternally supplied product of *daughterless* and the presence of the zygote of the wild-type *sisterless-a* and *sisterless-b* genes (see Cine 1985, 1988, Mainee et al 1985). Shortly after gastrulation, a non-sex-specific promoter for *Sxl* becomes active. If sufficient embryonic *Sxl* function has occurred, RNAs transcribed from this promoter are processed in a manner which allows translation of *Sxl* proteins. One of the functions of these proteins appears to be to maintain the female-specific *Sxl* splicing pattern necessary for production of the *Sxl* proteins. In males (X:A = 1:2) embryonic *Sxl* function appears not to be induced; post-embryonic transcription from the non-sex-specific promoter produces RNAs that are not translocated and are functionally dispensable (see Bell et al 1988, Maine et al 1985). Thus female and male cell states correspond to the presence or absence of functional *Sxl* proteins. The stability of the state is maintained by the continued induction of female-specific splicing by the active *Sxl* product in females or the continued lack of female-specific splicing caused by the lack of *Sxl* product in males. *Sxl* also controls the activity of the downstream gene *transformer* (*tra*) by inducing female-specific splicing of *tra* RNAs, leading to the production of active *tra* proteins in females. In the absence of *Sxl* function, *tra* RNAs are spliced in a manner that yields non-functional and dispensable RNAs (Boggs et al 1987, Nagoshi et al 1988). If *tra* activity is induced, it induces female-specific processing of the RNAs from the *doublesex* (*dsx*) gene. This process also requires the function of the *transformer-2* gene (Baker et al 1987, McKeown et al 1988, Nagoshi et al 1988). The female *dsx* RNAs are translated into active *dsx* female products which lead to female differentiation. In the absence of *tra* function, as in males, *dsx* RNA is processed to its default male mode, which produces male-specific *dsx* proteins, leading to male differentiation.

Scott: I have never troubled with defining cell states because I feel there is a danger in trying to impose something on the embryo that may or may not exist. Defining cell states can be a useful way to talk about things, but any definition other than that of a functional aspect of cells, a visible morphology of cells, or an observable aspect of cellular gene expression is arbitrary.

Lawrence: I didn't coin the word cell state, it is in the literature and I thought it needed a definition.

Scott: The trouble with definitions is that they are arbitrary. Why does a boundary between cells in different states have to be a straight line, for example? The cells in a *Drosophila* blastoderm are, roughly, hexagonally close packed, so most lines will only be approximately straight. I doubt the embryo cares whether such lines are straight or not. I think it is useful to talk either about gene expression—whether it be qualitative or quantitative differences—or about other observable aspects of cell morphology or function. The term 'cell states' can be used in a general sense to refer to these things. But a precise definition demands a reference to precise aspects of cell differentiation in individual cases. For cell determination, it may be important for a cell to have a gene product there transiently (and to lose the product may be an important event), or it may only be important what gene product, or combination of gene products, persists in a certain cell for a long time. Or it may only be important where a cell is in relation to another cell that has a certain gene product. I don't think we know the answers to these things well enough to make a rigorous definition of what a cell state is in terms of how that cell will behave subsequently in development. Also, because cells are adjusting frequently to changing regulatory inputs during development, the concept of a 'state', except in general terms, is uncomfortably static for me.

Janni (Nüsslein-Volhard) alluded to these other stripes—they aren't as well studied, although now there are antibodies for *hairy*. Eric (Wieschaus) and Peter Gergen, some time ago, as well as Pat O'Farrell and I, have discussed and reviewed some models in which the combination of pair-rule gene products in a cell determines how the cell will behave. For the pair-rule genes, the patterns of expression repeat roughly every eight cells along the anteroposterior axis of the blastoderm embryo. Each of those eight cells makes a different combination of pair-rule gene products. I could, arbitrarily, define each of those eight cells as having a different 'state', but I am not sure that this would be useful. The combinatorial model still seems to me a perfectly reasonable hypothesis, at least for pair-rule genes, for how cells decide what to do.

Certainly in some cases it's clear that segment polarity genes are expressed in finer stripes (one or two cells wide) within the three or four cell wide repeat units in which the earlier acting pair-rule genes are expressed. The cells in the anterior part of each segment almost certainly express different genes depending on their position within it. Why regard the anterior–posterior distinction as more primary than these other distinctions? If I was forced to define cell states in my arbitrary way, I would say any case in which you could distinguish cells on any basis, their behaviour, their gene expression or their position, would define a cell state.

Lawrence: That's useless!

Scott: I agree, but it is no more arbitrary than the other definitions. The real question we have to answer is which of these things, the gene expression patterns or the positions of the cells, are actually important for the behaviour of the cell. I would argue that for the blastoderm, we don't know that yet.

Lawrence: I have tried to argue that what evidence we have points to there being two cell states. We don't yet have convincing evidence for more, I am not saying that there couldn't be more.

Wolpert: What is the class of information that you would require to persuade you that you were wrong?

Martinez-Arias: First of all there is no argument in defining units within which pattern will be formed. The question is how do you pattern within those units? Peter seems to believe that there is a gradient, may be in the early embryo, and the cells are able to read their position in that gradient. I don't know whether the values that the cells get are fixed or not, in lineage terms. The argument that we would like to raise is that there is no such gradient—there might be a few reference points—and that information is then generated during proliferation by cells looking at their neighbours.

Concerning lineage, and this is an important aspect of Peter's definition, I cannot think of any experiments which would convince him.

Lawrence: If there were more than two cell states, then I would expect to see a gene, and it may be one of the genes that Janni referred to which has not yet been looked at, which would so define cells, with a sharp boundary, in such a way that you could allocate those cells to a third state. That gene would have a controlling role. It would then have to be maintained through lineage for a time, not necessarily right through development, otherwise the putative third cell state would be too transient to mean anything.

Martinez-Arias: Peter, do you think that the absence of *engrailed* expression therefore should affect only the cells that normally express *engrailed*? Is it part of your definition? As you know, it doesn't. We know that *wingless* and *engrailed* are not expressed in the same cells, and absence of *engrailed* results in the decay of *wingless* expression in the neighbouring cells.

Lawrence: I agree, if you remove *engrailed* from certain cells, it could have effects on other cells, but because of *indirect* consequences.

Wieschaus: The important difference between my view and the way Peter has talked about cell states is that at the blastoderm stage there are blastoderm cells which are different from each other with respect to many different gene products. Peter would regard that the central difference is this expression of *engrailed*. I believe that the different pair-rule genes also have effects on patterning independent of *engrailed*. My view could be disproved if altering the expression of these non-*engrailed* genes in an *engrailed*-background had no effect on the *engrailed* phenotype. If you did get effects on the patterning, then those genes must be working in some pathway independent of simply establishing the *engrailed* stripe. That is the kind of experiment that I would like to see done.

Nüsslein-Volhard: These genes all have phenotypes. Remove them and you see distinct phenotypes. You can't just say they don't matter.

Struhl: No one is saying that they don't matter, the question is in what way they matter.

Wieschaus: But they don't matter simply to regulate *engrailed*.

Struhl: Peter made the point, for example in the *Oncopeltus* experiments, that almost all the cells were derived from the anterior compartment. Yet this system could produce a number of different spatial signals which appeared in those experiments to be graded. The question is whether cell states is a useful concept to understand how that patterning occurred.

References

Baker BS, Nagoshi RN, Burtis KC 1987 Molecular genetic aspects of sex determination in *Drosophila*. Bioessays 6:66–70

Bell LR, Maine EM, Schedl P, Cline TW 1988 Sex-lethal, a *Drosophila* switch gene, exhibits sex-specific RNA splicing and sequence similarity to RNA binding proteins. Cell 55:1037–1046

Boggs RT, Gregor P, Idriss S, Belote JM, McKeown M 1987 Regulation of sexual differentiation in D. melanogaster via alternative splicing of RNA from the transformer gene. Cell 50:739–747

Cline TW 1985 Primary events in the determination of sex in Drosophila melanogaster. In: Halvorson HO, Monroy A (eds) Origin and evolution of sex. Alan R Liss, New York, p 301–327

Cline TW 1988 Evidence that sisterless-a and sisterless-b are two of several discrete 'numerator elements' of the X/A sex determination signal in Drosophila that switch Sxl between two alternative stable expression states. Genetics 119:829–862

Gergen JP, Coulter D, Wieschaus E 1986 Segmental pattern and blastoderm cell identities. In: Gall J (ed) Gametogenesis and the early embryo. Alan R Liss, New York, p 195–220

Lawrence PA 1973 A clonal analysis of segment development in *Oncopeltus* (Hemiptera). J Embryol Exp Morphol 30:681–699

Lawrence PA, Morata G 1977 The early development of mesothoracic compartments in Drosophila. Dev Biol 56:40–51

Lawrence PA, Johnston P 1989 Pattern formation in the *Drosophila* embryo: allocation of cells to parasegments by *even-skipped* and *fushi tarazu*. Development 105:761–768

Locke M 1959 The cuticular pattern in an insect—the intersegmental membranes. J Exp Biol 37:398–406

Maine EM, Salz HK, Schedl P, Cline TW 1985 Sex lethal, a link between sex determination and sexual differentiation in Drosophila melanogaster. Cold Spring Harbor Symp Quant Biol 50:595–604

Martinez-Arias A, Baker NE, Ingham PW 1988 Role of segment polarity genes in the definition and maintenance of cell states in the *Drosophila* embryo. Development 103:157–170

McKeown M, Belote JM, Boggs RT 1988 Ectopic expression of the female transformer gene product leads to female differentiation of chromosomally male Drosophila. Cell 53:887–895

Nagoshi R, McKeown M, Burtis K, Belote JM, Baker BS 1988 The control of alternative splicing at genes regulating sexual differentiation in Drosophila melanogaster. Cell 53:229–236

Steiner E 1976 Establishment of compartments in developing leg imaginal discs of *Drosophila*. Wilhelm Roux's Arch Dev Biol 180:7–30

Wright DA, Lawrence PA 1981 Regeneration of the segment boundary in *Oncopeltus*. Dev Biol 85:317–327

General discussion II

Kornberg: An interesting question that has been raised on a number of occasions is how the mechanism of segmentation in *Drosophila* relates to segmentation in a short germ band insect like the grasshopper. It appears that these insects develop with quite different programmes during these early stages of embryogenesis. Gary Struhl has described how in long germ band insects like *Drosophila*, spatial domains larger than a single segment are defined initially, and segment-sized regions are formed by partitioning of these larger domains. However, segmentation in short germ band insects involves cell proliferation and addition of segments with a rostrocaudal polarity.

We have investigated the mechanism of segmentation in short germ band insects by studying the expression of the grasshopper *engrailed* gene. In collaboration with Corey Goodman's lab, Kevin Coleman, a post-doctoral fellow in my lab, isolated a monoclonal antibody, 4D9, which is directed against the *invected* protein of *Drosophila*. The *invected* gene is closely related to *engrailed* in sequence and in pattern of expression. Enrique Martin in my lab and Nipam Patel in Corey Goodman's lab have characterized 4D9 further. 4D9 recognizes an epitope within the *invected* homeodomain that is conserved in the *engrailed* sequence. This portion of the homeodomain is also conserved in the *engrailed*-like gene of the grasshopper, and grasshopper embryos were stained with 4D9 to follow the process of segmentation.

In grasshopper embryos undergoing elongation, *engrailed* stripes were added one at a time, in a rostrocaudal direction; no intervals larger than a single segment were observed. This differs from early patterns of engrailed expression in *Drosophila*.

Klaus Sander raised the question of where the grooves form initially. Are they parasegmental, i.e. at the anterior edge of these new stripes, or are they segmental, that is at the posterior edge of these *engrailed* stripes? Unfortunately, our evidence does not allow us make a good judgement at this time.

Regarding the question of parasegments in the grasshopper embryo, it is relevant that the anterior border of the *engrailed* stripe is better defined, initially, than is the posterior border. Of course, in the mature germ band embryo, both of these stripes are equally well defined. This behaviour is consistent with the idea that the parasegmental side of the *engrailed* stripe is formed initially, and the segmental side is formed subsequently.

Scott: In the light of earlier discussion, and this idea that the head and thoracic segments would be formed all at once and the abdominal segments would be the budding ones, do you see the more anterior stripes in this case forming in synchrony and then the addition of posterior ones?

Kornberg: My understanding from Gary's model was that it applied to intermediate germ band insects.

Struhl: No, it also applies to short germ band insects. The general idea is that short germ band insects start with basically a head and then extend. In progressively longer germ band insects, something would have to remove the activities of head-determining genes from the posterior domain.

Kornberg: The question is what are the first stripes and where are they? What does the head look like initially? The first stripe to form in the grasshopper is T2, then T3, then A1 and then stripes begin to form in the head. So, at least with respect to *engrailed* as a molecular marker for segments, the head is not initially segmented.

Sander: The sequence of stripes appearing agrees with the results from the heat-shock experiments and X-ray damage experiments on grasshoppers and crickets. In these, the first responses occur more or less simultaneously in the thoracic and may be posterior gnathal regions, and then successively in the abdominal segments (see Sander 1988). This supports Gary and Janni's idea that the difference between anterior influences and posterior influences in *Drosophila* reflects the course of evolution.

Garcia-Bellido: There is an old proposal from Seidel of a differentiation centre in insect embryos, which consistently gives rise to segments T1 and T2. The position of this differentiation centre in the egg has no relation to the location or size of the embryo. I would be interested to hear your view of this differentiation centre.

Sander: Seidel conceived the idea for a differentiation centre from a primarily descriptive approach, which showed that gastrulation tends to start somewhere in the anterior thorax region, then spreads anteriorly and posteriorly. The same holds for visible segmentation and subsequent steps of differentiation. That description is not restricted to the short germ band insects; it also occurs in the honey bee, for example (Fleig & Sander 1986). In my view, that is the lasting value of the concept of a differentiation centre.

The problem arose about ten years later when Seidel connected this spatio-temporal sequence of differentiation with the mode of specification of individual segments. This was done under the influence of the concept of Spemann's organizer. Spemann earlier used the term differentiation centre, coined by Boveri for the echinoderm egg. Seidel's 'physiological differentiation centre' is much less well founded. I have looked at the examples where he (or others) claimed it and managed to contrive alternative explanation in every case, based on the assumption that the cells' capacity to replace eliminated neighbours is lost first in the differentiation centre (Sander 1976). So I believe Seidel's conclusion that the differentiation centre is handing out fates to the adjacent segments is not compelling, and it seems untenable in the light of later findings.

You also asked whether there is any pre-formation of this centre in the egg, as shown for the anterior and posterior centres. Until recently, I saw no indications of this, but now—with at least one gap gene showing a maternal rescue effect (Lehmann & Nüsslein-Volhard 1987)—I feel that some old results should be re-evaluated. I still would not expect to find maternally localized determinants in the middle region of the egg cell, but an ubiquitous maternal message might become restricted in its expression to the middle region, as a result of regulatory effects originating from the polar centres. Under certain experimental conditions this might yield results that could be mistaken for maternal localization.

Garcia-Bellido: There is another property added to that centre. Most of the homeotic mutations in selector genes (*Antennapedia* and *Bithorax* complexes) do transform segments into something that also corresponds to T1, T2. It is too much of a coincidence. The question then is whether this is a very important region in the embryo, possibly containing *knirps* or *Krüppel* or something which is being used as a reference in segmentation.

Sander: The explanations I gave as alternatives for Seidel's propositions (Sander 1976) involve the idea that cell fate is fixed at different times in different segments, starting from the T1/T2 region. Thus temporal priority might be linked to the respective gene activities, and this would account for the coincidence.

Garcia-Bellido: So you say that segmentation may use internal references (something in the presumptive T1 segment) and proceed anteriorly and posteriorly from them, rather than intercalating between ends defined by maternal products. But I disagree, because embryos in the polyembryonic *Hymenoptera* species are formed from random clusters of blastoderm cells with no relation to either the anteroposterior or the dorsoventral axes of the egg (Ivanova-Kasas 1972). They must self-organize! What spatial clues do they use to do so?

Nüsslein-Volhard: This theory of the thoracic state as the ground state is compatible with the finding that *Krüppel*, one of the gap genes, is independent of the terminal centres. When you remove the terminal influences it is expressed everywhere, whereas it is expressed in the middle of the egg in normal development.

Garcia-Bellido: That leads back to your comment that we don't have the inventory of all the genes that could be involved. *Krüppel* is obviously not one, but the fact that *Krüppel* does not affect all segmentation is not proof that the differentiation centre is not specified genetically.

Nüsslein-Volhard: There is this famous experiment where Seidel constricted the posterior pole and removed the posterior 10% and didn't get any development. Is that compatible with the idea that we start with a truncated embryo with head and thorax? That experiment should inhibit only abdomen formation, not the formation of head and thorax.

Wolpert: Can you tell us what the general issue is, because those of us who do not work with *Drosophila* simply don't understand?

Nüsslein-Volhard: We are back to the primitive insects—the notion of this differentiation centre and that the earliest segmentation is in the thorax and spreads from there towards the posterior and the anterior.

Wolpert: What is this telling me about segmentation? I am interested in the mechanism; where it starts is of little interest to me.

Garcia-Bellido: The issue is whether the reference for segmentation is at the ends or in the middle of the embryo. That is crucial in my opinion for how segmentation is occurring.

Cooke: I want to go back to the local mechanism of segmentation. Are there any somatic mosaic studies using a marker gene in conjunction with mutant states for the pair-rule genes and possibly the segment polarity genes to produce marked mutant clones in the blastoderm stages? These seem to be the genes where the gene products are intracellular proteins or nuclear proteins, or possibly molecules whereby cells put an instruction onto an adjacent cell. And these genes seem to be making the local pattern rather than being respondents. It is interesting to look at small mutant patches and ask questions about the autonomy of the response to pattern. If you have a blastoderm which has patches of various sizes that are not producing some subset of these genes at the right time, on how much of a cell by cell basis do the pattern effects occur?

Wieschaus: The mosaics that are easiest to make involve mutants on the X chromosome. When you make those kinds of mosaics, you see that large patches of mutant tissue are autonomous (Gergen & Wieschaus 1985). If you try to push the argument further and to look exactly at the mosaic boundary between mutant and wild-type cells, you are very hard pressed to find examples of non-autonomy. This is partially because even if you choose an amorphic pair-rule mutant like *runt*, the phenotype is still somewhat variable and the variability of the mosaic boundary is no greater than the variability in an animal which is fully mutant in all of its tissues—that's for pair-rule genes.

The next step is to look at the segment polarity genes. Experiments have been done using mitotic recombination to produce a very small patch, may be one or two cells in a clone (Wieschaus & Riggleman 1987). If you ask how the segment polarity genes produce their phenotypes, you can classify segment polarity genes into those which will produce an autonomous transformation and others which under the same experimental regime do not. For example, *armadillo* will produce a very small patch of denticles within an area of *naked* cuticle, just as you would see in mutants wholly *armadillo*. If you do the same experiments with *wingless*, you get a negative result; you don't find those clones.

One interpretation is that *wingless* is non-autonomous and *armadillo* is autonomous—that *armadillo* is in some way involved in internalizing information from the field and *wingless* is involved in generating that information. Those are the only published experiments, but I know that Peter (Lawrence) and other people have results on other kinds of mosaics.

Sander: One aspect of Lewis' question is how cell numbers relate to periodicity. Peter or Tom (Kornberg) might like to talk about haploid/triploid mosaic blastoderms and what you see in terms of *ftz* stripes. It is not cell counting that makes the periodicity (see Sullivan 1987).

Wolpert: There are published models of the reaction/diffusion type, that generate a succession of waves of decreasing wavelength which could give rise to the repeated pattern (Nagorcka 1988, Lacalli et al 1988). If those can be excluded, it would be enormously helpful to say so.

Struhl: I don't know if those could be excluded unequivocally. However, in the case of periodic patterns of expression of pair-rule genes, recent experiments argue strongly against such models. We know, primarily on the basis of studies of the gene *hairy*, that some pair-rule genes contain very large regulatory regions just upstream of the promoter, and that partial deletions which remove progressively larger chunks of this region inactivate the expression of particular stripes, one or two at a time (Howard et al 1988).

Wolpert: Randomly or from one end?

Struhl: Not progressively along the body, but nevertheless in a consistent order, e.g. first stripes 3 and 4, then stripe 7, then stripes 2 and 5, *etc*. That shows that the promoter for the *hairy* gene has distinct regulatory elements spread out over a large piece of DNA which seem to respond to singularities in the system. These singularities are presumably the products of the gap genes. You can compare that with a pair-rule gene like *fushi tarazu* (*ftz*), for which equivalent experiments been done trying to map the sites on the DNA which are responsible for the periodic response (Hiromi et al 1986). In this case, there does seem to be one key stripe element, which is a relatively small target. That, together with the fact that the function of the *hairy* gene seems to be critical for regulating the periodity of *ftz*, gives an interpretative system where there is a gene, *hairy*, which has an antenna on it that is looking for particular combinations of different gap gene products. It sees those singular signals which arise at regular intervals along the body and responds by being activated in a periodic pattern. Then its periodic expression helps to set up the periodicity of the active component, *ftz*, which subdivides the embryo into parasegments.

Wolpert: But what is the periodic signal?

Struhl: The periodic signal is essentially a set of singularities which are contributed by the gap gene functions.

Wolpert: That is not a periodic signal.

Struhl: Correct, but it is interpreted in the form of periodicity. The idea of some kind of wave function is very tempting, but that's not the only way in which you can generate periodicity. Hans Meinhardt pointed out that if you can generate enough singularities—you don't need that many—you can get periodicity.

Lawrence: Sullivan (1987) showed that the pattern of expression of *ftz* is independent of the cells within a stripe. Concerning all these models about how a stripe is formed, the people who make these models should remember that the stripe is asymmetric.

Kornberg: Given a very different mechanism of segmentation in the short germ band versus the long germ band insects, we can now directly test the role of these gap genes in a different setting. We are now doing experiments to isolate genes related to the *Drosophila* segmentation genes from the short germ band insects and ask, are they also involved in neurogenesis, are they also involved in segmentation?

References

Gergen JP, Wieschaus EF 1985 The localized requirements for a gene affecting segmentation in drosophila. Analysis of larvae mosaic for runt. Dev Biol 109:321–335

Hiromi Y, Kuroiwa A, Gehring WJ 1986 Control elements of the Drosophila segmentation gene fushi tarazu. Cell 43:603–613

Howard K, Ingham P, Rushlow C 1988 Region-specific alleles of the Drosophila segmentation gene hairy. Gene Dev 2:1037–1046

Ivanova-Kasas OM 1972 Polyembryony in insects. In: Counce SJ, Waddington CH (eds) Developmental systems: insects. Vol 1:243–271

Lacalli TC, Wilkinson DA, Harrison LG 1988 Theoretical aspects of stripe formation in relation to *drosophila* segmentation. Development 103:105–113

Lehmann R, Nüsslein-Volhard CH 1987 *Hunchback*, a gene required for segmentation of an anterior and posterior region of the *Drosophila* embryo. Dev Biol 119:402–417

Nagorcka BN 1988 A pattern formation to control spatial organization in the embryos of *Drosophila melanogaster*. J Theor Biol 132:277–306

Sander K 1976 Specification of the basic body pattern in insect embryogenesis. Adv Insect Physiol 12:125–238

Sander K 1988 Studies in insect segmentation: from teratology to phenogenetics. Development 104:111–123

Sullivan W 1987 Independence of *fushi tarazu* expression with respect to cellular density in *Drosophila* embryos. Nature (Lond) 327:164–167

In vivo competition identifies positive *cis*-regulatory elements required for lineage-specific gene expression in the sea urchin embryo

Roberta R. Franks, Roy J. Britten and Eric H. Davidson

Division of Biology, California Institute of Technology, Pasadena, California 91125, USA

Abstract. Several *cis*-regulatory elements within the 5′ regulatory region of the lineage-specific CyIIIa actin gene have been identified by *in vivo* competition. Sea urchin eggs were coinjected with a fusion construct in which the bacterial chloramphenicol acetyltransferase (CAT) gene is controlled by the CyIIIa regulatory domain, together with molar excesses of various DNA subfragments that are derived from this region. Each subfragment studied includes one or several known sites where highly specific interactions occur *in vitro* with nuclear DNA-binding proteins. Coinjection of excess molecules of some of these subregions results in a decrease in the activity of the CyIIIa-CAT fusion gene, as a function of the molar subfragment : CyIIIa-CAT ratio. This result implies that these sites compete with *cis* sequences linked to the CAT reporter gene for limited factors that positively regulate CyIIIa transcription in the embryo, and demonstrates the functional importance of a number of the DNA–protein interactions that have been observed *in vitro*.

1989 Cellular basis of morphogenesis. Wiley, Chichester (Ciba Foundation Symposium 144) p 156–171

Recent lineage studies, building on an extensive older literature, have shown that regularly developing sea urchin embryos have an invariant cell lineage (Cameron et al 1987, Davidson 1986, 1989). By 6th cleavage, the embryo has been divided up into territories, each of which consists of a set of contiguous clones, and each of which will give rise to one or more embryonic structures. Of these territories, those from which derive the aboral ectoderm and the skeletogenic mesenchyme will each produce only one differentiated cell type. A number of molecular markers (cloned genes and immunological probes) have been characterized, which serve as specific indicators of these very early embryonic programmes of differentiation (reviewed by Davidson 1986). The invariant spatial relations of the lineage founder cells, and the precise molecular criteria available for the differential gene functions expressed in their progeny, provide special opportunities for examination of one of the major conceptual issues in development. This is the mechanism by which differential programmes of gene expression are first instituted in space and time as embryogenesis begins. In this paper, we describe one facet of new evidence obtained in a continuing effort to understand at the molecular level the spatial and temporal regulation of the gene coding for the CyIIIa cytoskeletal actin. CyIIIa is expressed exclusively in the aboral ectoderm, and is here considered as an exemplar of an early, lineage-specific differential gene activation.

CyIIIa expression

The CyIIIa cytoskeletal actin gene is one of six functional actin genes in the genome of *Strongylocentrotus purpuratus* (Lee et al 1984). CyIIIa transcripts begin to accumulate 10–12 hours post-fertilization (Shott et al 1984, Lee et al 1986) and are confined throughout embryogenesis to the cell lineages that give rise to the aboral ectoderm of the completed pluteus (Angerer & Davidson 1984, Cox et al 1986, Cameron et al 1987). Gene transfer experiments have identified a regulatory region extending about 2.5 kb upstream of the CyIIIa transcription start site that promotes correct developmental expression. Thus, when a fusion gene construct consisting of the chloramphenicol acetyltransferase (CAT) reporter gene under the control of upstream CyIIIa sequences is microinjected into *S. purpuratus* eggs, CAT gene expression in the transgenic embryos is activated at about the same time and in the same cells as the endogenous CyIIIa gene (Flytzanis et al 1987, Hough-Evans et al 1987, Franks et al 1988, R. Franks & E. Davidson, unpublished results). A particularly interesting result was obtained on injection of the CyIIIa-CAT fusion into eggs of a relatively distant sea urchin species, *Lytechinus variegatus*. Correct temporal regulation was found to be preserved, but spatial regulation is abolished and expression occurs ectopically in all cell types of the recipient embryo (Franks et al 1988). This result implies that the *cis–trans* interactions responsible for spatial and temporal regulation are separable.

Some 15 different sites have been mapped where high specificity interactions occur *in vitro* between proteins extracted from embryo nuclei and DNA sequence elements of the CyIIIa regulatory region. These sites have been mapped by gel shift methods, DNase I footprinting and other procedures, and they are largely distinct from one another with respect both to their sequence and to the putative protein regulatory factors with which they interact (Calzone et al 1988, F. Calzone et al, unpublished results). The experiments briefly described here were undertaken to identify any such interactions that *in vivo* function positively in a regulatory sense. Without interfering with the structure of the regulatory domains, for example by deletions, we can test for positive function *in vivo* by competition, since it is possible to saturate the available positive regulatory factors by injecting a sufficient number of CyIIIa-CAT genes (Livant et al 1988, Flytzanis et al 1987). In such experiments individual subfragments of the CyIIIa regulatory region are coinjected in excess with the complete CyIIIa-CAT fusion gene. Coinjected DNA sequences that compete for binding of putative positive regulators of CyIIIa transcription are identified by the depression of CAT gene activity, the extent of which depends on the molar excess of competing fragment with respect to the CyIIIa-CAT reporter gene incorporated in the embryo.

Competition *in vivo* by CyIIIa regulatory sequences

DNA molecules injected in linear form into sea urchin egg cytoplasm are rapidly ligated into long concatenates that are stably incorporated into one or more blastomere nuclei during cleavage and are replicated together with host cell DNA throughout embryogenesis (McMahon et al 1985, Flytzanis et al 1985). Fusion gene constructs in which the upstream regulatory sequences of several different *S. purpuratus* genes were ligated to the CAT reporter gene are correctly regulated in time and space in the transgenic embryos derived from injected eggs (Davidson et al 1985, Flytzanis et al 1987, Hough-Evans et al 1987, Katula et al 1987, Sucov et al 1988). Thus the *cis–trans* interactions that developmentally regulate endogenous gene expression apparently function appropriately in the regulatory domains of these exogenous reporter gene constructs.

Flytzanis et al (1987) showed that over a range of a few hundred to several thousands of CyIIIa-CAT fusion genes injected per egg, the same average quantity of CAT enzyme is produced. It follows that these numbers of copies of CyIIIa regulatory sequence incorporated per blastomere nucleus are saturating for *trans*-acting regulatory factors. If the amount of CAT expression is indeed limited by the availability of regulatory factor, and the CyIIIa-CAT genes are present in excess, then coinjection of additional copies of the regulatory region together with the complete CyIIIa-CAT gene should competitively and stoichiometrically reduce CAT enzyme expression. Approximately this predicted behaviour was observed in the recent studies of Livant et al (1988). In these experiments, eggs were coinjected with 1500 molecules of linearized CyIIIa-CAT together with varying numbers of molecules containing the whole upstream regulatory region, and CAT enzyme activity was measured in the embryos that developed. An example of a competition experiment of this kind is reproduced in Fig. 1 (Livant et al 1988). Fig. 1a shows the restriction map of the 14 kb CyIIIa-CAT gene and the locations of two different fragments, each including all known elements of the regulatory region, that were used as competitor. The 9.5 kb *Sal*I fragment contains the entire 5′ flanking region and 5′ transcribed leader associated with the CyIIIa-CAT gene. The 2.7 kb *Sph*I–*Bam*HI fragment that was derived from this larger fragment includes all sites of *in vitro* interaction between nuclear factors and CyIIIa upstream sequence that have been detected (Calzone et al 1988). This fragment has been shown to be necessary and sufficient to promote correct ontogenic regulation of the CyIIIa-CAT gene in transgenic embryos (Flytzanis et al 1987, Hough-Evans et al 1987, Franks et al 1988, R. Franks & E. Davidson, unpublished results). The competitive behaviour of the CyIIIa 5′ flanking sequences is shown in Fig. 1b, in which the average number of CAT enzyme molecules produced per embryo is given as a function of the molar ratio of the competitor to CyIIIa-CAT genes injected. The results (solid line) indicate that the competition behaved as if about 40% ($\alpha = 0.4$; see legend) of the competing DNA were effective, by comparison to the ideal results (dashed line) that would have been obtained had reduction in CAT activity occurred with perfect stoichiometry with respect to the number of 5′ flanking sequences present. The average value of the parameter α for a large number of different experiments is 0.75, i.e. even closer to that expected for complete competition (R. Franks & E. Davidson, unpublished results). One explanation for these small differences between experimental results and ideal stoichiometry is that in some embryos of each batch concatenates containing competing DNA and little or no CyIIIa-CAT DNA were formed and incorporated in different cells from the bulk of the CyIIIa-CAT DNA. In any case, Fig. 1b provides clear evidence of systematic competition *in vivo*, and indicates that the concentration of CyIIIa-CAT reporter gene sequences is close to, or at, saturation with respect to at least some limiting nuclear factor(s) required for CAT expression.

To examine the competitive behaviour of subfragments of the whole regulatory region that contain only one or several sites of interaction, we introduced the appropriate small restriction fragments (or corresponding subclones) along with the CyIIIa-CAT reporter gene. However, when small DNA fragments are microinjected into sea urchin eggs they are not replicated measurably during embryogenesis, probably due to failure to concatenate and incorporate (R. Franks & E. Davidson, unpublished results). To circumvent this problem, competitor DNA subfragments much less than 1 kb in length were ligated *in vitro* to carrier genomic DNA molecules bearing restriction termini homologous with the competitor DNA termini, or blunt-ended if the competitor DNA molecule bore a blunt end. For injections, molecules ranging in size from 1 kb to approximately 15 kb were prepared from the ligated, high molecular weight carrier–subfragment concatenates by digestion with *Bgl* II, an enzyme for which none of the subfragments tested included a site. The actual concentration of the specific competitor DNA sequence in the injected preparation was then estimated by slot-blot hybridization.

In each competition experiment, eggs were coinjected with at least 1200 molecules of the CyIIIa-CAT reporter gene, together with increasing quantities of competing sequence that were always in molar excess with respect to the CyIIIa-CAT DNA. Genomic carrier DNA with an average length of approximately

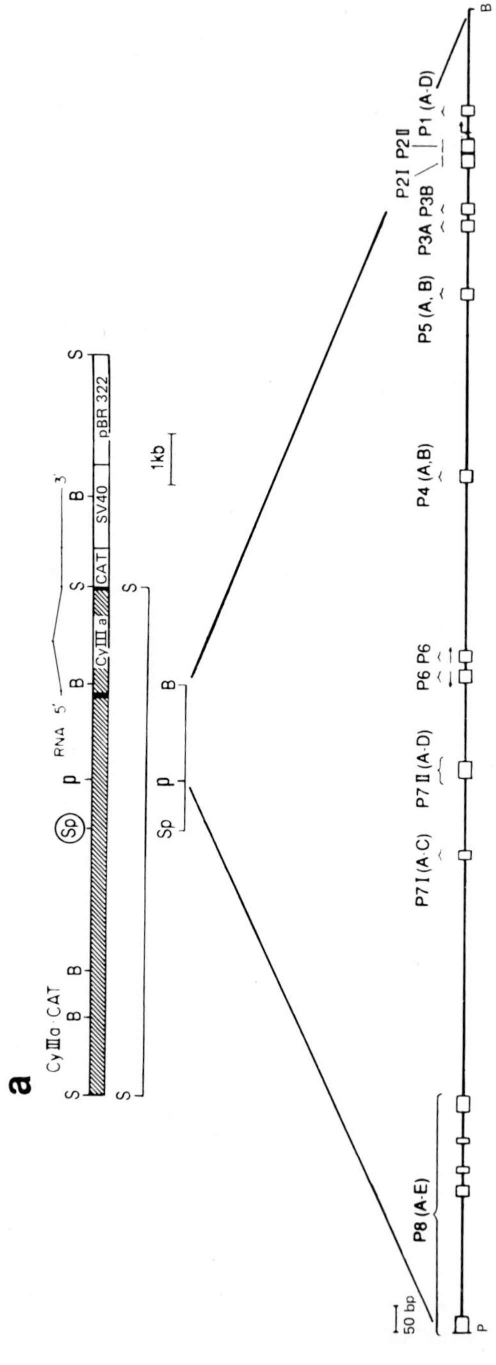

FIG. 1. (a) See legend on facing page.

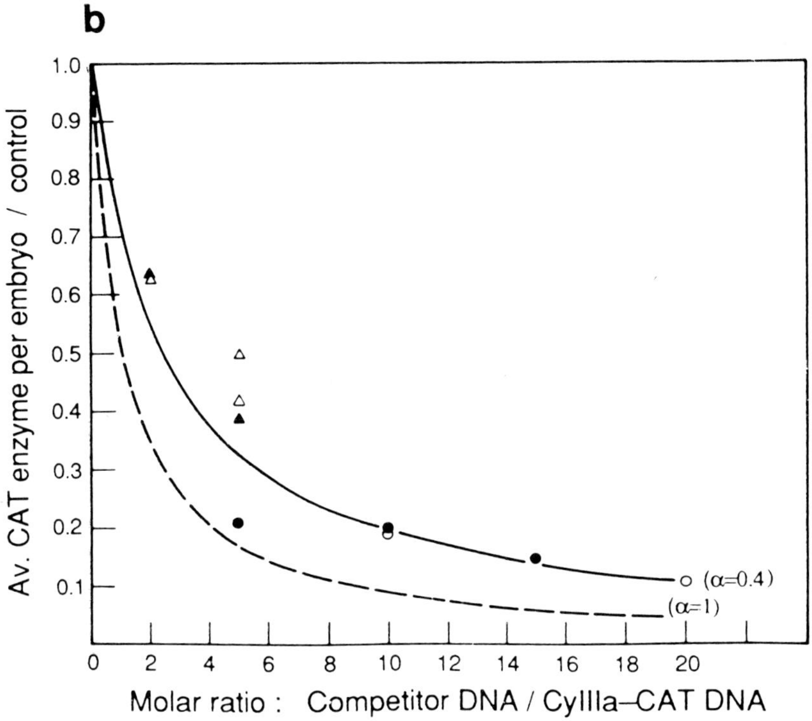

FIG. 1. (a) Map of the CyIIIa-CAT fusion gene with the two upstream fragments used as competitor DNAs. Also shown are the locations (open boxes) of all known sites of *in vitro* interaction between nuclear factors and CyIIIa 5′ flanking sequence. The protein–DNA complexes described by Calzone et al (1988) for each binding site are indicated (P1–P8). Restriction enzyme sites are: B, *Bam* HI; P, *Pst* I; S, *Sal* I; Sp, *Sph* I. SV40, simian virus 40; CAT, chloramphenicol acetyltransferase. (b) *In vivo* competition by excess regulatory DNA fragments decreases CAT production from coinjected CyIIIa-CAT reporter genes in transgenic embryos. Closed and open triangles denote experiments with the 9.5 kb *Sal* I competitor fragment. Closed and open circles denote experiments with the 2.7 kb *Sph* I–*Bam* HI competitor fragment. About 1500 molecules of CyIIIa-CAT were injected at each point, together with the indicated number of competitor fragments. The mass of DNA injected at each point was kept constant by the addition of varying amounts of sea urchin genomic DNA. Thirty to ninety embryos were pooled and assayed for CAT activity at each point. The relationship between CAT activity per embryo relative to the control in which no competitor DNA was included (y), and molar ratio of competitor DNA molecules to CyIIIa-CAT genes (x), can be approximated as $y=(1+\alpha x)^{-1}$, where α is the fraction of competitor fragments actively competing. Controls in these experiments averaged 5×10^6 CAT enzyme molecules per embryo. The dashed curve indicates the ideal form when $\alpha=1$. A least-squares-fit to the experimental data shown yields a value for α of 0.4. From Livant et al (1988).

5 kb was added to injection solutions as necessary, to ensure that all eggs were injected with the same total DNA mass (about 0.2 pg). The addition of a modest molar excess of carrier DNA has been found to increase CAT expression severalfold in embryos bearing the CyIIIa-CAT fusion, possibly by providing spacers between adjacent CyIIIa-CAT molecules in the incorporated concatenates (R. Franks & E. Davidson, unpublished results). Maximum expression is always attained at the molar ratio of carrier DNA to CyIIIa-CAT DNA greater than 10 utilized in these experiments. The competitive behaviour of several non-overlapping subfragments that were derived from the CyIIIa regulatory domain is shown in Fig. 2, in which the average CAT enzyme activity per embryo is expressed as a function of the average molar ratio of competitor to CyIIIa-CAT genes incorporated per embryo. This ratio was measured in pooled embryo samples by slot-blot hybridization using competitor DNA and CAT DNA sequence probes. Embryos that were coinjected with CyIIIa-CAT and an excess number of molecules of the entire 9.5 kb regulatory region were included in each experiment as controls. Thus, the competitive behaviour of a given subfragment was directly compared to that of the whole regulatory region in the same embryo batch. It can be seen that each of the non-overlapping subfragments labelled P4, P5 and P7 (a fragment that contains both the P7I and P7II binding sites) in Fig. 2(a)–(c) competes almost, if not exactly, as well as does the whole regulatory region. Since the activity of the CyIIIa-CAT reporter falls when the excess competitor sites bind the apparently limited factors specific for these sites, the interaction normally occurring at each such site must be necessary for full expression. Therefore, each of these subregions is likely to contain DNA sequence(s) that participate in positive regulatory interactions. Other fragments, such as that containing the binding site for P3A used in the experiment shown in Fig. 2(d), display relatively insignificant competitive activity, although highly specific DNA–protein interactions occur in these subregions as well (Calzone et al 1988).

Lineage-specific regulation of the CyIIIa gene

Our competition experiments have identified seven independent sites where positively acting *cis–trans* interactions apparently take place. Coinjection of molar excesses of fragments that contain either the P1, P2, P3B, P4, P5, P7I or P8 subregions, together with CyIIIa-CAT, severely depresses the activity of this fusion gene in transgenic embryos. Each of these DNA fragments displays one or more highly specific interactions with embryonic nuclear extracts *in vitro* (Calzone et al 1988). It follows that at least one of the DNA–protein interactions in each of these regions is required for CyIIIa transcription in the embryo. The regulatory interactions identified by *in vivo* competition are likely to function combinatorily in a quantitative, literal sense. Thus, for example, the results

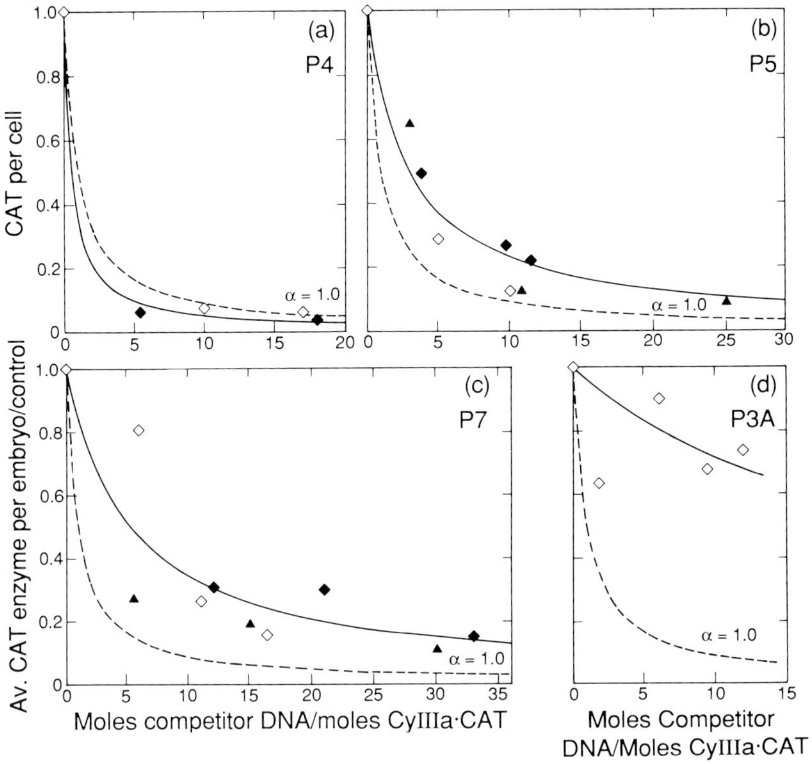

FIG. 2. *In vivo* competition by sub-elements of the CyIIIa 5′ regulatory region in transgenic embryos. At least 1200 molecules of CyIIIa-CAT were injected at each point together with varying numbers of competitor fragments that contain the binding regions for the indicated set of DNA-binding factors (Calzone et al 1988): P7 contains the P7I and P7II binding sites. The molar ratio of competitor sequence to CyIIIa-CAT sequence at each point was measured by slot-blot hybridization of DNA extracted from samples of the pooled embryos, using ^{32}P-labelled probes specific for each sequence. A second aliquot of embryos was pooled and assayed for CAT activity at each point. The relationship between CAT activity per embryo and the molar ratio of competitor DNA molecules to CyIIIa-CAT molecules is expressed as in Fig. 1(b). Embryos that were coinjected with CyIIIa-CAT and a molar excess of the entire 9.5 kb regulatory region were included in each experiment. A least-squares-fit to all of the pooled experimental data for the whole domain fragment yielded an average value for α of 0.75 (ideal stoichiometric competition would generate a value for α of 1.0). The values for α calculated for the whole domain in the experiments shown in (b) and (c) ($\alpha = 0.24$ and $\alpha = 0.59$, respectively) were used as normalization factors to correct the individual data points in these experiments so that the competitive effect of the individual fragments would be compared directly to the effect of the whole domain. In (a) and (d) the average value calculated for α ($\alpha = 0.75$) was used as a normalization factor. Thus the dashed curves in (a)–(d) indicate the competitive behaviour of the entire regulatory region. The solid curves indicate the behaviour of the individual subfragments. Competition is nearly stoichiometric in (a)–(c). There is little, if any, competition in (d).

shown in Fig. 2(a)–(c) demonstrate that competition for binding of the respective factor molecule(s) by any one of several protein-binding regions depresses transcriptional activity approximately as does competition by the entire regulatory region. It follows that the regulatory apparatus must be able to assess the states of all these different interactions in parallel, and in the absence of any one of them, the gene does not function. The spatial distribution of CyIIIa-CAT activity is also being monitored in transgenic embryos to determine if competition for factor binding by any of the sites of interaction might induce ectopic expression of the transgene, i.e. if any of these interactions are negative regulators of spatial expression. Preliminary evidence (B. Hough-Evans, R. Franks & E. Davidson) shows that embryos coinjected with CyIIIa-CAT plus excess molecules of two subfragments—those that contain binding sites for the factors P3A and P7II—display CyIIIa-CAT activity not only in aboral ectoderm cells, but also in gut, oral ectoderm and mesenchyme cells, all cell types that in transgenic controls (injected with CyIIIa-CAT alone) exhibit no CAT activity. It is interesting that for neither the P3A site (Fig. 2d) nor the P7II site (R. Franks and E. Davidson, unpublished results) was competitive depression of CAT synthesis observed by the method summarized here. This result suggests that spatial control of CyIIIa indeed includes negative regulatory interaction(s). There may be positive spatial regulation as well. In all, we can distinguish several different regulatory functions, each of which appears to be carried out by more than one specific interaction. Some interactions are probably responsible for positive temporal regulation; some may be general enhancer functions; and others are responsible for spatial control.

Conditional specification of cell lineage in the sea urchin embryo

The observation that regulation of an early lineage-specific gene such as CyIIIa depends on *cis–trans* interactions leads us back to the conundrum of how differential gene expression is first established at the beginning of the life cycle. Specification of the oral–aboral axis, and of the functional identity of the founder cells that give rise to the aboral ectoderm, are processes that occur after fertilization. This axis is not foreshadowed in the primordial organization of the egg; classical experiments show that in sea urchin embryos this axis, and hence cell specification along it, remains plastic throughout cleavage. An analysis presented elsewhere (Davidson 1989) concludes that cell specification along both animal–vegetal and oral–aboral axes depends largely on intercellular inductive events. A minimum of four such inductive interactions is required to explain the known evidence for lineage specification and lineage plasticity. One way to resolve this result with that discussed here is the proposal that the regulatory factors (positive and/or negative) that causally control spatial activation of lineage-specific genes are mainly globally distributed in the cleavage stage embryo, but that they are converted from an inactive to an active state (or *vice versa*) as a result of specific interblastomere inductions. These are envisaged as ligand–receptor interactions that would result in signal transductions which either covalently modify the embryonic regulatory factors or modify cofactors thereof. The consequence of their action would be conversion of given founder cells into specific gene regulatory domains, and specification of the differential patterns of gene expression that the progeny of these founder cells will display. In this way, a mechanism can be imagined that would account for the regulative plasticity of the sea urchin embryo, and its reliance on an invariant cell lineage through a conditional process of founder cell specification, but also for the convincing molecular evidence that even the earliest differentially expressed genes are controlled by specific DNA–protein interactions.

Acknowledgement

Research was supported by NIH grant HD-05753.

References

Angerer RC, Davidson EH 1984 Molecular indices of cell lineage specification in the sea urchin embryo. Science (Wash DC) 226:1153–1160

Calzone FJ, Thézé N, Thiebaud P, Hill RL, Britten RJ, Davidson EH 1988 Developmental appearance of factors that bind specifically to *cis*-regulatory sequences of a gene expressed in the sea urchin embryo. Genes & Dev 2:1074–1088

Cameron RA, Hough-Evans BR, Britten RJ, Davidson EH 1987 Lineage and fate of each blastomere of the eight-cell sea urchin embryo. Genes & Dev 1:75–85

Cox KH, Angerer LM, Lee JJ, Britten RJ, Davidson EH, Angerer RC 1986 Cell lineage-specific programs of expression of multiple actin genes during sea urchin embryogenesis. J Mol Biol 188:159–172

Davidson EH 1986 Gene activity in early development, 3rd edn. Academic Press, Orlando, Florida

Davidson EH 1989 Lineage-specific gene expression and the regulative capacities of the sea urchin embryo: a proposed mechanism. Development 105:421–445

Davidson EH, Flytzanis CN, Lee JJ, Robinson JJ, Rose SJ III, Sucov HM 1985 Lineage-specific gene expression in the sea urchin embryo. Cold Spring Harbor Symp Quant Biol 50:321–328

Flytzanis CN, McMahon AP, Hough-Evans BR, Katula KS, Britten RJ, Davidson EH 1985 Persistence and integration of cloned DNA in postembryonic sea urchins. Dev Biol 108:431–442

Flytzanis CN, Britten RJ, Davidson EH 1987 Ontogenic activation of a fusion gene introduced into the sea urchin egg. Proc Natl Acad Sci USA 84:151–155

Franks RR, Hough-Evans BR, Britten RJ, Davidson EH 1988 Spatially deranged though temporally correct expression of a Strongylocentrotus purpuratus actin gene fusion in transgenic embryos of a different sea urchin family. Genes & Dev 1:1–12

Hough-Evans BR, Franks RR, Cameron RA, Britten RJ, Davidson EH 1987 Correct cell type-specific expression of a fusion gene injected into sea urchin eggs. Dev Biol 121:576–579

Katula KS, Hough-Evans BR, Britten RJ, Davidson EH 1987 Ontogenic expression of a CyI actin gene fusion injected into sea urchin eggs. Development 101:437–447

Lee JJ, Shott RJ, Rose SJ, Thomas TL, Britten RJ, Davidson EH 1984 Sea urchin actin gene subtypes. Gene number, linkage and evolution. J Mol Biol 172:149–176

Lee JJ, Calzone FJ, Britten RJ, Angerer RC, Davidson EH 1986 Activation of sea urchin actin genes during embryogenesis. Measurements of transcript accumulation from five different genes in Strongylocentrotus purpuratus. J Mol Biol 188:173–183

Livant DL, Cutting AE, Britten RJ, Davidson EH 1988 An *in vivo* titration of regulatory factors required for expression of a fusion gene in transgenic embryos. Proc Natl Acad Sci USA 85:7606–7611

McMahon AP, Flytzanis CN, Hough-Evans BR, Katula KS, Britten RJ, Davidson EH 1985 Introduction of cloned DNA into sea urchin egg cytoplasm: replication and persistence during embryogenesis. Dev Biol 108:420–430

Shott RJ, Lee JJ, Britten RJ, Davidson EH 1984 Differential expression of the actin gene family of Strongylocentrotus purpuratus. Dev Biol 101:295–306

Sucov HM, Hough-Evans BR, Franks RR, Britten RJ, Davidson EH 1988 A regulatory domain sufficient to direct lineage-specific expression of a skeletal matrix protein gene in the sea urchin embryo. Genes & Dev 2:1238–1250

DISCUSSION

Melton: Has anyone looked for inducible factors which will cross a filter? Have you tried any heterologous growth factors to see what effects they might have?

Davidson: Yes, we hope to have results from those experiments in due course.

Cooke: Do you have positive evidence that the class of proteins you think are spatial specific regulators come to be present everywhere in the egg?

Davidson: There is some evidence which is not fit to discuss in public. Studies of the DNA-binding proteins suggest that DNA-binding *per se* is the most robust of the molecular functions that these proteins carry out. The process that is probably modified physiologically is the transcriptional activity. That is not tested in the DNA binding assays which we have done.

Heasman: Could you explain more about a negative space regulator factor?

Davidson: We know three things. If we compete out binding factors with excess quantities of either of two different DNA upstream sequences that bind nuclear proteins with very high affinity, we get ectopic expression of the gene. This says that a negative interaction is required to keep the gene switched off in gut or mesenchyme cells, for example.

Secondly, there are two additional aspects of the transpecific experiment. If you look for the DNA–protein complex which that particular probe sees in nuclear extracts of the second species, it is not there, although a different complex that may or may not be significant can be detected. In the second species the genes are correctly regulated temporally. If you make a species hybrid, there is proper spatial regulation, presumably because the sperm carries a zygotic regulatory gene which makes this product. Without these interactions (and there are probably two of them), there are enough activators ubiquitously present for this gene to be turned on. It's interesting that this is a gene that functions differentially across the oral–aboral axis, which is established after fertilization and can apparently be oriented anywhere in the plane perpendicular to the animal–vegetal axis, and which is plastic in that it can easily be respecified.

Cooke: Regarding your speculations about lithium, in amphibians, if you take the part of the egg which requires morphogenetic instructions to become anything other than ectoderm, before it has any of the natural inducers, then lithium will not act by itself as an inducer.

Davidson: On the other hand, lithium does have striking effects on amphibian embryos, from Elinson's work, which could be interpreted in similar ways (Kao et al 1986).

Cooke: A couple of labs have found that lithium greatly potentiates the thresholds of response to protein inducing factors. The ectodermal cells will make the structures they would normally have made in response to about 10 times as much inducing factor.

Davidson: In my opinion, it is not profitable to speculate too much on that aspect in the embryos we are discussing in the absence of further mechanistic studies, because most lithium biochemistry comes from investigations on why it's a good medicine for depression in humans. Lithium seems to intervene in at least two different signal transduction systems. How those balance out with one another or which, if either, is involved in this case, we really can't say.

Weisblat: Is the animal–vegetal axis preserved in the animal caps treated with lithium?

Davidson: It is and it can never be changed, according to experiments of Hörstadius and his school. If you take animal caps and turn them upside down, for example, the animal–vegetal axis won't regulate. If you take extra animal caps and pile them on top of each other on top of a vegetal region of an embryo, the result is the same as if the extra ones weren't there; there is no evidence for 'animal' action at a distance.

Sander: What are the criteria on which you base your statement that the polarity in the animal cap never changes except when you implant micromeres?

Davidson: In any experiments in which layers are taken from the middle of the egg and put together, the ectoderm always comes from the portion of the system which came originally from veg_1 or the animal layer, never from veg_2. I am simply quoting Hörstadius here, there is also less stringent evidence which led Hörstadius to make this statement based on morphological criteria.

Wieschaus: One of the differences between the effects of lithium and micromere transplantation is that micromeres are placed at one particular pole of the egg, whereas lithium presumably acts uniformly all over the surface.

Davidson: No, it is probably not acting uniformly over all the surface. One part of the animal cap surface was an intracellular surface. The outside of the sea urchin late cleavage embryo is a specialized structure, in that the cells are all polarized inside to outside at this point; it is covered with a hyaline layer and there are other surface specializations. It looks as if the site of action of lithium is the place where there was an intercellular and not an external surface.

Wieschaus: If you transplant the micromeres to another region of the embryo, what effect does that have?

Davidson: You can put them anywhere, as long as the cells are placed such that they can fuse. Wherever you put the micromeres, a second induction will occur. The lithium experiment I cited was done by von Ubisch in 1925 and has been repeated in Fred Wilt's lab recently. Some of these other early experiments are being repeated in various labs at present. The simplest and clearest of the experiments in the literature have all been reproducible so far.

Wolpert: You proposed an induction model, whereas the classical model is based on a double gradient. It would be very helpful to know your reasons.

My second question is related to the egg and the effect of lithium. Lithium will work on normal embryos in which you have made no cuts at all: it shifts the ectoderm/endoderm border. The obvious interpretation is that some gradient or threshold is being altered. We know there is polarity in the egg, the animal pole being different from the vegetal pole. The lithium experiment is one way of showing that there is some sort of graded property within the egg. So again, why have you chosen induction rather than a gradient mechanism?

Davidson: For several reasons. The best analysed inductive system where the inducing agent is diffusible so it can create a gradient is the amphibian system. John Gurdon's experiments suggest that the radius of diffusion from an inducing cell is approximately equal to the diameter of the whole sea urchin egg. If one has to account for these extremely sharp changes in cell fate which occur right at the borders of the cells (I appreciate there are, in principle, ways of getting sharp interfaces from intersections of gradients) and account for the fact that the cleavage planes can be altered by pressure and other things and still have the expected immediate neighbour effects, even in eggs that are deformed, it seems to me a less tortured interpretation.

Secondly, I am impressed with the fact that the double gradient argument that was adopted by Hörstadius has significant unsatisfactory aspects, even in the classical evidence for it. There is, in my opinion, no convincing evidence that there is a gradient originating at the animal pole. The whole argument for a double gradient is based on so-called quantitative effects. All of the experiments people have used to support that argument are subject to a trivial interpretation, which is that when you put less animal pole versus vegetal material you get incomplete transformation. When you use only a small amount of animal pole material, most of the cells in the recipient embryo are not in contact with what I would regard as the inducing cells. Furthermore, if you do things like piling extra sources of animal pole material on top of the egg, you would expect an additional effect, but nothing like that happens. So I don't think there is any good evidence for an animal-like gradient source. Furthermore, one of the restrictions in this system is that the micromeres will never produce any ectoderm. It is not a symmetrical, plastic system with equivalent animal and vegetal sources.

Wolpert: There are chemicals which can animalize the embryo.

Davidson: To my knowledge, none of those animalizing experiments have been done using molecular markers. So the outcome in terms of cell differentiation is not clear.

Wolpert: What about the lithium experiments and the gradient within the egg—you don't think there is a gradient within the egg either?

Davidson: I think that whether you describe the initial condition of the egg as a gradient or as a system in which some regulatory factors for micromere–veg_2 function are fixed to the cytoskeleton at the south pole of the egg, doesn't make a lot of difference. There could be a diffusion gradient earlier than that, but there is no requirement for one.

Wolpert: I would dispute that but I am going to resist.

Meinhardt: Although some people like the idea of mosaic eggs, it doesn't solve the problem. It only means that the pattern-forming reactions occur somewhat earlier. I like the lithium experiment, where you revitalize the pattern-forming system. You have shown there is a group of genes which suppress transcription in other cells. That's something we would expect from the models we proposed a long time ago. We would also expect that the enhancing element would be autocatalytic. Are there any indications for local autoregulation in your system?

Davidson: I can answer your speculation with another speculation. As Janet Heasman said, the time at which commitment appears in the amphibian is shortly after transcription resumes. The time at which it appears in the sea urchin system is much earlier in development, but perhaps not long after the initial specification, given that transcription is going on all the time. It may be that what is needed for commitment is transcription. One thing that happens in the cases we have looked at so far is that many of the regulators that we know about go from relatively low levels to much higher levels after the system has been committed. These things must be transcribed, but whether the effect is autocatalytic, we don't know yet.

Gardner: You said the lineage was invariant, but qualified it by saying that there are subtle differences in lineage in the animal region of embryos. Do any of those subtle differences lead to changes in the assignment of the fate of cells with respect to the major components of the larva?

Davidson: There are two kinds of differences. One is minor territorial geometrical differences without changes in cell fate. The other is that in some species the third cleavage plane, which is horizontal, can float up and down;

sometimes the border between veg_1 and the animal layer is altered. In the sea urchin we study, that is not a prominent phenomenon, but it is in some species.

Garcia-Bellido: How much information is there in the egg? What is the minimum amount of cytoplasm that will still generate a complete embryo, and from which region of the egg does it come?

Davidson: Morgan & Spooner in 1909 showed that you can centrifuge these eggs and move all the internal cytoplasmic inclusions to the opposite end of the egg without affecting either the cleavage pattern or the course of development. The polarization in the egg is either a diffusion source which continues to function or something that's tightly locked into the cytoskeleton in the cortex in the egg which is not disturbed at these centrifugal forces; it is the latter idea that I prefer.

Garcia-Bellido: If you sandwich micromeres between two animal caps, what happens? The question is whether the cells of the lower animal cap are recruited to form veg_2 progeny or whether they retain their own identity and give a double-headed embryo.

Davidson: My recollection from that experiment is that some of the embryos turned into monsters and some of them reorganized in such a way as to have one unified polarity. The monsters were certainly not bipolar or anything of this kind.

Garcia-Bellido: That is the question: to know if the polarity is there in the oocyte or whether it is the result of interactions between regions of the egg.

Wolpert: We have centrifuged sea urchin eggs (Mercer & Wolpert 1962) and found that all you shift is the mitochondria and the yolk granules, everything else is left intact. There was no evidence for a cortex at all. Where the polarity resides, we don't know.

I would contest three things at least. The first is that I find it very difficult to believe that veg_2 cannot give rise to ectoderm. In some embryos, if you divide them at the first horizontal cleavage plane (that is, the third cleavage) and throw away the animal half, a significant proportion of vegetal halves form a normally proportioned larva. I would like to see how an inductive model will ever give rise to that proportionality.

Davidson: In those cases, development is apparently delayed and there is a good bit of cell division.

Wolpert: There is no evidence for increased cell division, is there?

Davidson: I didn't say increase but there is cell division before the thing forms a blastula. I would say there is nothing in the literature inconsistent with the proposition that there are sufficient ectoderm cells deriving from veg_1 to re-create what's necessary. Those things usually don't give a perfectly normal embryo. They give stumpy looking embryos and what's missing is the elongation of the aboral ectoderm.

Wolpert: I think you should draw a distinction between a diffusion gradient of a morphogen and a fixed gradient. There is no evidence for a signal from the animal pole in Hörstadius' experiments; rather they suggest that there is some graded property within the animal region.

Davidson: Hörstadius himself commented that he was surprised that for some unaccountable reason when that experiment was done there was no strengthening of the animal signal, contrary to expectation from his own gradient interpretation. I specifically said that there are two ways of looking at it and I prefer the idea of cytoskeletally fixed regulators.

Meinhardt: Induction models can account for size regulation much better than gradient models.

Weisblat: Couldn't questions regarding the fate of specific cells in these various experiments be addressed using lineage tracers?

Davidson: We are undertaking a series of experiments with both lineage tracers and molecular markers. The right way to do it is using embryos into which you have injected a fusion gene so that you can be sure that changes in cell specification can be equated with changes mediated by the regulatory domains in given genes.

Gardner: Do you see any evidence for the presence of extracellular matrix components during cleavage? In the mouse, both laminin and collagen are found interstitially in the cleaving embryo, but it is not clear what role, if any, they play in specifying planes of cleavage.

Davidson: This has been studied extensively in sea urchins by David McClay. There is a battery of these substances which are differentially secreted on the apical and basal sides of the cells that constitute the blastula. Interference with their modification is extremely deleterious to subsequent morphogenesis. However, unlike amphibians, later development in sea urchins does not involve any multilayer inductions, because there aren't any multilayers in sea urchin embryos until late in embryogenesis. That's interesting, for example when considering the origins of nerve cells, it is more like delamination of neuroblasts in *Drosophila* than neurogenesis in a plate of cells overlying the mesoderm, as in *Xenopus*. In the sea urchin, there is no layer-to-layer induction during embryogenesis, except perhaps for stomodeal formation, where gut approaches the oral ectoderm wall. Of course during rudiment formation extensive and complex inductive interactions must occur.

References

Kao KR, Masui Y, Elinson RP 1986 Lithium induced respecification of pattern in Xenopus laevis embryos. Nature (Lond) 322:371–373

Livingston BT, Wilt F 1989 Proc Natl Acad Sci USA, in press

Mercer EH, Wolpert L 1962 An electron microscope study of the cortex of the sea urchin (*Psammechinus mulians*) egg. Exp Cell Res 27:1–13

Morgan, Spooner 1909 Arch Entwicklungsmech Org 28:104–117

von Ubisch 1925 Z Wiss Zool 124:361–381; 457–468; 469–486

von Ubisch 1925 Verh Phys-Med Ges Würzburg 50:13–19

Cell allocation and lineage in the early mouse embryo

R. L. Gardner

Imperial Cancer Research Fund, Developmental Biology Unit, Department of Zoology, University of Oxford, South Parks Road, Oxford OX1 3PS, UK

Abstract. In the early mammalian embryo, initially asymmetric cell contacts appear to induce blastomere polarization, elements of which can persist through cytokinesis. This leads to the generation of inner and outer populations of blastomeres which may subsequently diverge as a result of residing in distinct microenvironments. Similar processes may account for the generation of primitive endoderm versus primitive ectoderm, and that of trophectoderm versus inner cell mass. However, if this is the case, the response of cells to positional cues must change as a function of either their previous positional history or the number of cycles they have completed. Once these primary tissues have been established, specific interactions between them lead to further cellular diversification.

1989 Cellular basis of morphogenesis. Wiley, Chichester (Ciba Foundation Symposium 144) p 172–186

The early mammalian embryo appears to be an unusually unconstrained developmental system, in terms of the variability within species in temporal and spatial organization of its cleavage, as well as in its response to experimental perturbation (Gardner 1983, 1985a, Fleming 1987). This has led to the conclusion that specification of cell lineages in early mammalian development is unlikely to depend on either the segregation of cytoplasmic determinants or morphogenetic gradients set up in the egg around the time of fertilization. Attention has therefore been focused on identifying environmental factors which may direct the diversification of cells, and attempting to elucidate how they might do so. This paper briefly reviews the progress that has been made in this endeavour in recent years.

Specification of trophectoderm and inner cell mass cells

The trophectoderm of the blastocyst is generally thought to be composed of descendants of outer blastomeres of the late morula, and the inner cell mass (ICM) of inner blastomeres. The results of various studies in the mouse suggest that there is little or no exchange of cells between inner and outer populations once the blastocoel has begun to form, so that two discrete cell lineages are probably established by the early blastocyst stage (see Rossant 1987, Gardner & Beddington 1989). Whether these lineages remain discrete thereafter has been questioned recently by Pedersen and his colleagues, who have provided evidence based on intracellular injection of horseradish peroxidase that cells are recruited into the polar trophectoderm from the ICM during growth of blastocysts *in vitro* (Winkel & Pedersen 1988). However, using a non-invasive method of cell marking, Dyce et al (1987) found such cell movement to be exceptional and to occur no more frequently than that from trophectoderm to ICM. As noted elsewhere, these studies need to be extended to blastocysts which are allowed to continue their growth in the uterus rather than *in vitro* before their relevance to normal development can be assessed (Gardner & Beddington 1989).

Largely through the work of Johnson and his colleagues (reviewed in Johnson & Maro 1986), some progress has been made in recent years in understanding how the divergent differentiation of inner and outer blastomeres might occur during cleavage. The key finding is that all blastomeres of the embryo acquire an apical microvillous pole at the 8-cell stage, elements of which can persist through subsequent mitoses. Hence, depending on the orientation of cleavage in parent 1/8 blastomeres, one or both daughters may inherit a polar phenotype (Johnson & Ziomek 1981). Where polar–apolar pairs are produced by division of isolated 1/8 blastomeres the polar cell typically attempts to envelope its apolar sister (Johnson & Maro 1986). Furthermore, polar and apolar blastomeres partition efficiently to an outside and inside location, respectively, when allowed to do so in blastomere reaggregates (Johnson & Maro 1986). The polar–apolar difference would therefore seem to be sufficiently stable under normal circumstances to ensure partitioning of blastomeres to two distinct microenvironments in the morula, even though it can be erased under certain experimental conditions (Johnson & Maro 1986, Rossant 1987). Once partitioned, blastomeres clearly undergo further phenotypic divergence, presumably in response to local cues. This seems to be a gradual process, particularly in the case of inside cells, which can still be re-programmed to form trophectoderm at the early blastocyst stage. Viewed thus, the 'polarization' and 'microenvironmental' hypotheses can be regarded as providing complementary rather than alternative ways of accounting for the origin of trophectoderm and ICM cell lineages (Gardner 1983). This conclusion is supported by the finding that prolonged exposure of morulae to antisera that interfere with compaction without inhibiting the polarization or division of blastomeres can result in the formation of trophectodermal vesicles devoid of ICM tissue (Johnson et al 1986).

It is now clear that the allocation of blastomeres to inner and outer populations is not accomplished in a single step at the conclusion of 4th cleavage, but entails further recruitment of the presumably apolar daughters of outer polar 16-cell blastomeres to the inner population thereafter (Fleming 1987). Hence some ICM cells have resided inside the embryo for a shorter interval than others and may therefore become committed somewhat later. Possible implications of this for cellular differentiation within the ICM are considered below.

Diversification of cells within the trophectoderm

The minority of trophectoderm cells retaining contact with the ICM are the only ones that continue to proliferate beyond the blastocyst stage. These polar trophectoderm cells are therefore presumed to be progenitors of the various types of trophoblasts originating later in gestation (Gardner & Beddington 1989). The remaining mural cells begin to enlarge and become polytene once the blastocyst attaches to the uterine epithelium. It seems likely that the trophectoderm is composed initially of a population of equipotential cells, within which regional differentiation is specified later by the position of the ICM. Thus mural tissue can be made to form polar tissue simply by bringing it into contact with an ICM before it undergoes transformation into giant cells (Gardner et al 1973). Conversely, when polar trophectoderm is separated from the underlying ICM, it behaves like mural tissue, forming vesicles which can implant *in utero*, ceasing mitotic activity and beginning giant transformation both *in vivo* and *in vitro* (unpublished observations). Hence, in effect, the ICM prevents overlying trophectoderm cells from embarking on terminal differentiation, although how it does so is still obscure. Nevertheless, observations on blastocysts containing two physically separate ICMs suggest that a local interaction between the tissues is involved (R. L. Gardner, in preparation). As discussed elsewhere, ICM derivatives continue to play an essential role in sustaining proliferation of trophoblast well into the post-implantation phase of development (Gardner & Beddington 1989).

One step towards further characterizing this tissue interaction is to determine which ICM derivative(s) can promote trophoblast growth. Currently, the development of vesicles of mural trophectoderm containing microsurgically isolated primitive ectoderm or primitive endoderm tissue is being investigated with this aim. The great majority of such vesicles implant in the uterus, but typically give rise to morphologically aberrant conceptuses thereafter. Preliminary appraisal of the contents of resulting decidua on Day 8 of gestation suggests that sustained trophoblast proliferation is correlated particularly with the presence of visceral endoderm (unpublished observations). This is interesting because the visceral endoderm is not only closely associated with mitotically active diploid trophoblast in the early post-implantation embryo, but also because it secretes various types of molecules implicated in growth (Gardner 1983, Heath & Rees 1985). Recent evidence that binding to components of the extracellular matrix may be necessary for the activity of certain haemopoietic growth factors suggests how a regulatory molecule that is freely diffusible might nevertheless exert a local effect (Roberts et al 1988). A matrix is associated with the inner surface of proliferating trophoblasts from the blastocyst stage onwards.

Specification of endoderm versus ectoderm cells

Differentiation of cells within the ICM leading to formation of the primitive endoderm and primitive ectoderm has been examined rather less closely than the differentiation to trophectoderm or ICM discussed earlier. Nevertheless, since it also seems to entail production of enclosed and exposed populations of cells with distinct fates, it has been widely assumed to depend on recapitulation of processes occurring earlier in the morula. If there is, indeed, any basis for this assumption, it is necessary to postulate that the response of cells to positional cues changes—otherwise, additional trophectoderm rather than endoderm cells would be formed on the blastocoelic surface of the ICM. It is instructive to consider briefly what is known about differentiation of cells within the ICM in order to see how closely it corresponds with what happens earlier in the morula.

The primitive endoderm forms on the blastocoelic surface of the ICM where it is first clearly discernible when the blastocyst begins to implant. Despite earlier claims to the contrary, the time of appearance of this tissue is not delayed appreciably if implantation is postponed (Gardner et al 1988). Cell and tissue transplantation experiments show that the primitive endoderm is the sole progenitor tissue of the entire extra-embryonic endoderm of the post-implantation conceptus (Gardner 1985b).

Differentiation of the primitive endoderm is not accompanied by any obvious change in phenotype of the remaining ICM cells. These non-endoderm cells do, however, reorganize to form a more or less discrete core to the ICM known as the primitive ectoderm (Fig. 1). When transplanted between blastocysts, primitive ectoderm cells consistently colonize most, or all, tissues of the later conceptus, with the exception of trophectoderm and primitive endoderm derivatives (Gardner 1985b). Hence, differentiation within the ICM leads within 24 hours of the onset of blastulation to the production of two populations of cells, which clearly differ in developmental potential, as assayed by blastocyst injection.

There is no compelling evidence that all endoderm and ectoderm cells originate from precursors occupying exposed and enclosed positions, respectively, in the early ICM. However, this seems likely, because extensive rearrangement of ICM cells is not consistent with the coherent clonal pattern of growth observed throughout pre-implantation development (Gardner 1983, Rossant 1987). In addition, according to Enders and his colleagues (1978), cells showing early ultrastructural indications of endodermal differentiation are confined to the

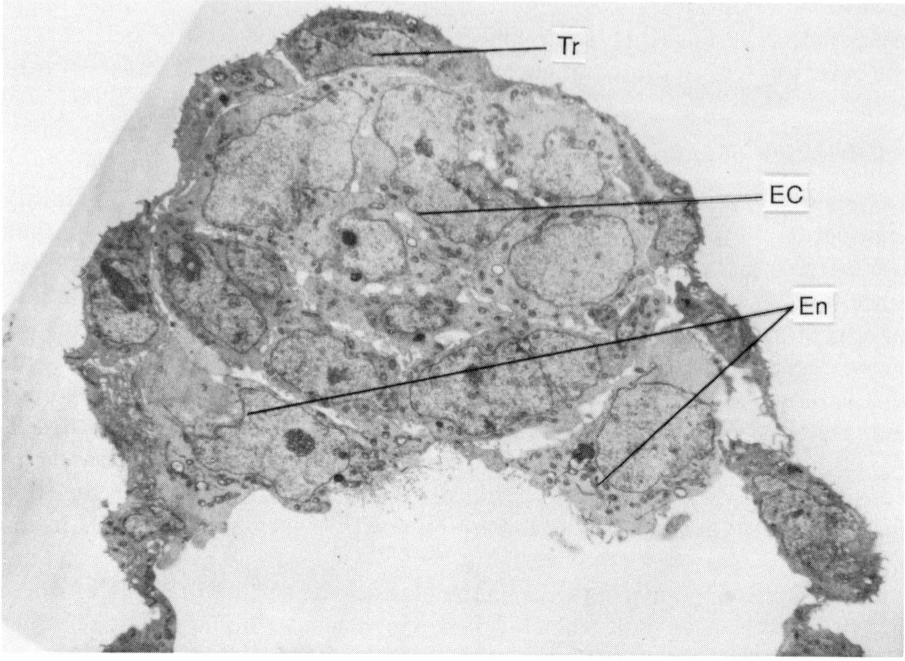

FIG. 1. Electron micrograph showing the inner cell mass region of a mature blastocyst with a distinct core of ectodermal tissue (EC) sandwiched between the polar trophectoderm (Tr) and juxtacoelic endoderm layer (En). Magnification: ×2430.

blastocoelic surface of the ICM. These workers also reported that the early ICM engages in a process akin to compaction which is partially reversed once endoderm is discernible. Whether this is accompanied by polarization of juxtacoelic ICM cells has not been addressed. Antibodies that interfere with compaction of morulae can also prevent intimate association between cells in cultured ICMs. According to Richa et al (1985), prolonged exposure of ICMs to such reagents prevents formation of an endoderm layer, which is the opposite of what would have been expected if differentiation within the ICM and morula depended on similar positional cues. However, the structures formed in these experiments were not examined closely enough to establish whether it was the actual differentiation of endoderm that was inhibited or its organization into an epithelium. If all cells in treated ICMs became endodermal, they might be unable to form an epithelium because of the absence of underlying ectoderm tissue.

While it is clear from studies on both early ICMs and aggregates of embryonal carcinoma (EC) cells in culture that primitive endoderm cells can differentiate on a free surface, it is by no means certain that they have to do so. The primitive endoderm appears to be already more than one layer thick at an early stage of its existence (Gardner 1985c); it remains to be established whether the deeper cells are mitotic descendants of those occupying a juxtacoelic position at the time of commitment. Furthermore, Fleming and his colleagues (1984) claim that trophectoderm processes cover most of the blastocoelic surface of the ICM until immediately before endoderm differentiation takes place. They argue that the role of these processes is to prevent the expression of totipotency and consequent trophectodermal differentiation of juxtacoelic early ICM cells, rather than simply to hold the ICM in place. This hypothesis implies that unlike trophectoderm, differentiation of endoderm is not dependent on asymmetric contacts between cells. However, the extent of coverage of the blastocoelic surface of the ICM varied considerably between blastocysts, a finding attributed to temporary withdrawal of processes by dividing trophectoderm cells (Fleming et al 1984). Movement of cells from polar to mural trophectoderm during growth of the blastocyst (Cruz & Pedersen 1985, Dyce et al 1987) might make an even greater contribution to periodic disruption of the continuity of this layer.

It is important to note that while early ICMs can regenerate an outer layer of trophectoderm and thereby assume the appearance of miniature blastocysts when cultured following isolation, by no means all of them do so (Fleming et al 1984). Some form endoderm; others yield non-integrated structures of uncertain cellular composition. One question posed by these findings is whether outside cells in a particular ICM behave uniformly, differentiating either as trophectoderm or endoderm, or, as might explain the occurrence of non-integrated forms, can yield both types of cell. Experiments in which phagocytosed melanin granules were used to label outer cells showed that those of a single ICM could indeed produce both trophectoderm and endoderm (see Fig. 2 and Nichols & Gardner 1984). Hence, early ICM cells can differentiate along either of two distinct pathways in response to external positional cues. This disparate behaviour of the cells could not readily be explained by microenvironmental heterogeneity in the cultures, and therefore seemed to reflect their own intrinsic heterogeneity, one obvious source of which is the marked asynchrony in cell cycles that occurs throughout the embryo by this stage of development. Attempts to test this possibility have so far yielded equivocal results, possibly because by the blastocyst stage the size of cells is no longer a reliable index of the number of cycles they have completed (J. Nichols & R. L. Gardner, unpublished data). However, a further possibility noted earlier is that the response of ICM cells to exteriorization depends on the length of time they have resided inside the embryo rather than the number of cycles they have completed before they are exteriorized. Thus, Fleming (1987) found that although all ICM cells derived from blastomeres which entered the morula after completion of the 4th cleavage reside in the periphery rather than the core of

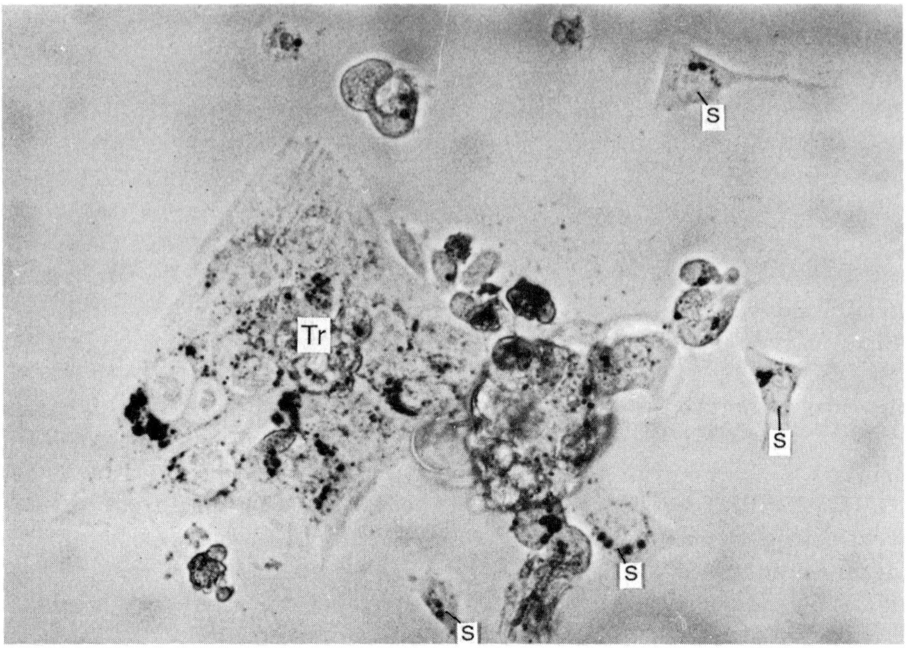

FIG. 2. Outgrowth formed by an isolated early inner cell mass whose external cells had been labelled by endocytosis of melanin granules. Note the presence of granules both in the sheet of trophoblast (Tr) and in solitary cells (S) shown to be endodermal by blastocyst injection. Magnification: ×370. (From Nichols & Gardner 1984.)

the ICM, only 10% of them lie on its blastocoelic surface. Hence, if totipotency is restricted to these cells, there may be no need to invoke special mechanisms to ensure that juxtacoelic cells form endoderm rather than trophectoderm. It is noteworthy that Fleming et al (1984) recorded very occasional juxtacoelic ICM cells in intact blastocysts which resembled trophectoderm cells ultrastructurally.

Specification of lineage within the primitive endoderm

Further lineage restriction probably does not occur in the primitive ectoderm until around the time of gastrulation, although little is known about when and where it takes place. One might expect segregation of the germ line to be a first step, but this is by no means certain. Later stages in the development of this tissue have been reviewed in Gardner & Beddington (1989); here attention is focused on differentiation within the primitive endoderm which results in the production of two very different but well characterized types of cell, the visceral and parietal endoderm. The visceral endoderm displays a typical epithelial organization and plays both a secretory and an absorptive role. The parietal endoderm is composed of cells which are more variable in morphology and arrangement, and which continue to mix extensively throughout growth of the tissue (Gardner 1983, 1985b).

Embryonal carcinoma cells, particularly those of the F9 line, have been used extensively as an *in vitro* system for studying aspects of endoderm formation, largely because of the ease with which their differentiation can be controlled. By exposing aggregates of these cells to certain drugs, it is possible to manipulate both the time course and direction of their differentiation (Hogan et al 1983). This is in marked contrast to the situation in the embryo, where critical events take place rather rapidly in a small population of cells at a stage that has so far proved refractory to both manipulation and culture. Nevertheless, the two systems have yielded closely comparable results.

The little that is known about differentiation within the primitive endoderm in the embryo can be summarized as follows. Cloning experiments have revealed that primitive endoderm cells are bipotential (Gardner 1985b). They nevertheless show a marked propensity for parietal rather than visceral differentiation which may be an artifact of the blastocyst injection assay (Gardner 1983). During the first day of its existence, the visceral endoderm resembles the primitive endoderm very closely in developmental potential (Cockroft & Gardner 1987). Cells can change from a visceral to a parietal phenotype, even beyond the stage at which markers of visceral differentiation are first expressed (Hogan & Tilly 1981, Gardner 1983), although there is no direct evidence that they do so during normal development. Such transformation seems to be precipitated by disruption of the relationship of visceral cells with underlying tissue. The reciprocal transformation of parietal into visceral cells has not been recorded. This is perhaps not surprising, because parietal cells appear to differentiate very early, possibly before they emigrate onto the mural trophectoderm from the ICM (Gardner 1985c).

Grover & Adamson (1986) have presented evidence that F9 aggregates express a similar biochemical phenotype at an early stage of differentiation, regardless of whether they are induced to form parietal or visceral endoderm. Furthermore, parietal differentiation can be switched to visceral, providing the change is made within approximately the first 24 hours of induction. Visceral differentiation can be switched to parietal up to a much later stage and, as appears to be the case in the embryo, probably involves an alteration in cellular relationships. However, no cloning experiments have been done so far to test the developmental potential of endoderm at various stages in the differentiation of EC cell aggregates. Hence it is not clear whether the different inducing conditions influence the direction in which bipotential precursor cells differentiate, or promote the preferential survival or proliferation of one or other type of endoderm cell.

Conclusion

Differentiation appears to take place *within* a population of initially equipotent cells in both the morula and ICM. In the morula, induction of cell polarization followed by its differential inheritance seem to be crucial early events. Whether this is also true for the ICM is at present uncertain because of the paucity of relevant data. If it does prove to be the case, studies on relatively tractable EC cells could contribute to a better understanding of the differentiation of trophectoderm, as well as that of endoderm. Clearly, once the three primary cell lineages are established in the blastocyst, interactions *between* them play a vital but as yet poorly defined role in initiating further cellular diversification.

Acknowledgements

I wish to thank Andy Copp, Tim Davies, Jonathan Slack and Jo Williamson for help in preparing the manuscript and both the Royal Society and the Imperial Cancer Research Fund for support.

References

Cockroft DL, Gardner RL 1987 Clonal analysis of the developmental potential of 6th and 7th day visceral endoderm cells in the mouse. Development 101:143–155

Cruz YP, Pedersen RA 1985 Cell fate in the polar trophectoderm of mouse blastocysts as studied by microinjection of cell lineage tracers. Dev Biol 112:73–83

Dyce J, George M, Goodall H, Fleming TP 1987 Do trophectoderm and inner cell mass cells in the mouse blastocyst maintain discrete lineages. Development 100:685–698

Enders AC, Given RL, Schlafke S 1978 Differentiation and migration of endoderm in the rat and mouse at implantation. Anat Rec 190:65–78

Fleming TP 1987 A quantitative analysis of cell allocation to trophectoderm and inner cell mass in the mouse blastocyst. Dev Biol 119:520–531

Fleming TP, Warren PD, Chisholm JC, Johnson MH 1984 Trophectodermal processes regulate the expression of totipotency within the inner cell mass of the mouse expanding blastocyst. J Embryol Exp Morph 84:63–90

Gardner RL 1983 Origin and differentiation of extraembryonic tissues in the mouse. Int Rev Exp Pathol 24:63–133

Gardner RL 1985a Origin and development of the trophectoderm and inner cell mass. In: Edwards RG et al (eds) Implantation of the Human Embryo. Academic Press, London (2nd Bourn Hall Meeting 1985) p 155–178

Gardner RL 1985b Clonal analysis of early mammalian development. Philos Trans R Soc Lond B Biol Sci 312:163–178

Gardner RL 1985c Regeneration of endoderm from primitive ectoderm in the mouse embryo: fact or artifact? J Embryol Exp Morph 88:303–326

Gardner RL, Beddington RSP 1989 Multi-lineage 'stem' cells in the mammalian embryo. J Cell Sci Suppl 10:11–27

Gardner RL, Papaioannou VE, Barton SC 1973 Origin of the ectoplacental cone and secondary giant cells in mouse blastocysts reconstituted from isolated trophoblast and inner cell mass. J Embryol Exp Morph 30:561–572

Gardner RL, Davies TJ, Carey MS 1988 Effect of delayed implantation on differentiation of the extra-embryonic endoderm in the mouse blastocyst. Placenta 9:343–359
Grover A, Adamson ED 1986 Evidence for the existence of an early common biochemical pathway in the differentiation of F9 cells into visceral or parietal endoderm. Dev Biol 114:492–503
Heath JK, Rees AD 1985 Growth factors in mammalian embryogenesis. In: Growth Factors in Biology and Medicine. Pitman Publishing Ltd, London (Ciba Found Symp 116) p 3–15
Hogan BLM, Tilly R 1981 Cell interactions and endoderm differentiation in cultured mouse embryos. J Embryol Exp Morph 62:379–394
Hogan BLM, Barlow DP, Tilly R 1983 F9 teratocarcinoma cells as a model for the differentiation of parietal and visceral endoderm in the mouse embryo. Cancer Surv 2:115–140
Johnson MH, Maro B 1986 Time and space in the mouse early embryo: a cell biological approach to cell diversification. In: Rossant J, Pedersen RA (eds) Experimental Approaches to Mammalian Embryonic Development. Cambridge University Press, Cambridge, p 35–65
Johnson MH, Ziomek CA 1981 The foundation of two distinct cell lineages within the mouse morula. Cell 24:71–80
Johnson MH, Maro B, Takeichi M 1986 The role of cell adhesion in the synchronization and orientation of polarization in 8-cell mouse blastomeres. J Embryol Exp Morph 93:239–255
Nichols J, Gardner RL 1984 Heterogeneous differentiation of external cells in individual isolated early mouse inner cell masses in culture. J Embryol Exp Morph 80:225–240
Richa J, Damsky CH, Buck CA, Knowles BB, Solter D 1985 Cell surface glycoproteins mediate compaction, trophoblast attachment and endoderm formation during early mouse development. Dev Biol 108:513–521
Roberts R, Gallagher J, Spooner E, Allen TD, Bloomfield F, Dexter TM 1988 Heparan sulphate bound growth factors: a mechanism for stromal cell mediated haemopoiesis. Nature (Lond) 332:376–378
Rossant J 1987 Cell lineage analysis in mammalian embryogenesis. Curr Top in Dev Biol 23:115–146
Winkel GK, Pedersen RA 1988 Fate of the inner cell mass in mouse embryos as studied by microinjection of lineage tracers. Dev Biol 127:143–156

DISCUSSION

Lawrence: Your lineage and cell affinity experiments in mammals fit very well with the ideas of compartments in insects. You have shown that there are several such putative compartments in mammals and there might be even more if you looked later in development. Do you think these are true compartments?

Gardner: It depends what one would emphasize as important about compartments. In the early work on *Drosophila*, emphasis was placed on geographical subdivision rather than division with respect to differentiated cell types. In the mouse embryo, you certainly reach points in time where distinct populations of cells, which can no longer be interconverted by experimental manipulation, emerge. At this stage, the number of cells in each of these populations is small, about 10 or 20, and all their mitotic descendants will be confined predictably to particular sets of tissues. But within those tissue derivatives, lineage is highly variable. For example, within the primitive endoderm, the clones vary considerably in both size and spatial distribution. What is happening is the carving up of the cell populations in terms of differentiated function, at this stage at least. What happens later in development, I don't know. You could argue that the stages before gastrulation are still basically a phase of pre-embryonic development, because these early differentiating lineages are there to establish conditions in which development of the embryo proper can occur.

If you look at the two early differentiating lineages, trophectoderm and endoderm, they show a number of peculiarities. They differ from the primitive ectoderm lineage in terms of the degree of methylation of certain genes, the pattern of X chromosome inactivation and the possible roles of the two parental genomes (Gardner 1988). They also show differences in the stability of switching off of genes. For example, you can much more readily reactivate X-linked genes on the inactive X chromosome in extra-embryonic tissues than in the embryo itself (Chapman 1986). So you could make the case that you are dealing with rather special circumstances throughout.

Davidson: What happens in the formation of the embryo itself? Could you review the results on lineage that have been obtained with retroviral and other kinds of markers with respect to diversification of cell type in the embryo?

Gardner: As far as I am aware, none of the retroviral studies yet undertaken relate to the embryo itself. Sanes et al (1986) were dealing with extra-embryonic lineages, showing in the mesoderm component of the yolk sac that while early induced clones could contribute to the mesothelium, endothelium and connective tissue, later clones contributed to either mesothelium or these other tissues.

Davidson: I thought there was an argument about the number of progenitor cells in the germline based on clones bearing retroviral insertions.

Gardner: The retrospective analysis, where you look at patterns in adults and try to infer what has happened earlier, is a very different game. It depends on various assumptions that it would be inappropriate to discuss now. Looking at those early post-implantation stages directly poses problems: the retroviral approach has presented a number of difficulties that haven't yet been solved for the embryo itself. The only way people have been able to look at cell fate in post-implantation embryos is to explant them into culture after grafting marked pieces of tissue between them. The severe limitation of this approach is that the maximum period for which development *in vitro* rivals that *in vivo* is 36–48 hours, so you have to do multiple rather than single-cell grafts. Rosa Beddington (1981) did a series of orthotopic grafts of anterior, mid and posterior pieces of ectoderm. She thereby obtained a fate map of the embryo that, with a little distortion, can be accommodated in the basic vertebrate plan that one sees in amphibia, in the chick, and even in fish embryos (Beddington 1983). When you then do heterotopic grafting, the process of commitment of cells, at least in a general regional sense, is found to occur first in the anterior end of the embryo and then to extend posteriorly (Beddington 1982). That's really all that can be said at the moment.

Davidson: Aren't there some horseradish peroxidase labelling studies that have been done on cell sorting?

Gardner: Kirsty Lawson (Lawson & Pedersen 1987) in particular has investigated the origin of definitive endoderm in the mouse embryo, using horseradish peroxidase as an intracellular marker. One problem with this enzyme is that it remains detectable only for up to 48 hours. This is often not long enough to ascertain the fate of cells definitively, particularly since cell death is such a conspicuous feature in the ectoderm during this phase of development. In addition, there is some evidence that enzyme injection may perturb the cycle and, consequently, the fate of cells in the early mouse embryo (Dyce et al 1987).

Davidson: So you are saying there are no lineage tracer data for the embryonic ectoderm post-implantation period.

Gardner: We need to get the retroviral approach working, because that would provide an analogous system to X-ray-induced somatic recombination in *Drosophila*. There are several problems: one is demonstrating that if the retroviral gene is integrated, it is actually expressed in all the descendants of a target cell, regardless of how they differentiate. Another is that there is so much cell mixing in mammalian development that, when you see multiple patches, you have the headache of deciding whether you are looking at dispersion of a clone or at multiple independent clones.

Jessell: If you use criteria other than eventual cell allocation to ask when individual blastomeres first acquire an identity, is there any information on that? Paul Wassarman has looked at a serine protease that he has named strypsin. My impression is that strypsin is expressed very early in a subset of blastomeres.

Gardner: That's in the blastocyst and it seems to be in certain trophectoderm cells. Again, without manipulation one doesn't know whether that's a trophectoderm response to proximity to the inner cell mass or an autonomous property of these cells. There is a very real difficulty in dealing with lineage during cleavage. Both the spatial allocation of cells and the timing of division are essentially unique for each embryo, so one can only formulate certain general rules. We know from the work of Graham & Deussen (1978) that in normal cleavage from the 2-cell to the 4-cell stage, the interval between division of the two blastomeres ranges from less than one minute to three hours. The first blastomere to divide, in those embryos in which there is a reasonable interval between division of the two, produces progeny which tend both to divide ahead of those of the later dividing blastomere and to form contacts with more neighbours, thereby making a disproportionate contribution to the inner cell mass.

Weisblat: The first cell cycle takes about 24 hours, so the variation is only 10% or 25% of the total, is that correct?

Gardner: Yes, from the 2-cell stage to the 4-cell stage it is up to 25% of the total.

Weisblat: In the leech, where cell cycles can be pretty long, there is also considerable variability in timing, so I wouldn't attach too much importance to this.

Gardner: It is just that it is a surprisingly unconstrained system, but one that still seems to yield a very reproducible end-point. It may mean that there is a lot of redundancy in developmental controls in mammals. Even if you strip off one layer, the results of, say, interfering with the expression of some regulatory gene may not be revealed in the way that it might be in more economical developmental systems.

Takeichi: Do you know when and how the dorsoventral axis is determined in the mammalian embryo? How is the position of the primitive streak determined?

Gardner: We don't know how it is determined. We know that it is specified as soon as a blastocoelic cavity has formed. The blastocyst always implants with the inner cell mass directed dorsally towards the mesometrium and in subsequent development the dorsal surface of the embryo is also mesometrial. An early idea was that the blastocyst attached to the uterus at random and then the inner cell mass migrated in response to some gradient within the uterus. We now know that if you mark the inner cell mass at a very early stage, it does not move relative to the trophectoderm. But its orientation does seem to be related to the uterus. Alden (1945) did the pioneering, very difficult experiment of taking an individual segment of the rat uterus, rotating it through 180° and suturing it back. When operated rats subsequently became pregnant, all the embryos that entered the segment of the uterus with reversed polarity had their inner cell masses pointing downwards. So orientation of blastocysts is definitely related to the uterus and once the inner cell mass position is fixed in that way the dorsoventral axis of the embryo is effectively specified. There is now some indication that during early implantation anteroposterior and left and right axes may also be specified (Smith 1985).

Takeichi: What kind of signal determines that dorsoventral axis?

Gardner: There is no information on that.

Weisblat: Did you say that the trophectoderm is maintained in the proliferative state by the primitive endoderm?

Gardner: No, by a derivative of the primitive endoderm, the visceral endoderm. I must stress, however, this is based entirely on observation of a strong correlation between continued growth of trophoblast and presence of visceral endoderm in an implantation chamber produced by trophectodermal vesicles injected with inner cell mass-derived tissue.

Weisblat: The part of the trophoblast that is actually dividing isn't normally in contact with the endoderm, is it?

Gardner: Yes it is. The dividing part of the trophoblast actually moves inside the embryo where it is lined by the visceral endoderm. Somewhat later, proliferating trophoblast cells are not in direct contact with the visceral endoderm but with other inner cell mass derivatives. If you isolate trophectoderm-derived cells from the post-implantation embryo, they immediately cease dividing and begin to endoduplicate DNA. That is why I raised this issue: the visceral endoderm or some other inner cell mass derivative might produce a molecule that promotes trophoblast growth but which may nevertheless have to act over some distance to exert a localized effect. Binding of a diffusible growth factor to extracellular matrix components is one way in which this might be achieved.

Wieschaus: This mechanism that counts cells, is it known whether it counts inner cell mass cells or trophectoderm cells?

Gardner: I think the view would be that the greater the packing of cells inside at the 16-cell stage, the more the outside cells are stretched and therefore tend to divide conservatively rather than differentially.

Wieschaus: But the more the outside cells are stretched, the more likely it is that they will divide in the direction of the stretch, that is produce two daughter cells that remain on the outside. If the outer cells are not stretched because there are a lot of outer cells, then they could more easily divide in the other direction.

Wolpert: I have never fully understood from Martin Johnson's work whether polarization is simply a way of getting some cells to become inside cells, or whether the polarization has anything to do with the specification of cells as inner cell mass.

Gardner: I think that's probably no more than a matter of opinion at this stage. My view would be that it is simply a way of establishing cells in two distinct micro-environments. The other part of this story is that continued development of the embryo beyond the blastocyst stage, when you have a cell population that is removed from its surface layer, may depend critically on the development of this transporting epithelium. So this early polarization of cells could also be related to developing a physiological surface for exchange by the time the embryo cavitates. I have never regarded it as more than a way of assorting cells whose differentiation depends subsequently on other, as yet unidentified, cues. If you prevent the embryo compacting by various manipulations, you don't interfere with cell division, but you nevertheless end up with a vesicle of trophectoderm (Johnson et al 1986).

Wolpert: So polarization alone does not directly lead to a different developmental pathway.

References

Alden RA 1945 Implantation of the rat egg. I. Experimental alteration of uterine polarity. J Exp Zool 100:229–235

Beddington RSP 1981 An autoradiographic analysis of the potency of embryonic ectoderm in the 8th day postimplantation mouse embryo. J Embryol Exp Morphol 64:87–104

Beddington RSP 1983 The origin of the foetal tissues during gastrulation in the rodent. In: Johnson MH (ed) Development in mammals, Vol 5. Elsevier Science Publishers, Amsterdam, p 1–32

Beddington RSP 1982 An autoradiographic analysis of tissue potency in different regions of the embryonic ectoderm during gastrulation in the mouse. J Embryol Exp Morphol 69:265–285

Chapman VM 1986 X-chromosome regulation in oogenesis and early mammalian development. In: Rossant J, Pedersen RA (eds) Experimental approaches to mammalian embryonic development. Cambridge University Press, Cambridge, p 365–398

Dyce J, George M, Goodhall H, Fleming TP 1987 Do trophectoderm and inner cell mass cells in the mouse blastocyst maintain discrete lineages? Development 100:685–698

Gardner RL 1988 Cell fate in the developing embryo. In: Jones CT (ed) Fetal and neonatal development. Perinatology Press, Ithaca, New York p 10–23

Graham CF, Deussen ZA 1987 Features of cell lineage in preimplantation mouse development. J Embryol Exp Morphol 48:53–72

Johnson MH, Maro B, Takeichi M 1986 The role of cell adhesion in the synchronization and orientation of polarization in 8-cell mouse blastomeres. J Embryol Exp Morphol 93:239–255

Lawson KA, Pedersen RA 1987 Cell fate, morphogenetic movement, and population kinetics of embryonic endoderm at the time of germ layer formation in the mouse. Development 101:627–652

Sanes JR, Rubenstein JLR, Nicholas JF 1986 Use of recombinant retrovirus to study post-implantation cell lineage in mouse embryos. EMBO (Eur Mol Biol Organ) J 5:3133–3142

Smith LJ 1985 Embryonic axis orientation in the mouse and its correlation with blastocyst relationships to the uterus II. Relationship from 4¼ to 9½ days. J Embryol Exp Morphol 89:15–35

Induction and the organization of the body plan in *Xenopus* development

Jonathan Cooke

Laboratory of Embryogenesis, National Institute for Medical Research, The Ridgeway, Mill Hill, London NW7 1AA, UK

Abstract. Various experiments are surveyed in this paper that may throw light on how the degree of spatial organization of the *Xenopus* embryo increases during development. The events of the 100 minutes or so that follow fertilization may do little more than orientate and give proportions, within the egg's yolky vegetal region, to the system that originates patterned inductive signals for organization of the mesodermal and ectodermal regions during blastula stages (10^2–10^4 cells). By onset of gastrulation, an outline plan for mediolateral and anteroposterior body organization has developed within induced tissue around the equator of the embryo, which seems to control subsequent development in two ways. It sets the spatial and temporal pattern of mechanical activities whereby the mesoderm rudiment drives the crucial shape changes that lay it and the neural rudiment out correctly. It is also the probable starting point for positionally specific gene transcription that begins immediately after gastrulation. Experiments with known inducing factors that explore the possible bases for the early 'pre-organization' in mesoderm are informative, but leave us far from a complete understanding. Evidence that inhibitory or modulating, as well as activating, signals are involved is surveyed.

1989 Cellular basis of morphogenesis. Wiley, Chichester (Ciba Foundation Symposium 144) p 187–207

In this paper I discuss results from two different kinds of experiments, which each bear on processes that control the orientation and proportions of the 'whole body' structural plan into which tissue is allocated during early development of the amphibian, *Xenopus*. This development differs from that of amniote vertebrates—those that form the embryo within a blastodisc or similar structure—in two ways. It employs more determinate processes, in the single giant egg cell, in the initiation and polarization of pattern (see also Melton, this volume), and it accomplishes essentially all the pattern formation for the primary body plan before growth begins. But the body plan arrived at is highly conserved across all the versions of vertebrate early development, and the features described above do facilitate a quantitative analysis that helps in assessing theories about fundamental mechanisms.

One type of experiment consists of perturbing, in a variety of surgical and other ways, the development of whole embryos up to around the 10^4 cell stage, i.e. during the first 8–10 hours and 13 cell cycles after fertilization. Up to this stage, no stable differences of cell phenotype have emerged beyond those involved in normal assignment to the three 'germ layers' (though see Heasman, this volume). The final results are analysed at a fixed, early larval stage, whose quantitative anatomy is well recorded (e.g. Cooke & Webber 1985, Cooke 1987). By this stage, 20 hours after any of the experimental perturbations, ability further to regulate or re-adjust the layout and proportions of body structures is lost, unless it occurs by a long process of adjusting relative growth which has not so far been assayed for. Such early larvae thus show either a normally proportioned, single integrated body plan (regulation, or systematically malproportioned and/or twinned plans (failure of regulation). Mesodermal and ectodermal (epidermal and central nervous system) patterns are closely and constantly co-ordinated: mesoderm, believed to be the leading layer in this pattern expression, is the one assayed. Pattern is assessed both in terms of proportions between the major tissue types or organ systems, and in terms of the normally equally well controlled distribution of tissue into segmental structures at different positions along the axis.

The other results discussed come from exposing competent tissue, in explants and in whole blastulae after microinjection, to a small, soluble protein that induces axial mesodermal tissue (Smith et al 1988). We believe that this molecule probably represents one 'activating' component of the regulated intercellular signalling system of mesoderm induction in this embryo, that operates *in vivo* between about the 10^2 and 10^4 cell stages. Mesoderm arises largely within tissue that would otherwise have formed only epidermis, in response to signals that are first produced in the part of the egg specified to form endoderm (though they may then propagate actively in responding tissue). Induction, and thus the original mesoderm territory that arises due to it, is significantly and crucially patterned from the outset. The experiments that I outline below have enabled us to begin to see how the principles of pattern formation might be embodied in the dynamics of cellular responses to inducer molecules.

Separable phases of development

In the larval structure achieved after the regulative phases for whole body pattern, we are seeing the combined effects of two successive and, we believe, functionally independent sets of processes. It is the first of these, the activation and polarization of the field for pattern within the egg, that is atypical in *Xenopus*, being completed in little more than an hour within the first one or two cell cycles. Instead of a process of progressive intercellular signalling, there is systematic reorganization between components of egg structure, which normally ensures formation of a single pattern with the proper terminal structures and proportions for the intervening regions. Certain structural correlates of the intracellular movements have been well studied, but the molecular physiology whereby these are converted into 'pattern values' around the material of the vegetal zone (future endoderm) is unknown (Vincent et al 1986). It is almost certainly *not* the gathering into particular locations of unique 'plasms' that contain the molecular precursors of tissue-specific inducer substances or determinants (Black & Gerhart 1986, Cooke 1987). Understanding of the fate map that describes the normal apportionment of egg material to body structure enables us to say that deviant forms, caused by perturbations of the egg, are such as we would expect from gradient mechanisms giving large scale positional information. That is, distortions of proportion, changes or doublings of polarity, and omissions of terminal sectors of pattern are seen, but never gaps or jumblings of structure. Certain dynamic features of the response of pattern to early perturbation, however, are not in accord with expectations from a gradient controlled by the generally modelled system of interacting, diffusing morphogen molecules (Meinhardt 1982, see later section).

This early phase of pattern establishment sets up the (normal or abnormal) initiating conditions for a second one, whose own regulative responses to disturbance seem far more likely to share biochemical mechanisms with a 'homologous' developmental phase in other types of vertebrates. This is the inductive system for preliminary mesoderm/ectoderm patterning. Over several hours, extending in *Xenopus* from around the 10^2 to the 10^4 cell stage, a system of intercellular inductive signals causes the mesoderm territory and perhaps even the origins of the central nervous system pattern to become organized in the more equatorial parts of the 'animal cap', a blastular cell population that would otherwise form only epidermal tissue (Gimlich & Gerhart 1984, Keller et al 1985). It now seems that the active molecules are extracellular, soluble factors which must function *via* membrane receptor/second messenger systems, and may be members of those protein families that include known 'growth factors', even though their function discussed here is distinct from this (Slack et al 1987, Kimelman & Kirschner 1987, Cooke et al 1987, Rosa et al 1988, Smith et al 1988).

The final result of this process, however many signal species it involves, is positively regulated in two respects. First, the partitioning of the animal cap into a mesodermal zone nearer the initiating source of induction and a remaining ectodermal and neural zone is subject to *active* proportioning. This normalizes the *relative* extents of the two zones, even when competent tissue is experimentally added or removed at the beginning (Cooke, work in progress). Secondly, by onset of gastrulation, the newly specified mesodermal territory is sufficiently firmly patterned to play an active autonomous guiding role in the complex mechanics of gastrulation. A more detailed discussion of this role of the gastrulating mesoderm rudiment, and of what we can mean by a specified and patterned mesoderm in relation to, for example, the results of single cell experiments at similar stages, occurs in a later section.

Before describing properties we have been able to attribute to either of these two sequential phases of pattern formation, a limitation on the analysis of whole body patterns at 5×10^4 cell stages must be pointed out. Normality or otherwise of pattern cannot be quantified before this differentiating stage. Even those fairly reliable clues that pattern *will* be abnormal, which occur in the timing and pattern of gastrulation movements (Cooke 1985, 1987), post-date both the crucial phases of spatial interaction we are describing. Thus, if experimental disturbances that we have good reason to believe are over by the 4-cell stage nevertheless permanently impair the development of pattern, we can diagnose a failure of the 'early' events to normalize themselves, plus a failure of the inductive system to compensate for this. On the other hand, when adjustments that correct pattern proportions occur only partially after early manipulations, we cannot be sure whether they were accomplished close in time to the one-cell stage or much later during the intercellular communications of mesoderm induction and patterning. The various intrusions that produce twinned body patterns provide a good example (see Cooke 1987).

When two dorsal/anterior, part-body patterns share development in one normal sized egg, they frequently exert a significant mutual effect upon the spatial scale with which structures are formed in each of them. This is not a necessary, mechanical consequence of twinned development; under some circumstances it does not occur. When it does operate, however, successive territories (zones of differentiation or segmental structure) within each of the axial patterns are allocated fewer cells than they would have been in a single pattern. Each set of body parts is thus more nearly complete than would otherwise be the case, given the limitation on total tissue. We suspect that contributions to such 'regulation' come from both the developmental phases. This is because perturbations to either the single-cell phase (tilting in gravity to produce supernumerary sets of egg re-organization movements) or the 10^3-cell blastula (surgical implantation of an extra dorsal blastoporal lip or 'mesodermal organizer' region) can produce such scaled-down twin patterns, but the phenomenon is most complete after the earlier type of interference. Only in such cases would the dynamic properties of the egg activation system, and then the entire time of the inductive mechanism, be available for interactions.

The early events and their relation to mesoderm induction

The re-organizations of the normal first cell cycle are the only early pattern-forming interactions that have full homeostatic, regulative ability in the *Xenopus* version of vertebrate development, albeit an ability that can quite readily be overridden by experimental manipulation. This system's main features are represented in Fig. 1. Failure to arrive at a normally distributed set of the unknown changes that are instituted around the egg by, say, two hours after fertilization is permanently recorded as incomplete or otherwise deviant body pattern (Cooke 1985). Undisturbed versions of such events lead to a remarkably constant quantitative balance of tissue type and structure *that is independent of natural variation in egg size*. Pattern-forming capacity in this early system could be registered as a simple profile of initial states for the inductive system whose signals will invade the marginal zone, such that pattern of particular character is initiated at various meridional positions around the blastula. Both whole eggs in which the early events have been disturbed or attenuated (Fig. 1C), and part-embryos from egg halves containing an imbalanced share of these initial states (Fig. 1D), reveal this imbalance in their subsequent body forms.

It seems that a fully 'activated' setting for the inductive system, normally restricted to an egg sector of particular, narrow angular extent, leads to development of head and dorsal midline structure, whereas a 'baseline' state, on meridians far from this position, gives rise by itself to mesoderm of a non-axial, ventrolateral character. Certainly, after pre-cleavage UV-irradiation has selectively prevented the egg movements that are normally associated with activation, a mesoderm of normal size but radially symmetrical, entirely ventrolateral, character is produced (Scharf & Gerhart 1983). Such 'UV' embryos show that a principle of spatial organisation in induction related to normal 'head-to-tail' patterning of the body is partially independent of the dimension of patterning set up by the early events, because it still operates in the radially symmetrical, non-axial version of development. Only the first-involuting mesoderm in gastrulation, due to occupy the anterior position in the body plan, includes haemoglobin-producing tissue in its specifications. Later-involuting mesoderm, originating from higher in the blastular marginal zone and thus further from the initiating source of inductive signals, has other specifications (Cooke & Smith 1987).

A simple working hypothesis for the organization that precedes mesoderm induction is that two species of initiating inducer each specify homogeneous mesoderms that would by themselves develop just two of the tissue types finally seen, and that in normal development these are produced, respectively, in larger and smaller blastular sectors of particular relative extents. Interactions interpolating the remainder of the pattern subsequently occur *between* these two territories *within* mesoderm, in a way that can be understood as a sequence of inductions or perhaps as the effects of a morphogen gradient. Dale & Slack (1987) formulate the problem in these terms. Such an interpretation encounters immense difficulties in matching the observed behaviour of the system, however, which must now briefly be described.

Polarized body patterns develop from whole eggs in which the structural rotation movements that activate development have been attenuated by low-dose or late UV irradiation. They also develop from isolated egg fragments that included less than a balanced quota from the minor, highly activated sector of the egg, and that omitted its centre. There is no return to complete radial symmetry of development, even though whole sectors of dorsal and anterior pattern of varying extent, such as head, part or all of the notochord and subsets of the normal complement of somite segments, have not been specified (Fig. 1A). The extent of structure missing varies widely whether defined in terms of absent states of mesoderm and ectoderm differentiation or in terms of 'levels' within a body plan of non-equivalent segments (Lewis & Wolpert 1971). One cannot think of one structure or tissue as being specified uniquely and directly by one of two initial states of the system for mesoderm induction, both of which must be present to give polarity. We therefore propose that the events ending after the first cell cycle set up a proportioned partition into two zones of mesoderm

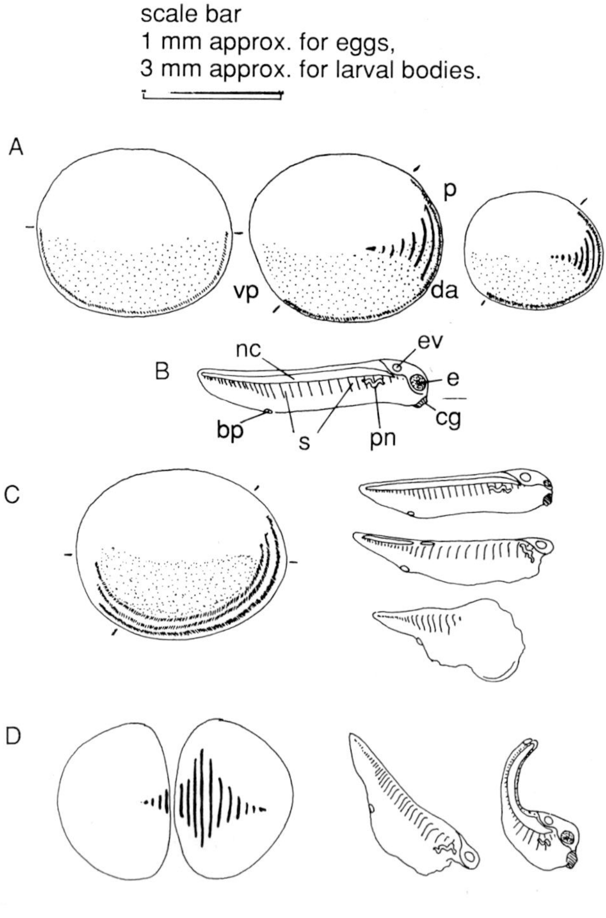

induction which specify a pair of states. These then interact to produce the final pattern of specifications but are not, themselves, any of those final specifications. Such a partitioning could be expressed via the relative concentrations or rates of release of a small number of initial inducers, rather than their presence or absence. The results of this kind of experiment certainly tell us that the final sequence of tissue types is not caused by the direct expression of the presence or absence of particular initial inducers, but rather by the extent and values of a gradient system in mesoderm, probably established by a cascade-like signalling system. This graded system, furthermore, is primarily interpreted by cells as what we will call *levels of body structure*, with several types of cell differentiation represented repeatedly at many of the levels, rather than as a geographical pattern of specific cell types (see Lewis & Wolpert 1971, also Driever & Nüsslein-Volhard 1988 for direct molecular evidence for this principle in *Drosophila*).

The above kind of evidence, from the behaviour of pattern in disturbed whole embryos and large fragments, is in accord with the responses of small cultured explants of competent blastular tissue to treatment with molecules that are likely to be the natural initiating inducers (see references in later section). Response, in terms of mesodermal tissue type and 'level' of structure produced, is progressive across a remarkably wide dose range above the lower threshold

FIG. 1. Dynamics of the pattern-activating system of the first cell cycle. A) Left and middle diagrams show fertilized egg with future midline of body in plane of page, before and after the reorganization (rotation of a thin outer component of structure around the cytoplasmic core) of the second half of the first cell cycle. This establishes a steep gradation with respect to some state of activation (vertical shading) that will control the character of future inducing signals, within the yolky endodermal region. Dorsoanterior (da), posterior (p) and ventroposterior (vp) extremes of pattern *as they will be specified in future mesoderm* are marked. Right diagram shows how this system is regulative in that a balanced extent and profile of activation is achieved regardless of individual egg size. B) The layout of principle structures and proportions in the normal body plan. nc, notochord; s, somites (relative but not real numbers of segments represented in this and following sketches); e, eyecup; ev, ear vesicle; cg, cement gland; pn, pronephros; bp, blastopore. C) Left diagram shows how a range of incomplete degrees of movement can be achieved, as after late or low-dose UV irradiation in the first cell cycle. These correlate in a general way with truncated profiles for the state of activation, which are duly reflected in various axial, polarized but dorsoanteriorly truncated body plans (right). D) Similar non-regulated but stable patterns for development are seen in separated half egg fragments where a cleavage that is oblique to the activation centre (head-on view in left diagram) gives reciprocal fragments that are imbalanced with respect to levels of activation. The body at extreme right has a complete segmental complement but is grossly biased towards head and dorsal structure, while that at middle is like a small version of the 'low-dose UV' body (C, right).

concentration, and the concentration thresholds which give rise to particular tissue types cannot easily be defined (Smith et al 1988, Cooke et al 1989).

The state of organization of mesoderm, and its role, at the onset of gastrulation

By the onset of gastrulation an outline plan for both dorsoventral and head-to-tail sequences of the body has developed within the recently induced mesodermal territory that is the marginal zone. This is a contentious statement in at least two ways. It might appear to conflict with results of single-cell implantation experiments (Heasman et al, this volume). There is also a widely held belief that the mesoderm only becomes further organized as a *result* of events that depend upon the movements of gastrulation, whereas we propose a reverse causal sequence, namely that mesodermal pre-organization itself organizes gastrulation. In reality, there are doubtless reciprocal interactions between the morphogenetic movements and increasing states of organization, but we feel that the degree of pattern existing in the mesodermal rudiment at the outset of gastrulation has been insufficiently recognized in recent years.

Our view of mesodermal pre-organization does not entail that individual cells within the marginal zone are committed to gene transcriptions proper to certain body positions or tissue types: single cells are known to be remarkably uncommitted at gastrula stages (Heasman et al, this volume). Their membership of future structures would be expected to be widely manipulable by dissociating them and allowing them to come to rest as members of different cell layers or presumptive body positions. Of late, research in development has tended to be directed at the mechanism of sequential restrictions of the potential of individual cells. This is understandable in view of the ascendency of technology for detecting synthesis of rare and transient gene products, and the striking success in portraying in molecular terms the geographical partitioning that precedes and accompanies gastrulation in the *Drosophila* embryo. In this search for the nature of determination seen as the transcription of specific gene combinations, the prior spatial diversifications that initially control the development of pattern have somewhat been forgotten. These are dynamically maintained gradations of physiological state across tissue, that serve as maps of what Wolpert has called 'positional value' (see also Child 1941), and have so far been harder to study in molecular/biochemical terms. Such variables in a morphogenetic field are embodied in states of the cells' metabolic/regulatory apparatus. A hypothetical example might be the degrees of phosphorylation, in particular systems of proteins, arranged by a small set of intercellular inductive signals whose varying concentrations are the positional information. We should not expect 'memories' of such information to survive cell dissociation, or even the explantation of small pieces of tissue into culture. The positional organization is a property of the

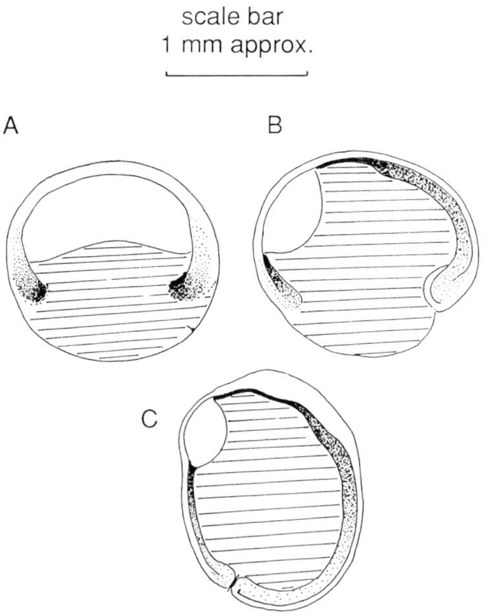

FIG. 2. The pre-organization of mesoderm related to gastrulation and to the body plan. A, B and C show longitudinal sectional views of initial, mid-gastrula and closing stages, respectively, in the process whereby mesoderm that was specified before gastrulation changes cell behaviour and actively involutes to form a new, middle cell layer. Graded density of shading for the mesoderm in each diagram represents the pre-organization, in relation to distance from initial sources of induction in the vegetal region, whereby mesoderm cells from successively higher in the marginal zone commence the change in their behaviour that causes involution at successively later times. Positions of cells within the pre-involution mesoderm are thereby significantly related to their expected final position in the body plan. In *Xenopus*, where involution spans 3–4 hours only, pre-organization of these properties has largely occurred before its onset. In slower-developing forms, the progressive organization of pre-involuted material may continue after gastrulation is underway.

entire tissue. Despite the pluripotent nature of individual cells within it, the mesodermal territory of an amphibian embryo that is about to begin gastrulation, observed whole or manipulated as large, gently handled parts, has long been known to be a spatially differentiated 'morphogenetic organ' (e.g. Cooke 1972a, Keller et al 1985, Keller & Danilchik 1988).

The pre-organization of mesoderm, which is depicted for *Xenopus* in Fig. 2, is necessary for subsequent normal development in two ways. It controls the mechanics of shape change by ordering the timing of involution (transition into the definitive mesodermal layer) by individual cells in the mesoderm of the gastrulating embryo, and it allows the first deployment of body position-specific gene activities in mesoderm very soon after its involution. The population of mesoderm cells provides, as Ray Keller has said, the motive engine of gastrulation. A variety of observations in the Laboratory of Embryogenesis at Mill Hill and elsewhere, and those from earlier decades, show that positional value at each location in pre-involuted mesoderm is first expressed during gastrulation. Each group of cells autonomously contributes local mechanical activity to the gastrulation process, by changing cell layer affinities (involuting) and commencing other schedules of mechanical change at a particular time within the time course of gastrulation as a whole. The nature of the geometric re-organization that is accomplished in involution, spreading and dorsal convergence of the mesodermal mantle requires that the participating cells change neighbours extensively. Positions of mesoderm as specified in the pre-involuted array nevertheless correlate significantly with positions in the final body plan, especially within precursor regions for the anterior extreme and the dorsal axis. It is clear that the pre-organization directs gastrulation, and it probably imparts to the forming mesodermal mantle a physiological positional 'map' that is sufficient for the activation of genes that may stabilize the regionalization of mesoderm very soon after its involution (e.g. Condie & Harland 1987).

The above contrasts with the widespread view that an ordering process, derived from global mechanical properties of gastrulation itself, organizes a mesoderm which originated with little intrinsic spatial gradation. A thumbnail sketch of such a view would be that not all mesoderm cells can make the layer transition of involution and migrate at once, because of general rules of cell behaviour (e.g. population pressure or contact inhibition controlling locomotion within a layer). The earliest-involuting tissue is somehow given 'head' specification by having been allowed to involute first; successively later involuting tissue derives more posterior specifications. A more extreme view would see the process of specifying tissue as mesoderm *per se* occurring only during gastrulation. Because of the extended nature of the gastrulation sequence, which brings successively more tissue into a special situation at the blastopore lip, a series of head-to-tail coded structures is specified.

All the above ways of conceiving how a body plan might be instituted at gastrulation involve the perception by cells of the biological equivalent of time, as an organizing principle. In the last described proposals, the 'ages' at which cells experience a new imposed situation cause particular positional specifications, whereas we propose that a pre-organized positional gradation is used by mesoderm cells to vary the 'age' at which they begin the autonomous new behaviours which actively orchestrate this phase of development. What is meant by biological 'age' in these ways of thinking about developmental organization? In amphibian embryogenesis the temporal sequence or, for want of a better word, 'clock', by which cells time the observed autonomous activities is determined from near fertilization. That is, sibling embryos fertilized within a few minutes of each other, if they are to produce normal body patterns, will commence and continue their schedules of gastrulation, 10–20 hours later, with less than 4% variability in timing at a given temperature. Individuals whose schedule of mesoderm involution lies outside these narrow time limits invariably exhibit corresponding distortions of, or omissions from, the normal body pattern.

A *prima facie* case exists for pre-gastrulation organization of specified mesoderm in amniote vertebrate embryos, but two uncertainties cloud the issue of whether there is an equally precise and necessary temporal organization. Synchronous fertilization followed by normal development of large samples is not readily observable. In addition, the organization of the embryo itself within the blastodisc or similar structure may be initiated later in the cell lineage history of development in such forms, at a time imperfectly controlled between individual zygotes. It is from this currently unobservable and possibly variable timepoint that the real 'countdown' of the clock of developmental age may begin.

We propose that of primary importance for body pattern is cells' ability to measure elapsed biological time in the early course of development, and to order their activities within the schedule of such elapsed time on the basis of locally perceived 'positional physiological' variables. The nature of the biochemical mechanisms for such temporal organization, currently quite obscure, represent a challenging and captivating problem for basic developmental biologists.

Experiments with known mesoderm-inducing signals: the principles of pre-organization in the marginal zone

At least three types of mesoderm participate in amphibian gastrulation as major elements of the spatial pattern, quite apart from such pre-organization as exists within each type. These are the prechordal mesoderm that organizes anterior head structure and involutes first, the dorsally converging axial-type mesoderm and the mesoderm of lateroventral type that does not spontaneously converge, elongate or differentiate as axial structures. Mesoderms of the latter two sorts involute over extended periods in a way that appears to reflect positive early anteroposterior organization in relation to those schedules (Cooke 1985, Cooke & Smith 1987). I now consider some experimental results obtained in our laboratory, mainly with a mesoderm-inducing protein (MIF) from *Xenopus*. This protein, XTC cell-derived MIF (Smith et al 1988) is a strong candidate to be a natural inducer of mesoderms of the first two above types that originate near the 'dorsal' side of the blastula. We have also used basic fibroblast growth factor, bFGF (see Slack et al 1987), a similar candidate inducer, but of lateroventral-type mesoderm. The experiments have been reported in detail (Cooke et al 1987, Cooke & Smith 1989) and only their implications are given here.

These mesoderm inducers belong to families of proteins that presumably act via appropriate membrane receptors and second messenger systems. When blastula animal cap tissue is experimentally diverted from ectodermal to mesodermal specification by encounter with the soluble factors, it shows the same autonomous timing of its first visible, cell behavioural response as does naturally induced mesoderm. That is, the 'age' (since start of development) when movement first occurs is invariant, characteristic of the part of the normal body plan for which the particular mesoderm would be appropriate and independent of the time at which the inducing signal was encountered. Where mesoderm has been ectopically induced across the entire blastocoel roof following injection of the factors into that cavity, these times of the 'involution' response, to XTC-MIF and mammalian bFGF, respectively, correspond to the beginning of the normal head-to-tail sequence of dorsal axial mesoderm involution and to a period when ventral, non-axial mesoderm is involuting. In normal gastrulation, involution actively continues for several hours within mesoderm of both these general types, and this is related to what we have called the pre-organization for head-to-tail regionality in mesoderm. By injecting XTC-MIF into the blastocoel, we have tried to gather evidence as to what principles, beyond the initial secretion of such an inducer at the (vegetal) edge of the field of tissue that is competent to respond, might set up and regulate such a pattern of pre-organization. The possibilities are many, since we do not know the precise role of this molecule or its *in situ* equivalent in the machinery of induction. We think it unlikely, however, that more than a few initiating signal molecules are involved.

Two simple hypotheses for the establishment of positional value are as follows. Passive diffusion and progressive sequestration of the molecule, within the restricted extracellular space in the marginal zone tissue, might set up a concentration gradient in relation to the vegetal source. This could be sufficiently steep for signal intensity, differentially responded to by the cells, to be the crucial positional variable correlated with original height of prospective mesoderm cells in the marginal zone. Alternatively, the signal from the initiating source could spread by active propagation, in that a very early part of cellular response might be the release of an identical or very similar signal. Since such signal propagation through tissue would have a particular rate per cell, intrinsic 'age' of cells at first encounter with (and response to) the signal could also provide information about their position in relation to the initiating source. We have studied in detail the precise time of the behavioural change corresponding to involution, but in the homogeneous field of ectopic mesoderm, after varying the blastular stage of injection and the concentration of the injected signal in samples from populations of synchronously fertilized embryos. The preliminary positional value acquired by the specified mesoderm does not vary significantly or consistently in relation to either of these variables, so that the normal organizing principles cannot be modelled adequately by either of the above simple principles. A concentration gradient of XTC-MIF and/or a systematic advance of this signal through tissue with 'age' may occur in normal development, but they do not by themselves give positional information.

A variety of evidence suggests that the natural patterning mechanism, imbuing mesodermal tissue with a variety of differentiation tendencies and other positional properties, involves both activating signals (for which XTC-MIF and FGF-like molecules are candidates) and further modulating or inhibitory signals that perhaps do more to create the spatial physiological gradations. This evidence is currently indirect, but worth recording in view of the long period likely to elapse before any direct identification of signals with such roles. It is surveyed in detail in Cooke (1989). The behaviour to be expected from such systems of pattern formation, i.e. of short range spreading activation followed by a dependent system of longer range inhibitory signals giving regulated spatial gradients, has been extensively modelled and described by Meinhardt (1982). That work should be consulted in order fully to grasp the implications of the dynamics that are seen in MIF-activated mesoderm formation (Cooke 1989, Cooke & Smith 1989).

Firstly, the sheet of mesoderm ectopically induced by MIF injection never shows its involution behaviour as early as does the embryo's 'own' prechordal mesoderm, and never goes on to develop as prechordal structure or even as notochord if left *in situ*. Even the highest concentrations of injected inducer, that saturate the response, give rise in this situation only to mechanically 'converging' but somite muscle-producing tissue. Small grafts of that blastocoel roof tissue, however, transferred to ventral sites in normal host blastulae less than an hour after initial exposure to the signal, organize whole supernumerary axial mesodermal patterns. These patterns do include notochord and prechordal, forebrain-inducing territories. By analogy with the normal dorsal blastoporal lip site, the mesoderm that lies closest to the natural initiating source of induction and the only other mesoderm that has these complete 'organizer' properties (Spemann & Mangold 1924, Cooke 1972b), tissue that causes formation of dorsoanterior structure should be found itself to participate in that structure. It is as if, after the abnormal, spatially homogeneous exposure of a large field to a high level of activating signal such as follows blastocoelic injection, a secondary downgrading or inhibiting influence follows. This can be circumvented if a smaller piece of recently activated tissue is removed to much less activated surroundings. 'Activation–inhibition' models for formation of spatial pattern in development often predict that maximal activation, or 'boundary' levels of positional value in the normal gradient, cannot be achieved when large fields are given a homogeneous, synchronous activating signal. Small, relatively activated pieces of tissue, however, removed to surroundings that act as a sink for inhibitors because they as yet are making none of their own, *can* break through to become fully activated and thus achieve boundary positional character without further exogenous activating signals.

Acknowledgements

It is a pleasure to acknowledge the many stimulating discussions of the various matters touched on in this paper with other members of the Laboratory of Embryogenesis at the National Institute for Medical Research, Mill Hill.

References

Black SD, Gerhart JC 1986 High frequency twinning of *Xenopus laevis* embryos from eggs centrifuged before first cleavage. Dev Biol 116:228–240

Child CM 1941 Patterns and problems of development. University of Chicago Press, Chicago

Condie BG, Harland RM 1987 Posterior expression of a homeobox gene in early *Xenopus* embryos. Development 101:93–105

Cooke J 1972a Properties of the primary organisation field in the embryo of *Xenopus laevis*. III. Retention of polarity in cell groups excised from the organiser region. J Embryol Exp Morphol 28:47–56

Cooke J 1972b Properties of the primary organisation field in the embryo of *Xenopus laevis*. II. Positional information for axial organisation in embryos with two head organisers. J Embryol Exp Morphol 28:27–46

Cooke J 1985 The system specifying body position in the early development of *Xenopus*, and its response to early perturbation. J Embryol Exp Morphol 89 (suppl):69–87

Cooke J 1987 Dynamics of the control of body pattern in the development of *Xenopus laevis*. IV. Timing and pattern in the development of twinned bodies after reorientation of eggs in gravity. Development 99:417–427

Cooke J 1989 The early amphibian embryo: evidence for activating and for modulating or self-limiting components in a signalling system that underlies pattern formation. In: Proc. NATO advanced research workshop; theoretical models for cell-to-cell signalling. Academic Press, in press

Cooke J, Smith JC 1987 The mid-blastula cell cycle transition and the character of mesoderm in UV-irradiated non-axial *Xenopus* development. Development 99:417–428

Cooke J, Smith JC 1989 Gastrulation and larval pattern in *Xenopus* after blastocoelic injection of a *Xenopus*-derived inducing factor; experiments testing models for the normal organisation of mesoderm. Dev Biol 131:383–400

Cooke J, Smith JC, Smith EJ, Yaqoob M 1987 The organisation of mesodermal pattern in *Xenopus laevis*: experiments using a Xenopus mesoderm-inducing factor. Development 101:893–908

Cooke J, Webber JA 1985 Dynamics of the control of body pattern in the development of *Xenopus laevis*. II. Timing and pattern in the development of single blastomeres (presumptive lateral halves) isolated at the 2-cell stage. J Embryol Exp Morphol 88:113–133

Cooke J, Symes K, Smith EJ 1989 Potentiation by the lithium ion of morphogenetic responses to a *Xenopus* inducing factor. Development 105:549–559

Dale L, Slack JMW 1987 Regional specification within the mesoderm of early embryos of *Xenopus laevis*. Development 100:279–295

Driever W, Nüsslein-Volhard C 1988 The bicoid protein determines position in the *Drosophila* embryo in a concentration-dependent manner Cell 54: 95–104

Gimlich RL, Gerhart JC 1984 Early cellular interactions promote embryonic axis formation in *Xenopus laevis*. Dev Biol 104:117–130
Heasman J, Snape A, Turner A, Wylie CC 1989 The establishment of regional identity in the *Xenopus* blastula. In: Cellular basis of morphogenesis. Wiley, Chichester (Ciba Found Symp 144) p 99–112
Keller R, Danilchik M 1988 Regional expression, pattern and timing of convergence and extension during gastrulation of *Xenopus laevis*. Development 103:193–210
Keller R, Danilchik M, Gimlich RL, Shih J 1985 The function and mechanism of convergent extension during gastrulation of *Xenopus laevis*. J Embryol Exp Morphol 89 (suppl):185–209
Kimelman D, Kirschner M 1987 Synergistic induction of mesoderm by FGF and TGF_b and the identification of an mRNA coding for FGF in the early *Xenopus* embryo. Cell 51:869–877
Lewis J, Wolpert L 1971 The principle of non-equivalence in development. J Theor Biol 62:478–490
Meinhardt H 1982 Models of Biological Pattern Formation. Academic Press, London
Melton DA, Ruiz i Altaba A, Yisraeli J, Sokol S 1989 Localization of mRNA and axis formation during *Xenopus* embryogenesis. In: Cellular basis of morphogenesis. Wiley, Chichester (Ciba Found Symp 144) p 16–36
Rosa F, Roberts AB, Danielpour D, Dart LL, Sporn MB, Dawid IB 1988 Mesoderm induction in amphibia: the role of $TGF\beta_2$-like factors. Science (Wash DC) 239:783–785
Scharf SR, Gerhart JC 1983 Axis determination in eggs of *Xenopus laevis*; a critical period before first cleavage, identified by the common effects of cold, pressure and UV irradiation. Dev Biol 99:75–87
Slack JMW, Darlington BG, Heath JK, Godsave SF 1987 Mesoderm induction in early *Xenopus* embryos by heparin-binding growth factors. Nature (Lond) 326:197–200
Smith JC, Yaqoob M, Symes K 1988 Purification, partial characterisation and biological effects of the XTC mesoderm-inducing factor. Development 103:591–600
Spemann H, Mangold H 1924 Uber induktion von embryonenanlagen durch implantation artfremder organisatoren. Roux's Arch Dev Biol 100:599–638
Vincent J-P, Oster, G, Gerhart JC 1986 Kinematics of grey crescent formation in *Xenopus* eggs: the displacement of subcortical cytoplasm relative to the egg surface. Dev Biol 113:484–500

DISCUSSION

Weisblat: If the degree of induction of head structures is dose dependent, why can't you just say that the height of the prospective mesoderm cells above the marginal zone determines how much of the growth factor they will 'see', and therefore gives them their appropriate fates?

Cooke: Because in the natural gastrulation there is this active temporal sequence whereby mesoderm cells involute early or late in some absolute time schedule. We know that is a necessary correlation, that it is part of the expression of belonging to a particular level of pattern within the normal body. However high you raise the concentration of injected inducing factor in the embryo, by injecting into the blastocoel and thus exposing cells to a homogeneous concentration, you never produce a mesoderm with the earliest-involuting, i.e. head, specification. If you allow such embryos to develop, that ectopic mesoderm never makes head differentiations. It will make them, however, if you then take an isolated piece of tissue in culture or graft one into very different surroundings in another embryo. It is that discrepancy, that when a piece of tissue is exposed to the factor in a way that is not spatially homogeneous it can achieve levels of body pattern which it cannot achieve if you leave it as a big sheet, which says that the concentration of this signal alone does not constitute positional information.

Weisblat: Is it not just that the material in the normal embryo goes by a different path—diffusing through the epithelium rather than coming from the blastocoel as in your experiments?

Cooke: We are saying that the sudden change of cell behaviour in mesoderm, whether induced either by these artificial means or in the natural sequence of events, is very sharp in time and very characteristic. Although it is not producing normal geometry because the things are in quite a different place, it is being presumptive dorsal structure. It is never being presumptive head, whether you give one unit per ml or 3000 units per ml.

Jessell: I am a bit confused by the literature on the relationship between XTC-inducing factor, FGF and TGF-β. Particularly in the relationship between commercially available TGF-β and whatever is equivalent to the XTC-inducing factor.

Cooke: bFGF and the TGF-βs belong to totally different families and they have different characteristics. An homologue of bFGF derived from the frog and present in the early embryo is known to be a powerful inducer, but of a different kind of mesoderm. TGF-β2, as obtained from mammals, is able by itself to induce dorsal axial, muscle-producing mesoderm in *Xenopus*. We don't know that it can induce notochord and head-producing mesoderm. Furthermore, the amount of this heterologous factor that is required is 30–40 times the concentration of this XTC-MIF that will have the same effect. The links are tenuous. The only reasons for assuming that XTC-MIF is a member of the TGF-β family are its biochemical purification characteristics and its characteristics when analysed on reducing and non-reducing gels. Then there are these various vegetal RNAs with very homologous sequences; I don't know what Doug Melton thinks the relationships are.

At the present state of our knowledge, it would be jumping the gun to say that a molecule which is the frog genome homologue of any of the currently named members of the TGF family is the natural inducer.

Melton: I agree with that. I would answer Tom's question in a slightly different way. The *Vg1* gene is unlikely to encode the XTC factor, from the results of nucleic acid hybridization tests. There is no evidence that the XTC factor exists in the embryo, although it does have an effect in the embryo. There is evidence that FGF is in the embryo.

It is too early, in my view, to say whether or not more than one growth factor is required to do any of this. Jonathan may have a different view, but one could propose that varying concentrations of one factor are sufficient to explain the different types of mesoderm induction.

Cooke: It is difficult to suppose that varying concentrations of the FGF class of factors would be sufficient to account for all the inductions, because of results from experiments using them *in vitro*. Varying concentrations of a member of the TGF-β/MIF family might be able to do it, in terms of tissue types. But there remains the other dimension of the head–tail (early to late involuting) organization, which is distinct.

Prima facie evidence that there are two different initiating inducers in natural embryogenesis comes from the 'UV embryo'. If this spoiling of the egg movements of the first cell cycle by UV irradiation is complete, the egg makes none of those movements that polarize the system of induction within it. The egg then develops in a radially symmetrical form and in those embryos there is almost as much mesoderm as usual. The mesoderm induction system is as extensive in terms of the number of cells that get recruited, but they are all of the lateroventral type, the type of mesoderm that is, on the whole, found *in vitro* after induction by FGF.

Gardner: Talking about the different types of mesoderm cell, I have always been confused about haemopoietic tissue in this context. Sometimes people observe what could be haemopoietic tissue and dismiss it as not necessarily being so. Haemopoietic tissue is presumably of mesodermal origin, but where does it lie in the scale of induction?

Cooke: It is at the bottom of the scale. The complication is that we think the proximity of endoderm is required in a permissive, rather than an instructive, way for tissue that's been specified to be blood-forming tissue to mature to the point where it makes haemoglobin. So when you do an animal cap assay, for example take isolated ectodermally specified tissue and expose it to these factors, we now think that differentiated blood is not formed, and believe this is because there has been no endoderm present in our experiments. The 'position value' for blood is at the ventral end of the scale; blood is greatly over-represented in 'UV' radially symmetrical embryos.

Gardner: Presumably, there is still ectoderm in these ectodermal caps, however much mesoderm induction you get. Is that important for the differentiation of the mesoderm?

Cooke: It is, in the situation in which we usually do it. One could probably engineer media in which mesoderm would be happy to do its thing uncapped for much longer, but for it to develop in simple media and go on to form bits of structure and body pattern, ectodermal capping is necessary. The pattern formation that is initiated, even by the soluble factor *in vitro*, includes leaving a large contingent of the cells behind as ectoderm—that in itself is very interesting. You can never take intact animal cap tissue and make it all form mesoderm, yet we can show that every single cell within the expanding cap tissue is competent to respond to the XTC mesoderm-inducing factor.

Gardner: In mammals, as the ectoderm differentiates and produces cells through the primitive streak, that complex of cells will form mesoderm and endoderm structures. But if you isolate the mesoderm on its own, it has only one very spectacular pattern of differentiation, it forms masses of brown adipose tissue (Skreb et al 1976).

Davidson: Jonathan, several years ago you carried out some experiments in which you took frontal partial embryos from early cleavage stages and claimed to observe a certain amount of preformation of pattern in that dimension. Would you still make the same conclusion in view of the current focus of the discussion on control of differentiation based on levels of these inducing factors?

Cooke: The point is that while the inductive events, which happen 'at right angles' to the early events in eggs, have yet to occur, even by the 4-cell stage those early events have partitioned the material into sectors in which for these purposes irreversibly different settings for the later induction have been established. There is no discrepancy between proposing a late, epigenetic character for inductions, and an earlier partitioning that cannot regulate complete pattern in two separated halves.

Davidson: Those are very early.

Cooke: But still too late in terms of regulation. These events finish in the first cell cycle. Frog embryos are not able to compensate for abnormalities of balance after that, whether material is removed surgically or simply not allowed to be set up normally, e.g. by manipulating eggs in gravity or by UV irradiation.

Wolpert: I believe you were saying that the pattern in the mesoderm was specified before gastrulation and that gastrulation was, in these terms, merely a permissive event to make the pattern manifest.

Cooke: Lewis, we go all around the world seeing the poster with a beautiful scanning electron micrograph of a sea urchin gastrulating and a quote from you that 'it isn't when you are born or when you are married that counts, it is when you gastrulate'. Well, I have to tell you that even by that stage, it is largely all over bar the shouting!

Wolpert: You are making a major statement. Most people have implied that the crucial feature in patterning of the mesoderm concerns involution and passing through the blastopore lip. You are saying it's not like that at all, but that the patterning is there before, and gastrulation is simply a permissive event driven by this pattern. That's a completely different way of looking at it.

Cooke: It is caused by a map of local autonomous mechanical tendencies and they are the first manifestations. It is obviously not a finely graded pattern, but the overall characteristics, that mesoderm is head, middle or tail, and dorsal or ventral in its developmental tendencies, are there. If an incomplete or wrongly proportioned body plan is going to be set up, it is visible at gastrulation as an abnormal sequence of absolute timings in gastrulation movement.

Wolpert: You are saying that no new specification occurs during gastrulation of the mesoderm.

Cooke: Not in this particular version of vertebrate development. Other types may turn out to be slower at specifying in relation to gastrulation.

Melton: In that context it will be of interest to see if the levels of *Xhox3* are uniform before gastrulation. Would Jonathan like to comment on that possibility?

Cooke: I am fascinated, because until I heard your paper, I hadn't fully appreciated that the onset of synthesis of some homeobox-containing gene products was a very early part of zygotic transcription after the mid-blastula transition. It might turn out that not only do we have these visual mechanical signs that mesoderm is pre-organized in a body plan in a typical way, but that the first deployment of genes which are going to stabilize that body plan has actually occurred. I wouldn't mind which way that result works out, in the sense that what we are talking about is a kind of organization.

Melton: Yes, but if it is uniform at that early stage, then Lewis Wolpert's view could still hold, and it could change during gastrulation.

Struhl: I am still confused about timing mechanisms. Jonathan is saying that the patterning has already largely occurred and yet at the same time you are saying that the patterning depends upon the timing.

Cooke: I said that the pattern was first shown up by timing, i.e. the timing is a reflection, not a cause, of the hidden pattern. When precisely the cells do what they do, is the first reliable way that they reveal to us, in the absence of molecular probes for other topographical differences, what they are.

Thinking about the possible active organizing role of timing at a still earlier stage, I described an experiment to find out whether it is the timing of receiving the inductive stimulus that is the patterning principle. It appears not to be. Cells do what they do at gastrulation because they are the right age to do it with respect to what part of the body they have been specified to belong to. But that specification is completely independent of when within the period of competence for induction in the blastula they received the stimulus telling them to do that.

Wolpert: In sea urchins, Eric (Davidson) would say that the patterning was prior to gastrulation. In the insects, once again it's prior to gastrulation. For amphibia we will have to rethink our ideas: it would certainly bring them into line with these other animals.

Cooke: You must remember we are talking about *Xenopus* and not even about

Axolotl. When you have different organisms, the mechanical events whereby they gastrulate are perhaps differently timed in relation to the cryptic events in the organization. I suspect that the answer if you did all the experiments on the *Axolotl* would be the same. In mammals, all this has to be put in the context that the events which play the same role in development as the first cell cycle events in *Xenopus* don't occur until there are probably several hundred cells. On the basis of these very weak symmetry-breaking stimuli, the clock-like nature of things might start much later.

Lawrence: The cells presumably invaginate in insects and in frogs because they are different, they have different surface properties. So they must be different *before* they gastrulate.

Wolpert: That is not necessarily the same thing. We are really talking about the setting up of the main anteroposterior axis. Jonathan is effectively saying that axis is set up before gastrulation, whereas most people had thought it was set up during gastrulation.

Lawrence: But why do the cells invaginate?

Cooke: Insects don't have to show their body position by their timing, because of the different geometry of gastrulation.

Weisblat: Have there been experiments where you invert ectoderm above the dorsal lip? According to the classical view that all determining events occur after cells have involuted, such an operation should make no difference, but according to the new view, you should get a piece of inverted axis.

Cooke: It depends at what time you do it. There are two kinds of experiments. This is hard to do in *Xenopus*, because it is developing so fast. If in the early blastula stage, you take a whole piece of tissue, which at the end of normal development would include mesoderm that was going to involute early and late, and was destined to form front and back of the body, respectively, and you turn that round, the result is normal development. That's presumably because events in the marginal zone haven't yet reached the degree of organization I am talking about.

Ray Keller has done many experiments on a later stage of organization than that, when gastrulation movements haven't yet begun but the marginal zone is showing anatomically that it is in a highly pre-organized condition. His results reveal that pre-organized condition in a beautiful way. He cut out little bricks of tissue and replaced them, either in different positions or keeping them *in situ*, and keeping the orientation normal or reversing the orientation of what he calls the pre-involuted marginal zone. If you make the polarity at variance with the polarity of the surroundings, you disrupt development enormously. The brick you have inserted keeps to the schedule by which the cells in it were due to perform their involution movements, but because it has been rotated in polarity, they do it in a completely reverse sequence and form a queue going the other way and therefore form a tissue barricade. Those kinds of experiments reveal that there is a big change of organization between four hours before gastrulation and one hour before gastrulation.

Reference

Skreb N, Svajger A, Levak-Svajger B 1976 Developmental potentialities of the germ layers in mammals. In: Embryogenesis in mammals. Elsevier, Amsterdam (Ciba Found Symp 40) p 27–45

General discussion III

Wolpert: It would be very nice to have a general discussion on comparative embryology. Not in the sense of just describing the morphology, but can we illuminate differences between closely related groups that distinguish what is just a clever mechanism for dealing with a special situation, like a very yolky egg, from fundamental mechanisms.

Cooke: When I was talking I was very aware that I was making some strong proposals about what are organizing principles of development. When you look at various other kinds of vertebrate development it is very hard to see these principles. But I would suggest that they might be there. For instance, in animals of a particular specialized form, like *Xenopus*, a lot of setting up is done in the very first cell cycle. One principle in organization of this body plan seems to be that cells have access to a clock. In the kind of development where the clock is started reliably a very few minutes after the start of development (which can be made synchronous in hundreds of embryos), it becomes very obvious when you look at them in the laboratory. In chick and mammalian embryos, it is very hard to ensure synchronous fertilization; superimposed on that their development may be differently organized, such that the clock which cells take notice of doesn't start in individual embryos in a tightly controlled way as a result of events in the first cell cycle, but starts by cells generating probabilistic mechanisms. Therefore the variation between sibling embryos that were fertilized at the same time is much greater. You get quite a different appearance, as if development is much less determined. Things can appear to be very differently organized on the surface because one isn't able to make particular kinds of manipulations and observations.

Davidson: That doesn't mean they are organized in the same way underneath.

Cooke: No, but you are entitled to stick to propositions that they might be. For instance, in mammals take the combination of Mike Snow's results with Rosa Beddington's results. There is a stage where the mouse blastodisc is organized into a primitive streak, which we would say is the stage where gastrulation is about to start. When you make appropriate dissections and explantations into culture within that stage you get an appearance which says that there is a considerable degree of pre-organization for developing the various geographical regions of the body at the very onset of gastrulation, whereas most people wouldn't think that was true.

Davidson: Is it true that the first diversification into, for example, the definitive endoderm or intra-embryonic mesoderm appears in either chick or mammal as the cells pass through Hensen's node or in the primitive streak?

Gardner: That seems to be the case, yes.

Cooke: All I am arguing is that the data on most types of vertebrate development don't yet allow one to insist on either viewpoint. Another thing that is being forgotten is that my proposals, even about frog development, relate to tissue-maintained properties and certainly not to properties that would survive isolation as single cells.

Wolpert: Claudio Stern would say that some of the primitive endoderm and mesodermal cells are present very early, well before gastrulation in the chick.

Cooke: We were talking about levels of pattern, not histologically defined cell type.

Davidson: I wasn't talking about pattern, I was talking about cells that you could identify as being this or that cell type.

Wolpert: You can't identify cells in the amphibian embryo as muscle before gastrulation, but we know that they are specified.

Davidson: In the amphibian embryo, initial specifications apparently occur much earlier, and are probably mediated by cell interactions before the resumption of transcription. My understanding is that in the mammalian embryo there is little evidence that any limitations on future cell fate occur prior to ingression of cells through the primitive streak.

Cooke: The distinction of which level of body patterning a group of cells is going to form is different from what histological type they are going to be.

If one considers pattern, I don't think the available results about mammals indicate whether or not they are different from amphibians.

Gardner: Not everything is environmentally induced or determined in the mammalian embryo; there seems to be an internal timing mechanism operating as well. There is evidence that blastomeres are polarized before the 8-cell stage, but before this stage polarity does not persist through cell division, as it does thereafter (Johnson et al 1988). It is not until about the 32-cell stage that blastocyst formation occurs. But as in other systems, if you interfere with cytokinesis by cytochalasin treatments, these events still occur on schedule.

Lawrence: What is this clock? Can somebody tell us what a clock is?

Cooke: In early development, I think it is a cell's ability to measure its age in an active and fairly accurate way and to respond to stimuli or to begin new patterns of behaviour at appropriate points.

Struhl: The question is whether there is a causal connection between the passage of time and embryonic events. In *Drosophila* we know that under certain temperature conditions the embryos will develop at a certain rate. But I don't think anyone working on *Drosophila* believes that timing plays a causal role. I don't see any reason to say that for any of the other systems that have been discussed.

Gardner: In mammals, activation of the timing mechanism may be related to breakdown of the germinal vesicle or the onset of meiosis, for example. Some very interesting information has emerged from studies on human *in vitro* fertilization. If eggs are inseminated *in vitro*, pronuclear development sometimes fails to occur on time. Apparently, fertilization can be postponed for up to 36 hours and the resulting embryo nevertheless reaches the 8-cell stage on schedule (Fishel et al 1984). This suggests that some event around the time of ovulation or even earlier sets development in motion. So long as the sperm enters before a certain point, development can continue without perturbation. If it does not, development fails.

Struhl: This is a bit extreme. In amphibians there is some evidence for oscillations or nuclear cell divisions early on. I was thinking more in terms of later developmental events, such as gastrulation.

Cooke: Richard, would you want to extend the response timing seen in early parts of mammalian development? Do you think it might run right through the prolonged sequence of events after that, until we get to stages of organization of body patterning that are equivalent to *Xenopus* late blastula or gastrula?

Gardner: It is very difficult to say. This problem of getting simultaneous fertilization is very difficult, particularly under *in vivo* conditions where it's so variable. As development progresses, you tend to get a greater and greater disparity; if the disparity is too much within a litter, the embryos that are behind can't, as in other systems, develop according to their own schedule. If they get too far behind, the maternal system has gone beyond the stage at which it can support them and they end up as a resorptions or failures in implantation. This enormously complicates looking at how far timing extends.

Wieschaus: One of the ways of getting a better handle on the clock would be to ask if there is any way that you could manipulate it or slow it down. You have told us about methods that block certain events but allow other events to occur at the normal time. These methods evidently do not affect the clock itself. Is there anything other than temperature that will affect the clock?

Cooke: There was an era when it was thought that all biological phenomena which gave evidence that the cell was measuring time could be perturbed by adding heavy water to the system, because it slows the free-running periodicity of biological oscillators. The mechanism for that remains obscure.

Garcia-Bellido: Has anyone looked at this clock mechanism in chimaeric embryos?

Davidson: There were classical experiments done in fish, amphibia and echinoderms. Species hybrids were made between organisms that had different rates of doing things. The events always adopted the timing of the maternal species until quite late in development. Then the rate becomes intermediate.

Garcia-Bellido: I don't mean a hybrid, I mean a chimaera.

Cooke: That kind of experiment has been done in amphibians, splicing together an old and a young amphibian embryo. The general result is that cells cannot be made to do things before they would have done them on their own time schedule. There is a great deal of cell autonomy: a clone of young cells in a tissue of older cells will not age faster.

Struhl: None of these results addresses the question as to whether or not the timing is causal.

Wolpert: There is something peculiarly attractive about time in the sense that those people who think about evolution always invoke heterochrony. It is worth bearing in mind how little we embryologists know about timing mechanisms.

References

Fishel SB, Edwards RG, Purdy JM 1984 Births after a prolonged delay between oocyte recovery and fertilization in vitro. Gamete Res 9:175–81

Johnson MH, Pickering SJ, Dhiman A, Radcliffe GS, Maro B 1988 Cytocortical organization during natural and prolonged mitosis of mouse 8-cell blastomeres. Development 102:143–158

Genetic control of cellular interactions in *Caenorhabditis elegans* development

Judith Kimble*† and Judith Austin†‡

**Department of Biochemistry, †Laboratory of Cell and Molecular Biology, ‡ Department of Genetics, University of Wisconsin, 1525 Linden Drive, Madison, Wisconsin 53706, USA*

Abstract. During development of the nematode, *Caenorhabditis elegans*, cell interactions play a significant role in controlling cell fate. Regulatory cells in the somatic gonad control proliferation in the germline and induce formation of the vulva in the hypodermis. In the early embryo, mesodermal cells are induced in a process similar to embryonic induction. In addition, interactions between precursor cells of equivalent developmental potential direct one cell to adopt one fate and the other to adopt a second fate. Two genes have been identified in *C. elegans* that appear to mediate cell interactions. The *glp-1* gene is required for embryonic induction and continued germline proliferation; *lin-12* is required for cells of equal developmental potential to adopt different fates. Genetic mosaics indicate that *glp-1* may be part of the receiving mechanism in controlling germline proliferation.

1989 Cellular basis of morphogenesis. Wiley, Chichester (Ciba Foundation Symposium 144) p 212–226

Interactions between cells are critical to the control of growth and differentiation of multicellular organisms. Such interactions were first observed in classical experiments with sea urchin and frog embryos (e.g. Driesch 1891, Spemann & Mangold 1924). More recently, cellular interactions that regulate development have been described in the nematode, *Caenorhabditis elegans*. In this brief review, we describe several cellular interactions that influence development in *C. elegans* and two genes, *glp-1* and *lin-12*, central to the control of these interactions. Remarkably, these two genes appear to encode similar proteins, indicating that they arose during evolution by duplication and that diverse regulatory interactions in development may rely on a similar underlying biochemical mechanism.

The three types of regulatory interactions that we discuss are summarized in Table 1. They include control of germline proliferation by the distal tip cell (Kimble & White 1981), induction of pharyngeal mesoderm in the embryo (Priess & Thomson 1987), and regulation among precursor cells of equivalent developmental potential to adopt different fates (Sulston & White 1980,

TABLE 1 Regulatory cell interactions in *Caenorhabditis elegans* development[a]

Signalling cell	*Receiving cell*	*Normal fate of receiving cell*	*Fate of receiving cell after removal of signal*	*Interaction deduced*
Distal tip cell (dtc)	Germline	Germ cells in mitosis near dtc	Germ cells enter meiosis near dtc	Dtc signals germline to continue mitosis
P_1 blastomere	AB blastomere	AB generates cells of anterior pharynx	AB does not generate anterior pharynx	P_1 signals AB to produce anterior pharynx
1° cell[b] e.g. anchor cell precursor, AC	2° cell e.g. ventral uterine precursor, VU	2° fate e.g. ventral uterus	2°→1° e.g. VU becomes AC	1° inhibits 2° from becoming 1°

[a]Only cell interactions discussed in this review are listed.
[b]Cells in equivalence groups are primary (1°), secondary (2°), or tertiary (3°). Only one example is provided in this table: the anchor cell precursor is 1° and the ventral uterine precursor is 2°.

Kimble 1981, Sternberg & Horvitz 1986). The discovery of each of these interactions emerged from one basic type of experiment. In each case, when one cell (the signalling cell) was physically removed, either by ablation with a laser microbeam or by needle puncture, the fate of a neighbouring cell (the receiving cell) was dramatically altered. For example, in the gonad, a somatic cell is located at the distal end of a tube of germline cells. [The germline tissue is actually a syncytium. However, each germline nucleus occupies its own membrane-bound alcove of cytoplasm located at the edge of a common anuclear cytoplasm (Hirsh et al 1976). Each germline nucleus and its cytoplasm is called a germ cell for simplicity.] Normally, germ cells located close to the distal tip cell are in the mitotic cell cycle, whereas more proximal germ cells enter meiosis. When the distal tip cell is killed, in either sex and at any time during post-embryonic development, germ cells leave the mitotic cell cycle, enter meiosis and undergo gametogenesis. Thus, the distal tip cell must signal to cells of the germline to continue mitosis. The mitogenic influence of the distal tip cell is required for growth of the germline during post-embryonic development and for maintenance of a stem cell population of germ cells in the adult. Other experiments suggest that the distal tip cell signal acts over a distance and that its position, usually at the distal tip of the testis or ovotestis, establishes the polarity of the gonad.

Two genes have been identified that are involved in control of the interactions outlined in Table 1. The mutant phenotypes of these genes are summarized in Table 2 and are described briefly below. One of these genes, *glp-1*, was identified in a specific attempt to find genes that control cellular interactions (Austin & Kimble 1987). When the distal tip cell is ablated, germ cells that would normally be mitotic enter meiosis and differentiate. Therefore, the mutant phenotype we sought was entry into meiosis by germ cells that normally would be in mitosis. In a screen of approximately 20 000 mutant chromosomes, six mutations were isolated with this phenotype. All fell in a single complementation group on chromosome III. This locus was named *glp-1* for *g*erm*l*ine *p*roliferation defective. The *glp-1* locus was identified independently in a general screen for maternal effect mutants that result in defective embryogenesis (Priess et al 1987). The *lin-12* locus was identified in a general screen for mutants that were defective in the cell lineages of the vulva (Ferguson & Horvitz 1985).

The *glp-1* gene is necessary for control of continued mitoses in the post-embryonic germline (Austin & Kimble 1987, Priess et al 1987). Mutant animals lacking *glp-1* (but born of a heterozygous *glp-1*/+ mother) produce only 4–8 germ cells (Table 2). These few germ cells enter meiosis and undergo spermatogenesis. The processes of meiosis and gametogenesis are normal in *glp-1* mutant animals. Experiments using a temperature-sensitive allele of *glp-1* indicate that the *glp-1* product is required for germline growth from larval stage L2 throughout adulthood. It is thus central to proliferation of the germline during larval development and to maintenance of a stem cell population in the adult. Similarly, experiments in which the distal tip cell was removed at various times of development showed that this regulatory cell must signal to the germline throughout larval development and adulthood for continued germline mitoses.

TABLE 2 Mutant phenotypes of *glp-1* and *lin-12*

Gene	*Genetic descriptor*	*Mutant phenotype*	*Temperature-sensitive period*	*Defective interaction*
glp-1[a]	m(+/−); z(−/−)	Germ cells enter meiosis near dtc	From L2 through adult	Distal tip cell/germline
glp-1[b]	m(−/−); z(−/−)	AB does not generate anterior pharynx	Early embryo (4–28 cells)	P_1/AB
lin-12[c]	recessive (lf)	2°→1°	For VU→AC: L2/L3 moult	1°/2°
lin-12[c]	dominant (gf)	1°→2°	—	1°/2°

m, maternal genotype; z, zygotic genotype; lf, loss-of-function; gf, gain-of-function; dtc, distal tip cell; VU, ventral uterine precursor; AC, anchor cell precursor. In all cases, the temperature-sensitive period of gene function corresponds to the time of development at which the relevant interaction is taking place.

[a] Austin & Kimble (1987). [b] Priess et al (1987). [c] Greenwald et al (1983).

In addition to this germline phenotype, *glp-1* mutants have an embryonic phenotype that indicates the need for maternal *glp-1* product in embryogenesis (Austin & Kimble 1987, Priess et al 1987). This embryonic phenotype was observed in conditional mutants of *glp-1*. At permissive temperature, *glp-1(ts)* homozygotes produce a normal number of germ cells, but when shifted as adults to the restrictive temperature, progeny do not survive. *glp-1(ts)/glp-1*(+) heterozygous cross-progeny of a *glp-1(ts)* mother do not survive; therefore the *glp-1* product must be contributed by the mother for survival of her progeny. The lethal phenotype of *glp-1* embryos includes defects in hypodermal morphogenesis and pharynx development (Priess & Thomson 1987, Austin & Kimble 1987). The embryos have a near normal number of cells, but do not possess the anterior half of their pharynx (Table 2) and do not change from a ball of cells to an elongated worm. Since they do possess the posterior part of the pharynx, differentiation of pharyngeal cells *per se* is not affected. Using reciprocal shift experiments, the stage of embryogenesis at which the maternal *glp-1* product is required has been found to be from 4–28 cells of embryogenesis. This temperature-sensitive period corresponds extremely well with the time at which anterior pharynx development is induced (Priess & Thomson 1987). This induction event appears to be an interaction between the AB blastomere (or its descendants) and the P_1 blastomere (or its descendants). Although no interactions are known to be required for hypodermal morphogenesis, such interactions may exist.

What role does *glp-1* play in regulating cellular interactions? Is it a component of the signalling mechanism or the receiving mechanism? To address this question, we examined genetic mosaic animals in which either the distal tip cell or the germline was defective for *glp-1* (Austin & Kimble 1987). Our results are summarized in Table 3. They indicate that the *glp-1* activity necessary for germline mitoses is produced by the germline. This implies that *glp-1* encodes a component of the receiving machinery for the distal tip cell–germline interaction. The mosaic analysis also shows that *glp-1* activity in the distal tip cell is neither necessary nor sufficient for germline development. Thus, *glp-1* does not encode the distal tip cell signal.

The effects of *lin-12* mutations on development have been extensively reviewed (Sternberg & Horvitz 1984, Greenwald 1987). Here, I emphasize one point: *lin-12* is involved in the regulation of interactions between cells in equivalence groups. Two types of *lin-12* allele have been isolated. Recessive alleles (loss-of-function, *lf*) transform cells of equivalent potential to one fate, and dominant alleles (gain-of-function, *gf*) transform the same cells to a second fate. Thus, in wild-type animals, where two equivalent cells can adopt either a primary or secondary fate, in *lin-12(lf)* mutants both cells adopt the primary fate, and in *lin-12(gf)* mutants both adopt the secondary fate. Opposite effects on development by *lf* and *gf* alleles have been observed in other genes that regulate cell fate (e.g. Sternberg & Horvitz 1984).

TABLE 3 Genetic mosaic experiments with *glp-1*

Animal	*Genotype*		*Phenotype*	
	Distal tip cell	*Germline*	*Distal tip cell*	*Germline*
Wild-type	*glp-1*(+)	*glp-1*(+)	Present	2000
Mutant	*glp-1*(−)	*glp-1*(−)	Present	4–8
Mosaic	*glp-1*(+)	*glp-1*(−)	Present	4–8
Mosaic	*glp-1*(−)	*glp-1*(+)	Present	2000

The results summarized here are from Austin & Kimble (1987).

By genetic analysis, *glp-1* was shown to map 0.02% to the right of *lin-12* (Fig. 1) (Austin & Kimble 1987). This region of the *C. elegans* genome is represented by a large series of overlapping cosmids (Greenwald et al 1987). It was therefore possible to use the appropriate cosmids as hybridization probes in an attempt to identify the *glp-1* gene. We found that the ZK506 cosmid, which neighbours *lin-12* (Fig. 2), detects alterations in two different *glp-1* mutations on Southern blots (Austin & Kimble, in preparation). One EMS-induced (ethyl methanesulphonate) mutation, *q172*, is associated with a small, 300 bp deletion in a 1.5 kb *Eco*RI fragment of this cosmid; a gamma-ray induced mutation, *qDf2*, is associated with a large deletion that breaks in the same 1.5 kb *Eco*RI fragment and removes the rest of the DNA in ZK506 to the right of that fragment (Fig. 3). The right end of *qDf2* has not yet been determined. From this region, we have identified one major transcript of 4.8 kb that is likely to encode the *glp-1* product (Fig. 3).

Fortuitously, in a search of the *C. elegans* genome for *lin-12* homologues, a gene in the ZK506 cosmid was identified (J. Yochem & I. Greenwald, personal communication). When a phage bearing this *lin-12* homologue was used as a

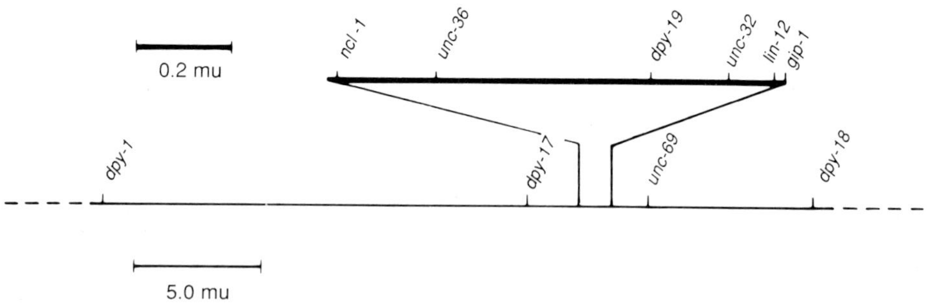

FIG. 1. Genetic position of *lin-12* and *glp-1*. *glp-1* maps 0.02% to the right of *lin-12* on chromosome III. Reproduced with permission (and minor modification) from Austin & Kimble (1987).

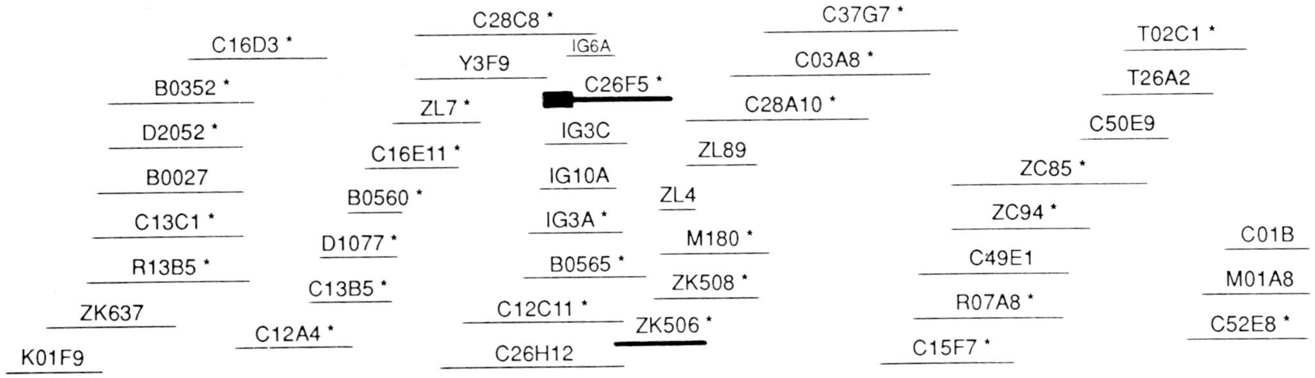

FIG. 2. The *lin-12* contig. Numerous overlapping cosmids have been identified in the region of *lin-12*. The *lin-12* gene, carried on C26F5, is indicated by an enclosed black rectangle. The *glp-1* gene maps to a neighbouring cosmid, ZK506 (Austin & Kimble, in preparation). This figure was prepared from a cosmid map provided by A. Coulson and J. Sulston (unpublished data).

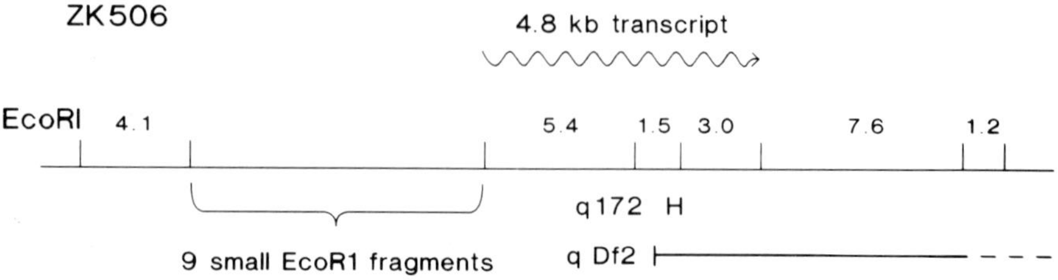

FIG. 3. Molecular analysis of *glp-1*. The *glp-1* gene has been identified by two deletions that are associated with *glp-1* mutants. *q172* is a 300 bp deletion in a 1.5 kb *Eco*RI fragment in ZK506; *qDf2* is a larger deletion that breaks in the same *Eco*RI fragment and that extends to the right. A major transcript of 4.8 kb has been identified. (Austin & Kimble, in preparation.)

hybridization probe to *q172* DNA, the 300 bp deletion associated with this *glp-1* mutation was detected on a Southern blot (J. Austin & J. Kimble, in preparation). Therefore, *glp-1* is likely to be identical to the *lin-12* homologue. Another gene homologous to *lin-12* has been identified in *Drosophila* (Wharton et al 1985). This gene, *Notch*, regulates the decision between differentiation as an epidermal precursor cell and a neuroblast. Thus, all three homologous genes, *glp-1*, *lin-12* and *Notch* influence a developmental decision of cell fate. Presumably, all do so by mediating a cellular interaction that regulates this decision. Since the deduced amino acid sequences of *lin-12* and *Notch* have the molecular characteristics predicted for membrane proteins (Yochem et al 1988, Greenwald 1985, Wharton et al 1985), *glp-1* may also encode a membrane protein. The functions of *glp-1*, *lin-12* and *Notch* in mediating cellular interactions are not known. A simple possibility is that each encodes a receptor. For *glp-1*, the receptor would bind the signal emitted by the distal tip cell (or P_1). Alternatively, it may encode a component of that receptor, or some other membrane protein that is essential to transduction of the distal tip cell (or P_1) signal.

Fig. 4 presents one model for the molecular function of *glp-1* in the germline. We show the *glp-1* product as a receptor located in the membrane of the germline syncytium. Upon binding of the signalling molecule produced by the distal tip cell, this receptor transduces the signal to direct continued mitotic divisions in the germline. We suggest that the *glp-1* protein may be present throughout the germline tissue and that it is the position of the distal tip cell and its signal which determines where *glp-1* will actively direct germline mitoses. This model is supported by results from genetic mosaic analysis that place *glp-1* function in the germline, and by the homology of *glp-1* with *lin-12*, which suggests that *glp-1* may encode a membrane protein. Although this model is extremely speculative, it serves as a useful starting point for understanding how the *glp-1* gene product functions.

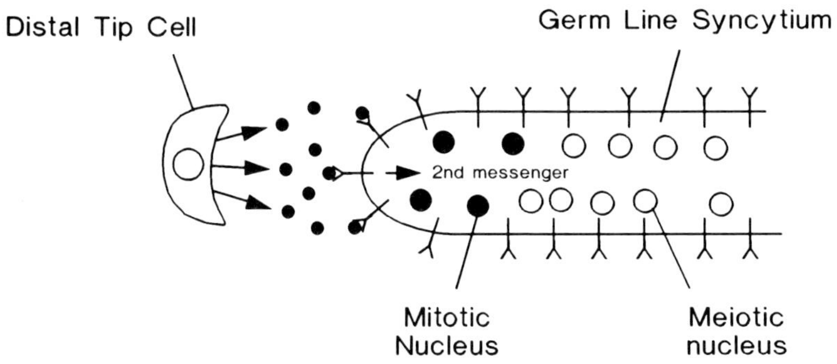

FIG. 4. Model for *glp-1* function. The distal tip cell is a somatic cell that signals to the germline syncytium. Germline nuclei at the distal end of the syncytium remain in mitosis; germline nuclei further proximal enter meiosis. In this figure, we have drawn the distal tip cell at some distance from the germline to permit drawing of signal molecules in the intercellular space. In the animal, the membrane of the distal tip cell is closely juxtaposed to the membrane of the germline. No basement membrane separates the distal tip cell from the germline; instead a basement membrane encapsulates both and separates the gonad from the surrounding pseudocoelom (Kimble & Ward 1988).

Based on the genetic mosaic experiments of Austin & Kimble (1987) and the predicted *glp-1* molecular structure of J. Yochem and I. Greenwald (personal communication), we propose that *glp-1* encodes a component of the membrane receptor for the distal tip cell and P_1 signals. In this figure we show the *glp-1* product as a receptor (Y) that is present throughout the germline. This distribution of *glp-1* product is suggested for simplicity; there is no evidence to date of its localization. We propose that the distal tip cell emits a signal (●) that binds the *glp-1* receptor locally. Since the distal tip cell signal appears to act over a distance (Kimble & White 1981), we propose that the ligand-activated *glp-1* generates a second messenger that diffuses in the germline syncytium to exert its mitogenic influence on germline nuclei more proximally.

Several major questions remain unanswered. Does the same *glp-1* product mediate both its germline and embryonic functions? If so, what are the signals and are they the same? Is the *glp-1* protein a component of the receptor itself as depicted in Fig. 4 or does it serve some other function that is critical to transduction of the signal? What is the distribution of the *glp-1* protein in the germline? And how is *glp-1* function limited to the AB cell in the embryo? Is the *glp-1* protein inherited on the membrane of the oocyte and distributed to all blastomeres? If so, a localized signal may activate *glp-1* to induce pharyngeal differentiation in AB. The answers to these questions are now accessible. Starting with the cloned *glp-1* gene, we can analyse the nature of the *glp-1* products and their distribution during development and we can investigate the regulation and function of *glp-1* after its reintroduction into the *C. elegans* germline and embryo.

Acknowledgements

We are grateful to John Yochem and Iva Greenwald for sharing their unpublished results on a *lin-12* homologue in ZK506. This research was supported by a U.S. Public Health Service grant GM31816 and a Research Career Development Award HD00630 to J. K. J. A. is a trainee of the NIH Predoctoral Training Program in Genetics GM07133.

References

Austin J, Kimble J 1987 *glp-1* is required in the germ line for regulation of the decision between mitosis and meiosis in *C. elegans*. Cell 51:589–599

Driesch H 1891 Entwicklungsmechanische Studien. I. Der Werth der beiden ersten Furchungszellen in der Echinodermenentwicklung. Experimentelle Erzeugung von Theil und Doppelbildungen. II. Uber die Beziehunger des Lichtes zur ersten Etappe der tierischen Formbildung. Z Wiss Zool 53:160

Ferguson EL, Horvitz HR 1985 Identification and characterization of 22 genes that affect the vulval cell lineages of the nematode *Caenorhabditis elegans*. Genetics 110:17–72

Greenwald I 1985 *lin-12*, a nematode homeotic gene, is homologous to a set of mammalian proteins that includes epidermal growth factor. Cell 43:583–590

Greenwald I 1987 The *lin-12* locus of *C. elegans*. Bioessays 6:70–72

Greenwald IS, Sternberg PW, Horvitz HR 1983 The *lin-12* locus specifies cell fates in *Caenorhabditis elegans*. Cell 34: 435–444

Greenwald I, Coulson A, Sulston J, Preiss J 1987 Correlation of the physical and genetic map in the *lin-12* region of *Caenorhabditis elegans*. Nucleic Acids Res 15:2295–2307

Hirsh D, Oppenheim D, Klass M 1976 Development of the reproductive system of *C. elegans*. Dev Biol 49:200–219

Kimble J 1981 Alterations in cell lineage following laser ablation of cells in the somatic gonad of *Caenorhabditis elegans*. Dev Biol 87:286–300

Kimble J, Ward S 1988 Germ-line development and fertilization. In: Wood WB (ed) The nematode *Caenorhabditis elegans*. Cold Spring Harbor Press, Cold Spring Harbor, New York, p 191–213

Kimble J, White JG 1981 On the control of germ cell development in *Caenorhabditis elegans*. Dev Biol 81:208–219

Priess JR, Schnabel H, Schnabel R 1987 The *glp-1* locus and cellular interactions in early *C. elegans* embryos. Cell 51:601–611

Priess JR, Thomson JN 1987 Cellular interactions in early *C. elegans* embryos. Cell 48:241–250

Spemann H, Mangold H 1924 Uber induktion von Embryonalanlagen durch Implantation artfremder Organisatoren. Roux's Arch Dev Biol 100:599

Sternberg PW, Horvitz HR 1984 The genetic control of cell lineage during nematode development. Annu Rev Genetic 18:489–524

Sternberg PW, Horvitz HR 1986 Pattern formation during vulval development in *C. elegans*. Cell 44:761–772

Sulston JE, White JG 1980 Regulation and cell autonomy during postembyronic development of *Caenorhabditis elegans*. Dev Biol 78:577–597

Wharton KA, Johansen KM, Xu T, Artavanis-Tsakonis S 1985 Nucleotide sequence from the neurogenic locus *Notch* implies a gene product that shares homology with proteins containing EFG-like repeats. Cell 43:567–581

Yochem J, Weston K, Greenwald I 1988 The *Caenorhabditis elegans lin-12* gene encodes a transmembrane protein with overall similarity to *Drosophila Notch*. Nature (London) 335:547–550

DISCUSSION

Garcia-Bellido: The genetic mosaic analysis you presented has been done with only one allele. Have you tried any other alleles?

Kimble: No, the alleles we have available in my lab are very similar to each other, and we have no reason to believe one would be different from another in a mosaic analysis. We prefer to concentrate on the molecular analysis of *glp-1* for the moment and to generate appropriate nucleic acid and antibody probes for localization studies.

Weisblat: One of the *glp-1* alleles reported by Priess et al (1987) does not affect the germline, but does alter embryos. Is it known where the mutation in this allele is located?

Kimble: No, that is not yet known.

Weisblat: The residual signal that you see on Northern blots of *glp-1* RNA obtained from animals with no germline, could this be somatic?

Kimble: That is one possibility. The mutant we used with a reduced number of germ cells has about 20 instead of the normal 2000. The small amount of RNA seen in that mutant might be present in those few germ cells, or it might be present in a somatic tissue.

Meinhardt: How do you explain the mutant phenotype of *lin-12*, where regions of germline proliferation are found at both distal and proximal ends of a single germline tube? Do you think that some sort of inhibition doesn't work any more?

Kimble: Greenwald and her colleagues showed some years ago that many different cells are transformed from one fate to another in *lin-12* mutants. We postulate that one of the somatic gonadal cells in the proximal area has been transformed so that it now has distal tip cell activity. This newly transformed cell would then influence neighbouring germ cells to remain mitotic.

Nüsslein-Volhard: Is this the lack-of-function phenotype of *lin-12*?

Kimble: Yes.

Nüsslein-Volhard: According to your model, one expects that there is a group of genes which should have the same phenotype as *glp-1*.

Kimble: That is right, but we did not isolate other mutants with a phenotype like that of *glp-1*. There are several possible explanations for failure to detect other genes involved in this process. There may be redundant genes; if so, elimination of one would not cause a mutant phenotype. Such genes may be used for more than one cell interaction, in which case elimination of one might cause a different mutant phenotype such as embryonic lethality. The maternal rescue of *glp-1* embryos permitted us to see the germline phenotype. If other genes are involved in the same processes as *glp-1* but are not rescued, they might be zygotic lethal.

Wieschaus: The *glp-1* product is used for cell interactions in the early embryo. It is then not used until the animal grows up and has a gonad. Are there other cell interactions that occur in the worm outside those two time points?

Kimble: There are several other cell interactions that occur during *C. elegans* development. The best known is anchor cell induction where a single regulatory cell in the somatic gonad, the anchor cell, induces underlying hypodermal cells to generate a vulva. Although there are genes that are candidates for mediating this interaction, the evidence is not as compelling as it is for *glp-1* and *lin-12*. There are many genes known that are required for normal proliferation of the germline and for normal development of the vulva. Any or all of these genes may mediate the cell interactions that influence these developmental events.

Garcia-Bellido: Are there other regions in the genome homologous to *lin-12* or *glp-1*?

Kimble: John Yochem is characterizing *lin-12* homologues of *C. elegans* in the laboratory of Iva Greenwald at Princeton University, but has only preliminary data at the moment.

Davidson: In that syncytial gonad, what are the real dimensions over which the second messenger is supposed to act?

Kimble: That is a difficult question to answer because we cannot detect cells as they make the decision to leave mitosis and enter meiosis. As an estimate of the order of magnitude, I would say about 100 μm.

Garcia-Bellido: Have you looked at where the *lin-12* dominant is expressed?

Kimble: Genetic mosaics of *lin-12* are currently being examined in Greenwald's lab. I would like to point out one important difference between cell interactions influenced by *glp-1* and those influenced by *lin-12*. Cell interactions that depend on *glp-1* activity occur between two very different types of cells (e.g. the somatic gonadal distal tip cell and germ cells). The decision regulated is between two fates (e.g. mitosis and meiosis) in only one of the interacting cells. In contrast, interactions that depend on *lin-12* occur between equivalent cells. In mutants lacking *lin-12*, both cells follow one fate; in mutants bearing gain-of-function dominant mutations in *lin-12*, both cells follow the alternate fate. We have not found a gain-of-function mutant of *glp-1*; in such mutants, we would predict that germ cells normally in meiosis might remain in mitosis instead.

Struhl: What is the phenotype of the *glp-1 lin-12* double mutant?

Kimble: We are constructing that double mutant now, so the result is not yet known.

Lawrence: Could you tell us more about *Notch*.

Kimble: Notch is one of the neurogenic loci in *Drosophila*. In wild-type development there is a sheet of cells from which neuroblasts delaminate. The remaining cells generate epidermal cells. Activity of the *Notch* gene is required in the epidermal cells to prevent them from delaminating and becoming neuroblasts; if *Notch* is absent, all cells delaminate and enter the nerve lineage.

Nüsslein-Volhard: Only in the ventral ectoderm, whereas *Notch* is expressed everywhere.

Struhl: There is one critical similarity: the laser ablation experiments of Doe and Goodman (1985) showed that if you remove the cell that would normally delaminate, then another cell from the epidermis would delaminate. This suggests that the cells somehow decide which is going to go out and which is going to stay. That's very similar to the *lin-12* situation, where only one of the two cells makes the decision and the other makes the alternative decision.

Kimble: Yes; and in addition, the cell interactions influenced by *Notch* are more similar to those influenced by *lin-12* than to those influenced by *glp-1*.

Cabrera: You presented evidence that *glp-1* functions in the germline and is therefore part of the receiving mechanism in the distal tip cell–germline interaction. This contradicts the result obtained in *Drosophila* where *Notch* is clearly non-autonomous.

Wieschaus: One of the issues that complicates autonomy or non-autonomy of *Notch* is that the result depends on the size of the mosaic patch. If you produce very large patches of *Notch*, generally you get the *Notch* phenotype. If you produce a single *Notch* cell in the epidermis, that cell can remain epidermal, even though *Notch* would normally cause it to become a neuroblast. Anything bigger than one cell, even two adjacent cells, and you begin to see an autonomous effect of *Notch*.

McKeown: That just tells you the distance over which the signalling molecule works.

Wieschaus: If there are two adjacent cells, each of those cells is surrounded on most of its interfaces by wild-type cells, yet that is not enough for a *Notch* cell to be rescued by this type of non-autonomy. It must be totally surrounded by wild-type cells. I can't think whether that's going to influence how you want to interpret the cellular basis for the non-autonomy but it seems that it should.

McKeown: Gerhard Technau and Jose Campos-Ortegas (1987) have done experiments in which they have taken *Notch*$^-$ blastomeres and transferred them into wild-type embryos. Rather than developing as only neural cells, they are capable of developing as both ectodermal and neural cells. The implication is that *Notch*$^-$ embryos are not missing a receptor: they are missing the signal. That's exactly the opposite of what you are saying here.

Kimble: The difficulty interpreting the *Notch* mosaic results suggests to me that we may not understand some important facts about these loci. One possible complication is that the genes may produce more than one product. Perhaps, the same gene can produce both signal and receptor. For example, the entire *glp-1* gene might encode a membrane protein that acts as a receptor, whereas a protein made from the 5′ half of *glp-1* would consist of EGF-like peptides and might generate the signal.

How do I accommodate our genetic mosaic results with such a suggestion? I must postulate that the mutation used for mosaic analysis is located at the 3′ end and does not affect production of the 5′ transcript. We have preliminary results that most of our *glp-1* mutations map to the 3′ region of the gene. This includes the mutation used for mosaic analysis.

Heasman: We have done some preliminary experiments on isolated mouse germ cells. Can we draw some parallels here? We take germ cells at the time that they are just beginning to migrate; they are mitotic cells and in culture those cells divide for a while and then slow down and appear to become meiotic cells and have a different phenotype. However, when we use a medium conditioned by the germ ridges from embryos, we can maintain the mitotic state in culture. This activity is blocked by antibodies against the EGF receptor. Furthermore, in the *W* mutants where there are no germ cells, conditioning of the medium by germ ridges does not stimulate mitosis. I wonder whether a similar EGF-type growth factor is involved in both embryonic systems.

Kimble: It was recently reported that the *W* locus encodes a protein similar to the c-*kit* proto-oncogene, a tyrosine kinase receptor (Geissler et al 1988). To answer your question, I believe that the maintenance of stem cell populations of mitotically dividing cells may rely on similar mechanisms in both *C. elegans* and in higher vertebrates and in both germ cells and other tissues. For example, there is evidence that haemopoietic stem cells depend on the 'microenvironment,' so they too may require interaction with other cells to continue proliferation.

Kornberg: You said initially that the allele you did the mosaic analysis with was a null allele. How did you define it as a null?

Kimble: Our proposal that this allele is a null allele is based on our characterization of multiple *glp-1* alleles. These include one small internal deficiency and one large deficiency that breaks in the middle of the gene. All strong alleles, including the deficiencies and the allele used for mosaic analysis, are fully penetrant and have identical phenotypes. However, it is certainly possible that these mutants knock out only one function of the *glp-1* gene.

Cabrera: Does this putative intracellular domain of *glp-1*, *lin-12* or *Notch* have homology to a tyrosine kinase domain?

Kimble: No. These genes are homologous in their putative extracellular domains to EGF; they are not homologous to the EGF receptor, which is a tyrosine kinase.

Wolpert: I find it somewhat ironic that you are using *Caenorhabditis* to talk about cellular interactions and receptors and so forth, and very little about lineage. Is there really no autonomous generation of diversity? I am not talking about the very early stages where there are obviously cytoplasmic differences.

Kimble: A mechanism of cell determination relying strictly on cell lineage is much more difficult to prove experimentally than a mechanism relying on cellular interactions. Alterations in cell fate after laser ablation of individual cells can be used convincingly to delineate cell interactions, but the lack of an effect in the same experiment does not provide equally compelling evidence that cellular interactions have no influence.

Davidson: Nevertheless, is it not fair to say that the experiments done by Priess & Thomson (1987), which demonstrate the requirement for cellular interaction in the anterior pharynx, do not give the same kind of result for the development of the posterior pharynx? There are other analogous situations where one part of the given structure is made one way and the other part is made in another way.

Kimble: That is possible, but again relies on interpretation of a negative result.

Davidson: It's a little stronger than just a negative result. The same experimental protocol results in one thing in one region and another in another place.

Wieschaus: You showed this terribly abnormal embryo which lacks *glp-1* activity. Then you showed Jim Priess' embryo which had only an abnormal pharynx. That suggests that the *glp-1* product must control more than just pharynx development and therefore that cellular interactions must be playing roles in other events in embryogenesis. Can you use your temperature-sensitive mutant to investigate this under precisely controlled conditions?

Kimble: We have shown that both embryonic morphogenesis and anterior pharynx development require maternal *glp-1* product between the stages of four and 28 cells. If we shift these *glp-1* mutants from the permissive to the restrictive temperature at the 8-cell stage, larvae are formed, but the anterior pharynx does not develop. This result suggests that embryonic morphogenesis depends on some event occurring between four and eight cells, while pharynx induction depends on an event taking place between eight and 28 cells.

Wieschaus: Have you done pulses of permissive or restrictive temperature?

Kimble: No.

Nüsslein-Volhard: Have you done *in situ* hybridization or antibody staining?

Kimble: We have begun *in situ* experiments, but have not yet developed antibodies. We have no results to report yet.

Morata: A few years ago you said that the equivalence group could be defined as units of interactions in the ventral nerve cord in the nematode. You implied that they could be equivalent to compartments in *Drosophila*. Have you identified genes which specifically recognize these equivalence groups?

Cabrera: I think there is a substantial difference between equivalence groups and compartments. In the former, cellular interactions take place to specify different cell lineages. In the latter, boundaries are created between cells. Within these boundaries, different building strategies are implemented, such as anteroposterior or dorsoventral polarity and proximodistal differential cell growth.

Kimble: The analogy I drew between equivalence groups in *C. elegans* and compartments in *Drosophila* was based on the idea that, in each case, distinct groups of cells might be set aside during embryogenesis to establish specific regions of the animal's anatomy.

References

Doe CQ, Goodman CS 1985 The role of cell interactions and cell lineage in the determination of neuronal precursor cells. Dev Biol 111:206–219

Geissler EN, Ryan MA, Housman DE 1988 The dominant-white spotting (*W*) locus of the mouse encodes the *c*-kit proto-oncogene. Cell 55:185–192

Priess JR, Thomson JN 1987 Cellular interactions in early *C. elegans* embryos. Cell 48:241–250

Sulston JE, White JG 1980 Regulation and cell autonomy during post-embryonic development of *Caenorhabditis elegans*. Dev Biol 577–597

Technau GM, Campos-Ortega JA 1987 Cell autonomy of expression of neurogenic genes of *Drosophila melanogaster*. Proc Natl Acad Sci USA 84:4500–4504

Structure and function of the bithorax complex genes of *Drosophila*

Ana Busturia, Jordi Casanova*, Ernesto Sánchez-Herrero† and Ginés Morata

Centro de Biología Molecular, Facultad de Ciencias, Universidad Autónoma, Cantoblanco, Madrid 28049, Spain

Abstract. The bithorax complex consists of three genes, *Ubx*, *abd-A* and *Abd-B*, which together specify the characteristic development of parasegments 5 to 13 of *Drosophila*. These genes are structurally homologous; they are of similar size, are transcribed in the same orientation and they all have a homeobox near the 3′ end of their transcription unit. Genetic and molecular analyses of *Ubx* suggest that the gene contains one transcription unit encoding the protein products and at least three *cis*-regulatory regions. Two of these, *abx* and *bxd*, promote the activity of the *Ubx* transcription unit to the levels appropriate for parasegments 5 and 6, respectively. A third regulatory element, called *Cbx*-like, prevents the expression of *Ubx* anterior to parasegment 5. The gene *abd-A* is not as well known, but genetic and molecular studies indicate at least one *cis*-regulatory region downstream of the 3′ end of the transcription unit. In the gene *Abd-B* there are two distinct *trans*-acting elements, called *m* and *r*. The *m* element is a conventional homeotic function, which specifies the identity of parasegments 10 to 13. The *r* element is specific for parasegment 14 where it suppresses a number of homeotic functions (including *m*). Molecular analysis indicates that *Abd-B* contains two transcription units with a common 3′ end which correspond to the *m* and *r* elements defined genetically.

1989 Cellular basis of morphogenesis. Wiley, Chichester (Ciba Foundation Symposium 144) p 227–242

When describing the evolution of insects, it is currently accepted (Snodgrass 1935) that they derive from some form of annelid-like worms. The body of these ancestors was formed by a chain of homologous units—metameres—most of which were morphologically similar or identical. This original situation then altered; some metameres in the head, and probably some in the telson, fused, obscuring the initial metameric organization. The rest of the metameres were gradually modified so that each of them came to have a specific pattern, usually

**Present address*: Neurobiology and Behavior Center, Columbia University 722, West 168th Street, New York NY 10032, USA
†*Present address*: Genetics Department, University of Cambridge, Downing Street, Cambridge CB2 3EH, UK

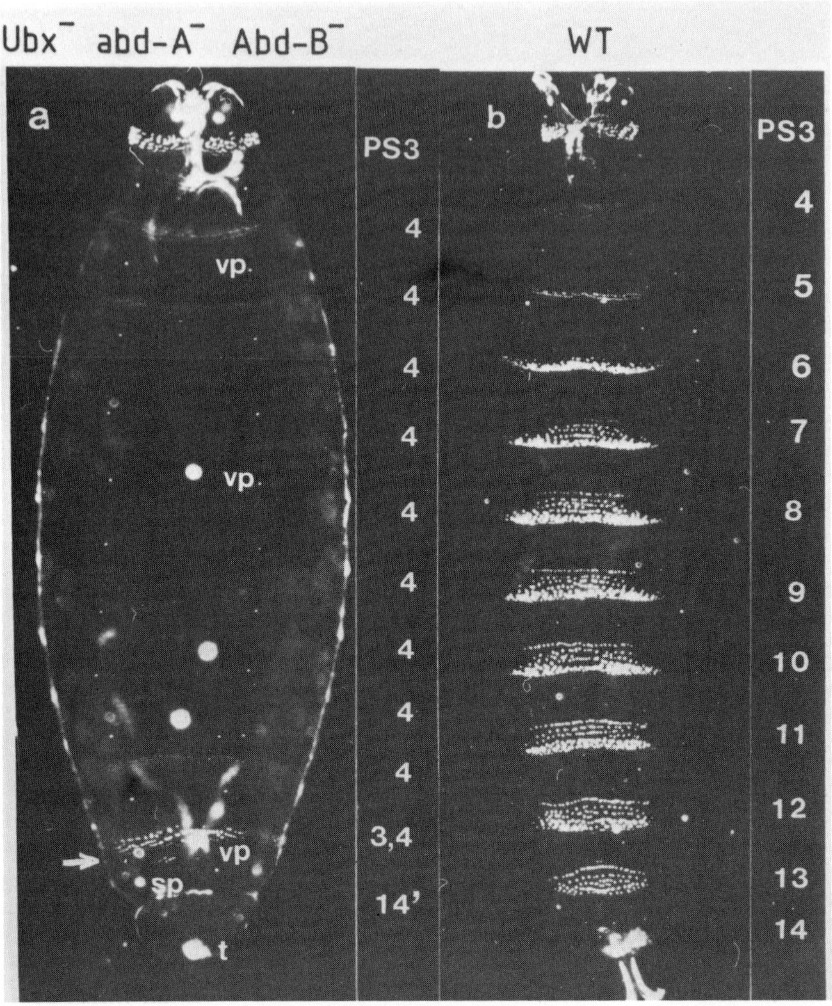

FIG. 1. Ventral aspect of larvae lacking (a) or containing (b) all the BX-C functions. In b) larvae, the head, parasegment 3 (PS3) and 4 are normal. Parasegments 5–12 develop as replicas of parasegment 4; their thoracic features are indicated by the presence of ventral pits (vp) in each metamere. The metamere corresponding to parasegment 13 (arrow) exhibits a mixture of the identities of parasegments 3 and 4. Parasegment 14 (14′) is abnormal and presents sclerotic plates (sp). More posterior structures like the tuft (t) are not affected. (Reproduced with permission from EMBO Journal.)

referred to as identity (thoracic or abdominal), of its own. This process of diversification must have been correlated with modification of the genetic machinery responsible for the generation of the new identities.

Although the nature and mechanisms of this genetic change are not understood, in *Drosophila* some of the relevant genes have been identified and characterized. These genes are called *homeotics*, and their principal feature is that their function is restricted to one or several metameres, where they play a critical role in establishing the identity. When a homeotic gene is inactivated by a mutation, a metamere or a group of metameres acquires the characteristic identity of another (Morata & Lawrence 1977).

Because the homeotic genes act only on specific metameres, these must be formed before homeotic function is activated. The process of metamerization is a complex one (see Nüsslein-Volhard & Roth, Struhl, this volume) and requires the cooperation of several categories of genes.

Many of the homeotics are clustered in two groups, the Antennapedia (ANT-C) and bithorax (BX-C) complexes. In this article we deal mainly with the structure and function of the BX-C genes.

The developmental role of the BX-C genes

At the time the BX-C genes become active (blastoderm stage, see review by Akam 1987), maternal and segmentation genes have fulfilled their function and the body is already subdivided into a chain of parasegments (Martínez-Arias & Lawrence 1985), each formed by a posterior and an anterior compartment. There are 15 of these parasegments.

The best way to assay for the function of the BX-C genes is to compare the phenotypes of embryos lacking the BX-C and normal ones (Fig. 1). In BX-C$^-$ embryos parasegments 1–4 develop normally, but parasegments 5–13 develop as a chain of reiterated parasegments 4. Parasegment 14 is abnormal but not transformed into 4 (Lewis 1978, Struhl 1981, Casanova et al 1987). This phenotype defines the developmental role of the BX-C: it determines the identity of parasegments 5–13. This region of the body is referred to as the bithorax domain and is constituted by 18 compartments which develop independently from the blastoderm stage (Lawrence & Morata 1977). It suggests that the BX-C is complex in both structure and function to account for all this diversity.

The number of BX-C genes

The classical model of BX-C genetic structure and function (Lewis 1978) proposed that the complex contains a large number of genes, at least one per metamere; a total of nine genes. This and related models (Struhl 1984, Hayes et al 1984) were based on the phenotypes of viable mutations, many of which affect predominantly one segment or parasegment.

However, detailed mutagenesis experiments (Sánchez-Herrero et al 1985 a,b, Tiong et al 1985) indicate that the BX-C contains only three complementation groups, i.e. three genes: Ultrabithorax (*Ubx*), abdominal-A (*abd-A*) and Abdominal-B (*Abd-B*). Mutations completely inactivating any of these genes are lethal, but mutant embryos are able to secrete larval cuticle so the identity of the larval segments can be studied. In every case, several metameres are affected. In Ubx^- embryos, parasegments 5 and 6 are transformed into 4 and there is a minor effect in parasegments 7–12. In $abd\text{-}A^-$ embryos, parasegments 7–9 are transformed to 6 and parasegments 10–13 are abnormal. Finally, in $Abd\text{-}B^-$ mutants, parasegments 10–13 develop as duplicates of 9 and parasegment 14 is abnormal.

These three genes appear to be completely independent of each other, although their specific products cooperate to determine the different metameric identities. The expression of any of them is not affected by the presence or functional state of the others (Casanova et al 1987). Furthermore, by using appropriate chromosomal rearrangements, it is possible to generate flies with normal segment patterns in which the genes *Ubx* (Struhl 1984) or *Abd-B* (Tiong et al 1987) are physically separated from the others. This indicates that the clustering of the BX-C genes is not a requisite for normal function. It probably reflects the evolution of the BX-C genes from a common ancestor (Lewis 1981), but it has little functional significance.

Structure of the BX-C genes

The three BX-C genes appear to be large (Bender et al 1983, Karch et al 1985) and complex in structure (Casanova et al 1985a, 1986, Karch et al 1985). One principal common feature is that each of them contains a homeobox (McGinnis et al 1984, Scott & Weiner 1984, Karch et al 1985). This and the observation that they are all transcribed in the same direction (Rowe & Akam 1988, Kuziora & McGinnis 1988) support a common evolutionary origin.

Ubx is the best known gene of the complex (Lewis 1963, Casanova et al 1985a). Genetic analysis indicates that there are three classes of mutations which eliminate totally or in part the function of *Ubx* (Casanova et al 1985a). The majority of the mutations belong to the *Ubx* class. These are lethal and transform parasegments 5 and 6 into 4. These mutations are breakpoints or small deletions and map anywhere in a region of DNA of about 75 kb (Bender et al 1983). Then there are two infrequent classes of mutations: the *abx* class includes the *abx* and *bx* mutations, which are equivalent (Casanova et al 1985a). They are viable and their effect is limited to parasegment 5, which develops like 4. There is little or no effect on parasegment 6. These mutations are caused by insertions (the majority) or deletions (Bender et al 1985) and are coincidental with the *Ubx* mutations in a region of approximately 25 kb (Fig. 2).

The *bxd* class includes the *bxd* and *pbx* mutations. They are viable but parasegment 6 exhibits the identity of 5. Parasegment 5 is not affected.

Mutations of this class are insertions, breakpoints or deletions, and all map distal to the *Ubx* and *abx* genes (Fig. 2).

Mutations of the *abx* class complement those of the *bxd* class (Lewis 1978). However, *Ubx* mutations fail to complement both the *abx* and *bxd* classes. Thus, though the *abx* and *bxd* functions can be considered as independent, they both need to be linked in *cis* with *Ubx*; there is a requirement for *abx* and *Ubx* to establish the identity of parasegment 5 and of *bxd* and *Ubx* for the identity of parasegment 6.

In addition, there is a heterogeneous group of dominant mutations that modify the expression of *Ubx*: the *Cbx*-like group. They cause an ectopic expression of *Ubx*, which is normally manifested as a partial transformation of wing into haltere. This affects the anterior wing compartment which belongs to parasegment 4, anterior to the normal domain of *Ubx* expression. Each of these mutations is probably a unique event altering the spatial regulation of *Ubx* by a different mechanism, as suggested by their disparate location within the gene (Bender et al 1983). However, they pinpoint regions of the gene involved in defining *Ubx* spatial constraints (see below).

Consideration of the molecular structure of the gene and the nature and location of its mutations helps to identify the different functional components of *Ubx*. As shown in Fig. 2, there is a large transcription unit extending for 75 kb (Beachy et al 1985, Hogness et al 1985) in the region where the *Ubx* and *abx* mutations map. This is called the *Ubx* transcription unit and is responsible for the morphogenetic functions of *Ubx*. The elimination or inactivation of the protein products encoded by this transcription unit results in completely null phenotypes (Beachy et al 1985, Wienzierl et al 1987). The spatial and temporal distributions of these *Ubx* protein products have been extensively studied (White & Wilcox 1984, Akam & Martínez-Arias 1985). What is relevant here is that the amounts of *Ubx* product in parasegments 5 and 6 are very different, being higher in the latter.

The *bxd* mutations define a DNA region located upstream of the *Ubx* transcription unit (Fig. 2). This is about 30 kb long (Bender et al 1985) and although it gives rise to several transcripts, these seem to have little or no functional role (Hogness et al 1985). Therefore we have ignored the *bxd* transcription unit when considering the functional aspects of *Ubx*. The *bxd* mutations reduce the level of expression of the *Ubx* transcription unit in parasegment 6, probably to that in parasegment 5 (Cabrera et al 1985). Normal parasegment 5 is not affected. The mechanism of the *cis* interaction between *bxd* DNA and the *Ubx* transcription unit is not known, but its role is clearly regulatory: to promote *Ubx* transcription to the level appropriate to parasegment 6.

The *abx* mutations map in the long intron of the *Ubx* transcription unit (Fig. 2) and are small deletions or more frequently insertions (Peifer & Bender 1986).

These mutations reduce or eliminate *Ubx* products in parasegment 5, but have no effect in parasegment 6. Therefore the transcriptional capacity of the *Ubx* unit is not affected by *abx* mutations. This suggests that, as in the case of *bxd*, the role of the *abx* region is regulatory: to promote the level of transcription of *Ubx* to that appropriate to parasegment 5. Both genetic (Casanova et al 1985a, 1987) and molecular studies (Peifer & Bender 1986, Bender et al 1987) point to the same conclusion.

There may be several other regulatory components in the *Ubx* gene, but so far this has been little explored. These are indicated by the *Cbx*-like mutations mentioned above. Cbx^{3} and Cbx^{Twt} are inversions with breakpoints located just downstream of the *Ubx* transcription unit, which is not altered by these mutations (Bender et al 1983). It has been shown that the phenotype of Cbx^{3} is due to ectopic expression of the *Ubx* transcription unit (White & Akam 1985). These observations suggest the existence of a negative regulatory sequence preventing the expression of *Ubx* anterior to parasegment 5 (Fig. 2). The interpretation of the phenotype of the three other members of the *Cbx*-like group is less clear. Cbx^{1}, Cbx^{2} and *Hm* are all rearrangements affecting the *Ubx* transcription unit (Cbx^{1}) or the *bxd* region, and are associated with gain as well as loss of some bithorax functions (Lewis 1982, Casanova et al 1985b). In the rearrangements Cbx^{2} and *Hm*, there is the possibility that expression of the *Ubx* transcription unit may be directed by the new DNA sequences to which it is fused (Bender et al 1985). This may produce an ectopic expression and result in a *Cbx*-like phenotype. We can therefore consider the *Ubx* gene to be formed by one transcription unit and at least three *cis*-regulatory regions (Fig. 2); *abx* and *bxd* function as positive regulators, promoting transcription of the gene in parasegments 5 and 6, respectively. The region proximal to the 3′ end would act differently, preventing the expression of the gene anterior to parasegment 5.

The gene *abd-A* is the least known of the BX-C. It is about 60 kb long (Karch et al 1985). There is a transcription unit with exons spread over more than 20 kb (Rowe & Akam 1988). At least two mutations, *iab-2*^{K} and *abd-A*^{S3} have been located downstream of the transcription unit. These mutations are viable and, unlike other *abd-A* mutations, affect only the second abdominal segment, which is partially transformed into the first. The effect of these mutations on the distribution of *abd-A* protein is not yet known, but by comparison with the *abx* and *bxd* mutations of *Ubx*, it is likely that they reduce *abd-A* expression in the second abdominal segment. This would indicate the existence of a *cis*-regulatory region downstream of the transcription unit.

There are recent results that suggest a functional and perhaps structural homology between *Ubx* and *abd-A*. A mutation called *Cl* is a partial deletion of both *Ubx* and *abd-A* genes (Rowe & Akam 1988). Since the two genes are transcribed in the same direction and the two breaks of *Cl* are in introns, the result is a hybrid gene with an uninterrupted open reading frame having the

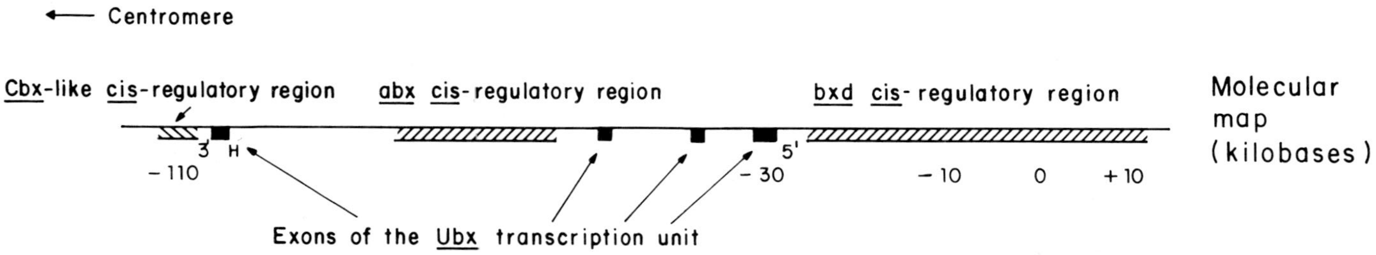

FIG. 2. Organization of the *Ubx* gene. There is one transcription unit spanning about 75 kb from the 5′ to the 3′ end, which encodes the protein products executing *Ubx* functions. All the *Ubx* mutations that completely inactivate the gene map within the transcription unit. The *abx cis*-regulatory region is defined by the position of the *abx* and *bx* mutations and lies between the second microexon and the 3′ exon. The *bxd cis*-regulatory region is defined by the location of the *bxd* and *pbx* mutations and extends more than 30 kb upstream of the transcription unit. The *Cbx*-like regulatory region is located just downstream of the transcription unit and is defined by the mutations Cbx^3 and Cbx^{Twt}. The nature of the regulatory role of the latter is different from the others (see text).

5′ end of *abd-A* and the 3′ end of *Ubx* (Rowe & Akam 1988). This hybrid product is functional and determines a mixed thoracic and abdominal identity (Casanova et al 1988). The hybrid gene contains the *abx* regulatory region intact and this results in the *abd-A* promoter initiating transcription in parasegment 5 and 6 where *abd-A* is normally not active (Casanova et al 1988). This suggests that the *abx* region can regulate the activity of the *abd-A* promoter, in turn suggesting a functional homology of *Ubx* and *abd-A*.

The *Abd-B* gene possesses some features that make it quite different from the other two BX-C genes. Genetic analysis has shown that the gene contains two distinct elements, called *m* and *r* (Casanova et al 1986). The *m* element is a conventional homeotic function that specifies the identities of parasegments 10, 11, 12 and 13. One class of *Abd-B* mutations, (m^-r^+) affects only this function. The *r* element is specific to parasegment 14 where it suppresses the function of *m* and other head-forming gene(s). There is another class of *Abd-B* mutations (m^+r^-) that inactivates only this function (Casanova et al 1986). A third class (m^-r^-) inactivates both functions at once and therefore fails to complement either of the two previous classes.

The molecular analysis of the *Abd-B* gene is not yet complete but there is very good agreement between the results of genetic and molecular studies. The gene is about 100 kb long (Karch et al 1985). The m^-r^+ and the m^+r^- mutations map to different regions of the gene, suggesting specific functions within it. There are two classes of RNAs encoded by *Abd-B* (Sánchez-Herrero & Crosby 1988, DiLorenzi et al 1988, Kuziora & McGinnis 1988), according to their spatial expression as defined by *in situ* hybridization; one (4,3/4,8 kb) appears in parasegments 10–13 and the other (3,2/3,7 kb) in parasegment 14 (and possibly part of 15). This correlates very well with the phenotypic domains of *m* and *r* mutations (Casanova et al 1986) and suggests that these transcripts encode the *m* and *r* functions, respectively. The evidence for the 4,3/4,8 kb class has been supported by the *in situ* hybridization studies performed by Kuziora & McGinnis (1988) in Pc^- embryos. In those embryos, thoracic and abdominal parasegments acquire the identity of parasegment 13 (Lewis 1978, Duncan & Lewis 1982) and this transformation depends exclusively on having an intact *m* element. ($Pc^-m^+r^+$ and $Pc^-m^+r^-$ embryos exhibit the same phenotype in the parasegment 1–13 region, Casanova et al 1986.) Kuziora & McGinnis have shown that only the α-class of *Abd*-B transcripts (corresponding to the 4,3/4,8 kb class of Sánchez-Herrero & Crosby 1988) is ectopically expressed early enough to affect the identity of larval parasegments. The two classes of transcripts have exons in common in the 3′ ends which map around the position + 153 kb of the BX-C DNA map (Karch et al 1985). The 5′ ends of the transcripts have not yet been located, but that of *m* must be proximal to the position + 170 kb while that of *r* is further than + 180 kb (Sánchez-Herrero & Crosby 1988).

The structure and disposition of the transcripts explains the existence of three classes of mutations in *Abd-B* (Casanova et al 1986). The m^+r^- mutations map distal to +170 kb and therefore affect only the *r* transcripts. The m^-r^+ mutations affect the *m* transcripts or their expression. One of these mutations, *iab-7*^{D14}, eliminates a specific exon of the *m* transcripts (Sánchez-Herrero & Crosby 1988, DiLorenzi et al 1988, Kuziora & McGinnis 1988) but leaves *r* transcripts unaffected. However, many of the m^-r^+ mutations map downstream of the 3′ end of the transcription units in the interval defined by the positions +95 kb to +140 kb in the DNA map (Karch et al 1985). This suggests the presence of a *cis*-regulatory region necessary for the proper expression of *m* products in parasegments 10–13. Finally, m^-r^- mutations would map to the region common to the two classes of RNAs. Only one mutation of this class has been mapped, *iab-8*^{65}, which is located at +165 kb (Karch et al 1985). This mutation is atypical in that, unlike the other m^-r^- alleles, the *m* function is defective but also abnormally expressed in the *Abd-B* domain. This may mean that it affects some regulatory region located near the 5′ end of the *m* transcription unit.

Thus, unlike *Ubx* and *abd-A*, the *Abd-B* gene contains two distinct, but structurally related, products. The *m* product is functionally homologous to the *Ubx* and *abd-A* products. Its spatial expression, like that of *Ubx* and *abd-A*, is probably controlled by *cis*-regulatory sequences. The *r* product is special in that its function is principally regulatory and it is active in a region, parasegment 14, where there is no other BX-C activity.

Functional characteristics of the BX-C genes

The finding that the BX-C contains only three genes (Sánchez-Herrero et al 1985a,b, Tiong et al 1985) and the phenotype of null mutations for each of them suggested a new way to examine the functional aspects of the complex. There are three distinct major functions which overlap in certain body regions. The identities of parasegments 5 and 6 are determined exclusively by the gene *Ubx*. Parasegments 7, 8 and 9 require both *Ubx* and *abd-A*, whereas the identities of parasegments 10–13 are specified by *Ubx*, *abd-A* and *Abd-B* (see Morata et al 1986 for a review). This model is strongly supported by the phenotype of triply mutant embryos *Ubx*$^-$ *abd-A*$^-$ *Abd-B*$^-$ (Casanova et al 1987), which is identical to that of embryos lacking the entire BX-C, more than 300 kb (Bender et al 1983, Karch et al 1985).

The phenotype of the different mutant combinations is shown schematically in Table 1, which illustrates functional aspects of and interactions among the BX-C genes. In the case of combination 2, in which *Ubx* is the only active BX-C gene, the mutant embryos develop with a normal set of parasegments 1–6 followed by a chain of reiterated parasegments 6. Thus the morphogenetic ability

TABLE 1 Sense and nonsense BX-C codewords

Combination	*Identity of parasegments*										
1) Ubx$^-$ abd-A$^-$ Abd-B$^-$	1–4	4	4	4	4	4	4	4	4	3–4	14′
2) Ubx$^+$ abd-A$^-$ Abd-B$^-$	1–4	5	6	6	6	6	6	6	6	6	14′
3) Ubx$^+$ abd-A$^+$ Abd-B$^-$	1–4	5	6	7	8	9	9	9	9	9	14′
4) Ubx$^+$ abd-A$^+$ Abd-B$^+$	1–4	5	6	7	8	9	10	11	12	13	14
5) Ubx$^-$ abd-A$^+$ Abd-B$^+$	1–4	4	4	mp	mp	mp	mp	mp	mp	mp	14
6) Ubx$^+$ abd-A$^-$ Abd-B$^+$	1–4	5	6	6	6	6	mp	mp	mp	mp	14
7) Ubx$^-$ abd-A$^+$ Abd-B$^-$	1–4	4	4	mp	mp	mp	mp	mp	mp	mp	14′
8) Ubx$^-$ abd-A$^-$ Abd-B$^+$	1–4	4	4	4	4	4	mp	mp	mp	mp	14

of *Ubx* is limited to the generation of identities 5 and 6. In normal parasegments 5 and 6, *Ubx* is the only BX-C gene active and therefore the functional state of *abd-A* and *Abd-B* is immaterial in those parasegments (compare combinations 2,3 and 6 in Fig. 3). The combined functions of *Ubx* and *abd-A* (combination 3) determine the identities of parasegments 7,8 and 9. The activity of *abd-A* establishes the distinction of identities 7,8 and 9 with respect to 6 (compare 2 and 3), whereas *Ubx* is necessary to provide a basal identity 6 on which *abd-A* can act (compare combinations 4 and 5).

Similarly, the identities of parasegments 10–13 are determined by the combined activities of the three genes (combination 4). *Abd-B* specifies the distinct identities of 10–13, but requires a ground identity 9 determined by *Ubx* and *abd-A* (compare combinations 3,5,6 and 8).

Parasegment 14 is special in that this combinatorial mechanism does not operate there. Only *Abd-B* is active in this parasegment and it does not perform a conventional homeotic function.

Thus, excepting parasegment 14, the entire BX-C domain, parasegments 5–13, can be subdivided into three subdomains, according to the local requirements for BX-C genes: the *Ubx* domain, constituted by parasegments 5 and 6, is defined by the + − − formula (codeword). The *abd-A* domain, parasegments 7,8 and 9, is defined by the + + − codeword. Finally the *Abd-B* domain, parasegments 10–13, is defined by the + + + codeword. These three combinations (plus the − − − determining identity 4) can always be interpreted correctly by the cells ('sense' codewords, Struhl 1982), so that mutant combinations (1,2 and 3) producing them develop normal identities in inappropriate metameres. Other mutant combinations (5,6,7 and 8) create unusual 'nonsense' codewords, − + +, + − +, − + −, − − +, which cannot be correctly interpreted by the cells and produce mixed patterns (mp, Table 1) unlike any other found in the fly.

Acknowledgements

We thank the Spanish Comision Asesora de Investigación Científica y Técnica (CAICYT) and EMBO for financial support.

References

Akam M 1987 The molecular basis for metameric pattern in the Drosophila embryo. Development 101:1–22

Akam ME, Martínez-Arias A 1985 The distribution of Ultrabithorax transcripts in Drosophila embryos. EMBO (Eur Mol Biol Organ) J:4:1689–1700

Beachy PA, Helfand S, Hogness DS 1985 Segmental distribution of bithorax complex proteins during *Drosophila* development. Nature (Lond) 313:545–551

Bender W, Akam M, Karch F et al 1983 Molecular genetics of the bithorax complex in Drosophila melanogaster. Science (Wash DC) 221:23–29

Bender W, Weiffenbach B, Karch F, Peifer M 1985 Domains of cis-interaction in the bithorax complex. Cold Spring Harbor Symp Quant Biol 50:173–180

Bender M, Turner FR, Kaufman TC 1987 A developmental genetic analysis of the gene regulator of postbithorax in Drosophila melanogaster. Dev Biol 119:418–432

Cabrera CV, Botas J, García-Bellido A 1985 Distribution of Ultrabithorax proteins in mutants of Drosophila bithorax complex and its transregulatory genes. Nature (Lond) 318:442–456

Casanova J, Sánchez-Herrero E, Morata G 1985a Prothoracic transformation and functional structure of the Ultrabithorax gene of Drosophila. Cell 42:663–669

Casanova J, Sánchez-Herrero E, Morata G 1985b Contrabithorax and the control of spatial expression of the bithorax complex genes of Drosophila. J Embryol Exp Morphol 90:179–196

Casanova J, Sánchez-Herrero E, Morata G 1986 Identification and characterization of a parasegment specific regulatory element of the Abdominal-B gene of Drosophila. Cell 47:627–636

Casanova J, Sánchez-Herrero E, Busturia A, Morata G 1987 Double and triple mutant combinations of the bithorax complex of Drosophila. EMBO (Eur Mol Biol Organ) J 6:3103–3109

Casanova J, Sánchez-Herrero E, Morata G 1988 Developmental analysis of a hybrid gene composed of parts of the *Ubx* and *abd-A* genes of Drosophila. EMBO (Eur Mol Biol Organ) J 7:1097–1105

DiLorenzi M, Ali N, Saari G, Henry C, Wilcox M, Bienz M 1988 Evidence that the Abdominal-B *r* element function is conferred by a transregulatory homeoprotein. EMBO (Eur Mol Biol Organ) J 7:3223–3232

Duncan I, Lewis EB 1982 Genetic control of body segment differentiation in Drosophila. In: Subtelny S (ed) Developmental order: its origin and regulation. Alan R Liss, New York, p 533–554

Hayes PH, Sato T, Denell RE 1984 Homoeosis in Drosophila: the Ultrabithorax larval syndrome. Proc Natl Acad Sci USA 81:545–549

Hogness DS, Lipshitz HD, Beachy PA et al 1985 Regulation and products of the *Ubx* domain of the bithorax complex. Cold Spring Harbor Symp Quant Biol 50:181–194

Karch F, Weiffenbach B, Peifer M et al 1985 The abdominal region of the bithorax complex. Cell 43:81–93

Kuziora MA, McGinnis W 1988 Different transcripts of the Drosophila *Abd-B* gene correlate with distinct genetic sub-functions. EMBO (Eur Mol Biol Organ) J 17:3233–3245

Lawrence PA, Morata G 1977 The early development of mesothoracic compartments in Drosophila. Dev Biol 56:40–51

Lewis EB 1978 A gene complex controlling segmentation in Drosophila. Nature (Lond) 276:565–570

Lewis EB 1981 Developmental genetics of the bithorax complex in Drosophila. In: Brown DD, Fox CF (eds) Developmental biology using purified genes. ICN–UCLA Symp Molec Cell Biol, New York Academic, New York 189–208

Lewis EB 1982 Control of body segment differentiation in Drosophila by the bithorax gene complex. In: Embryonic development Part A: genetic aspects. Alan R Liss, New York p 269–288

Martínez-Arias A, Lawrence PA 1985 Parasegments and compartments in the Drosophila embryo. Nature (Lond) 313:639–642

McGinnis W, Levine MS, Hafen E, Kuroiwa A, Gehring WJ 1984 A conserved DNA sequence found in homeotic genes of Drosophila Antennapedia and bithorax complexes. Nature (Lond) 308:428–433

Morata G, Lawrence PA 1977 Homeotic genes, compartments and cell determination in Drosophila. Nature (Lond) 265:211–216

Morata G, Sánchez-Herrero E, Casanova J 1986 The bithorax complex of Drosophila: an overview. Cell Differ 18:67–78

Peifer M, Bender W 1986 The anterobithorax and bithorax mutations of the bithorax complex. EMBO (Eur Mol Biol Organ) J 5:2293–2303

Rowe A, Akam M 1988 The structure and expression of a hybrid homeotic gene. EMBO (Eur Mol Biol Organ) J 7:1107–1114

Sánchez-Herrero E, Casanova J, Morata G 1985a Anatomy and function of the bithorax complex of Drosophila. Cold Spring Harbor Symp Quant Biol 50:165–172

Sánchez-Herrero E, Vernos I, Marco R, Morata G 1985b Genetic organization of Drosophila bithorax complex. Nature (Lond) 313:108–113

Sánchez-Herrero E, Crosby MA 1988 The Abdominal-B gene of Drosophila melanogaster: overlapping transcripts exhibit two different spatial distributions. EMBO (Eur Mol Biol Organ) J 7:2163–2173

Scott MP, Weiner AS 1984 Structural relationships among genes that control development: sequence homology between the *Antennapedia*, *Ultrabithorax* and *fushi-tarazu* loci of *Drosophila*. Proc Natl Acad Sci USA 81:4115–4119

Snodgrass RE 1935 Principles of Insect Morphology. McGraw-Hill, New York

Struhl G 1981 A gene product required for correct initiation of segmental determination in Drosophila. Nature (Lond) 293:36–41

Struhl G 1982 Genes controlling segmental specifications in the Drosophila thorax. Proc Natl Acad Sci USA 79:7380–7384

Struhl G 1984 Splitting the bithorax complex of Drosophila. Nature (Lond) 308:454–457

Tiong S, Bone LM, Whittle JRS 1985 Recessive lethal mutations within the bithorax complex in Drosophila. Mol Gen Genet 200:335–342

Tiong S, Whittle J, Gibbin M 1987 Chromosomal continuity in the abdominal region of the bithorax complex of Drosophila is not essential for its contribution to metameric identity. Development 101:135–142

Weinzierl R, Axton JM, Ghysen A, Akam M 1987 Ultrabithorax mutations in constant and variable regions of the protein coding sequence. Genes & Development 1:386–397

White RAH, Wilcox M 1984 Proteins products of the bithorax complex in Drosophila. Cell 39:163–171

White RAH, Akam ME 1985 Contrabithorax mutations cause inappropriate expression of Ultrabithorax products in Drosophila. Nature (Lond) 318:567–569

DISCUSSION

Martinez-Arias: You said that in the heat shock *Ubx* experiment, the reason there is no effect in the abdominal region is because *abd-A* represses *Ubx*. It is difficult to see how the *abd-A* gene can repress the heat shock *Ubx* construct, because there are no *Ubx* control elements on which it could act.

Morata: The issue is, what is the relative role of each *bithorax* gene in the patterning process. With the heat shock system, *Ubx* can be expressed everywhere and one can see where it has an effect and where it does not. In the abdominal segments, we see no morphological consequences of over-expressing *Ubx*. This is probably because in the abdomen *abd-A* and *Abd-B*, which play a dominant role with respect to *Ubx*, are expressed. In experiments with *Antennapedia* under heat shock control, Gibson & Gehring (1988) obtained a similar result; the over-expression of *Antennapedia* in the metathorax and abdomen has little or no effect, presumably because of the dominant role of the *bithorax* genes.

Martinez-Arias: Do you or Gary have any ideas as to how?

Struhl: I think the confusion comes from the use of the term suppressed. There are two possible mechanisms, both of which you see normally, apparently. One is that, for example, an *abd-A* or *Abd-B* gene can suppress *Ubx* by directly inhibiting its transcription. I think from what you have described of the *r* element, it is possible there that the suppression is of a slightly different type, which is that the *r* protein interferes with the function of the *Ubx* or *abd-A* or *Abd-B* gene products. Both types of mechanism can occur at the same time. In the heat shock experiment, you are affecting the downstream function of the *Ubx* protein, whereas in wild-type *abd-A* and *Abd-B* change the pattern of expression of *Ubx* by affecting the synthesis of the *Ubx* protein. The confusion arises I think because you use one word to describe both types of event.

Davidson: But aren't you adding to a possible confusion? These are proteins which can interact on the control regions of the DNA. One protein may be competing with another protein for a *cis*-regulatory site.

Struhl: That's true, except that in the heat shock experiment, *Ubx* protein is present at high levels throughout.

Davidson: That excess *Ubx* protein and the *r* protein could be competing for a binding site of a factor which is normally present abdominally.

Morata: In the case of the *r* product, there is evidence that other homeotic gene products are being suppressed. For example, in parasegment 14 of *esc* embryos, the *r* product will eliminate *Antennapedia* protein. The *Ubx* protein is also eliminated in the same parasegment (Casanova & White 1987). We don't know what happens in this respect in the heat shock experiment.

In *esc*$^-$ embryos, which contain no *bithorax* function but do contain the *Antennapedia* gene, if they are *r*$^-$, the whole central nervous system labels very strongly with anti-*Antennapedia* antibodies. But if the embryo is *r*$^+$, the

Antennapedia protein is essentially eliminated in parasegment 14 (J. Wirz et al, unpublished work). The regulation is at the level of the protein.

Garcia-Bellido: It is difficult to see how such a large amount of transcription of *Ubx* can be prevented. An alternative view is that it is very difficult to get deformation in the adult by heat shock, because the system has a lot of inertia in its genetic interactions. Even if a lot of protein with a short half-life is produced, the system recovers from that. Obviously, the prevalent signal does not initiate a cascade.

I have difficulty in understanding how, in the same population of cells, products of different homeotic genes, selector genes, can co-exist if the posterior-acting selectors do repress the transcription of the anterior-acting ones. How is it that the anlage is a heterogeneous population in terms of the amount of gene expression per cell? I propose that to account for this there have to be extra control elements in the very system that the homeotic genes are specifying. I propose that there are genes downstream, which regulate the expression of the upstream genes. (In our terminology, the reactivator genes modulate the expression of the selector genes.)

That brings me to the problem of the expression of selector genes in parasegments in the embryo and in segments in the adult. It may be that this is due to new genetic clues to which the selector genes respond. For example, you can provoke *Contrabithorax* (*Cbx*) mutations, which are specific to parasegment 4, to express *Ubx-P* in the head by simply changing, by independent mutations, the growth of the head anlage. If you make a double mutant of *opthalmoptera*, which produces wings in the head, and *Cbx*, the mesothoracic cell wing in the head is converted to a haltere. You have to deduce that the expression of *Ubx* has been elicited in the head, which is confirmed by the use of anti-*Ubx-P* antibodies (J. Botas, unpublished work). That would explain the transgression of the parasegment, the growing together of cells expressing different homeotic genes, and the observation that although you get large amounts of *Ubx* protein in the abdominal segments, you still don't get the pattern you would expect if *Ubx* were the only selector gene active there.

Wolpert: It comes as a bit of a shock that within one segment, gene expression in the individual cells is heterogeneous. The global pictures of the activity of the homeotic genes really don't give a fair picture of what is happening on a cell to cell basis.

Lawrence: Some cells are definitely expressing more than others—whether within the particular domain or parasegment in question there are cells that express none is not certain.

Struhl: It is a heterogeneous but stereotypic pattern.

Morata: Concerning the expression of heat shock genes in the adult, you can have expression of *Antennapedia* in adult cells and that causes ectopic legs.

Garcia-Bellido: If you do that with a similar *Ubx* construct, the same is true;

it is extremely difficult to get homeotic transformation by heat shock pulses at the end of development.

Morata: That is the point I wanted to make. We have not tried hard enough to obtain transformation of adults with heat shock, so I don't really know whether it is difficult or not. But we have to consider that the expression of homeotic genes in imaginal cells lasts a long time, these genes are active through the whole of larval development. You cannot heat shock a larva for long enough perhaps to have the required effect.

The trick may be to give pulses of heat shock at different times. I agree that some of the *Antennapedia* legs one gets with heat shock are partial, but others are more complete. So one can transform adult cells as well as larval cells.

Lawrence: One of the points you made quite strongly is this transformation of the unit of action of homeotic genes from parasegment to segment. I don't think you are using quite the same criteria as you might insist upon in early development. For instance, in early development you want to show that the gene is both expressed and required in parasegmental units. When you try to convince us that this has changed to segments later on, you don't use those same criteria. You look at mutant phenotypes, which are difficult, and often use weak alleles. Could you give a good example of expression of and requirement for one of these homeotic genes at the segmental level later in development?

Morata: I agree that there is no conclusive proof of this transition. I believe there is circumstantial evidence in a number of cases. There are some significant data, for example, from *Antennapedia* dominant mutations. These are gain-of-function mutants, some of them are very strong. In these cases, what you observe is an ectopic leg in the place of the antenna. This leg contains an anterior and a posterior compartment of mesothoracic type, that is, the homeotic transformation appears to be segmental, not parasegmental. In the experiments of Gibson & Gehring (1988), ectopic expression of *Antennapedia* is induced by the heat shock system. The result is an antennal leg with a mesothoracic pattern—there you have genuine ectopic expression of the normal *Antennapedia* product and the effect looks segmental.

Lawrence: But that's only suggestive, because you don't know about the other genes.

Morata: I agree, I am not convinced myself but I like the idea. Consider the *engrailed*[1] mutation, although I realize this is a weak mutation. You can interpret the *engrailed*[1] wings as having been formed from an anterior compartment (parasegment 4) with normal pattern and a posterior compartment (parasegment 5) which, because of the mutation, also develops an anterior pattern. That is, in the absence of *engrailed*, both compartments develop the same pattern, even though they belong to different parasegments. This observation suggests that the set of selector genes specifying the pattern of the entire wing, which is a segmental structure, is the same. The implication is that these genes are deployed segmentally in adult structures.

Martinez-Arias: I would like to give a concrete example, which is what prompted us to notice these changes. Gary Struhl will recall that a few years ago we had an argument about *Sex combs reduced* (*Scr*). He had shown quite conclusively that in the adult the requirement for *Src* genetically was very clearly delimited by segmental boundaries. The interpretation of the embryonic phenotypes was open for discussion. We looked at the expression of *Scr* in the embryo with DNA probes and later with antibodies. It is quite clear that in the embryo very early on you see a well defined parasegmental domain that slowly, as development proceeds, is shifted—first, when the mouth parts appear, towards segmental, and then in the discs it is clearly segmental.

Lawrence: I accept your evidence, but it is important to decide whether the gene is working in segmental units or in sets of compartments. It might be a set of compartments which just happens to be limited by a segmental boundary, at least at one end.

References

Casanova J, White RAH 1987 Trans regulatory functions in the abdominal-B gene of the bithorax complex. Development 101:117–122

Gibson G, Gehring WJ 1988 Head and thoracic transformations caused by ectopic expression of antennapedia during Drosophila development. Development 102:657–675

Cadherin-mediated specific cell adhesion and animal morphogenesis

Masatoshi Takeichi, Kohei Hatta, Akinao Nose, Akira Nagafuchi and Mayumi Matsunaga

Department of Biophysics, Faculty of Science, Kyoto University, Kitashirakawa, Sakyo-ku, Kyoto 606, Japan

Abstract. Cadherins are a family of transmembrane glycoproteins which are responsible for Ca^{2+}-dependent cell–cell adhesion. Each member of the family displays a unique pattern of tissue distribution. The expression of cadherin molecules in an embryo is spatiotemporally regulated so as to be associated with a variety of morphogenetic events. Antibodies against cadherins perturb the morphogenesis of tissues, indicating their importance in tissue formation. When cells were transfected with cDNAs encoding cadherins, they expressed the exogenous cadherin molecules and these were functional. Using these transfected cells, we tested whether each member of the cadherin family has a binding specificity, and found that cells preferentially adhere to those expressing the identical cadherin type. The cadherin-mediated specific adhesion also occurred in the attachment of neurites to the surface of other cells. These observations suggest that cadherins are crucial for the selective adhesiveness of cells, and thus for the embryonic morphogenetic processes in which specific adhesive interactions of cells are involved.

1989 Cellular basis of morphogenesis. Wiley, Chichester (Ciba Foundation Symposium 144) p 243–254

Animal cells are known to recognize particular cell types when making intercellular connections (Moscona & Moscona 1952, Townes & Holtfreter 1955, Roth & Weston 1967). A key factor in the recognition process is thought to be the specific adhesiveness of cells, the property by which cells preferentially adhere to particular cells. To elucidate the molecular basis of the specific cell adhesion is thus an important issue in understanding animal morphogenesis.

Many kinds of cell adhesion molecules have been identified: three molecular families, the immunoglobulin superfamily (Cunningham et al 1987), the integrin superfamily (Hynes 1987) and the cadherin family (Takeichi 1988), have been best characterized at the molecular level. Most members of these families are not ubiquitous, but are expressed in restricted cell types or restricted regions of an organ, and interact with specific receptors or ligands. Any of these molecules, therefore, could be a candidate for the factor regulating specific cell adhesion. In this paper, I focus on the cadherin family and discuss its role in specific cell adhesion and animal morphogenesis.

Cadherins

Cadherins are a group of functionally and structurally related glycoproteins which are responsible for Ca^{2+}-dependent cell–cell adhesion (Takeichi 1988). They are divided into different types such as E-, P- and N-cadherin, and L-CAM (liver cell adhesion molecule). These types differ in their tissue distribution as well as in their immunological specificities.

We and others isolated cDNA clones encoding the mouse E-cadherin (Nagafuchi et al 1987, Ringwald et al 1987) and P-cadherin (Nose et al 1987), and also the chicken N-cadherin (Hatta et al 1988) and L-CAM (Gallin et al 1987). Predicted amino acid sequences showed that they have a similar primary structure of 723–748 amino acids, all with a single putative transmembrane domain (Fig. 1). The molecules are identical in 45–58% of amino acids. The sequence of the cytoplasmic domain is highly conserved; that of the N-terminal region of the extracellular domain is also well conserved, suggesting the importance of these regions for the function of cadherins (Hatta et al 1988).

We introduced cadherin cDNAs, attached to the SV40 or β-actin promoter, into L cells which have little cadherin activity (Nagafuchi et al 1987). The cDNA-transfected L cells expressed functional cadherin molecules, acquiring the Ca^{2+}-dependent aggregating activity. This expression of cadherins induced morphological changes in L cell colonies. Normal L cells formed dispersed colonies, showing a typical fibroblastic morphology, while the transfected cells formed epithelioid colonies in which cells were connected to each other. These results provided direct evidence that cadherins are crucial for tight intercellular connections.

Cadherins in histogenesis

Inhibition of cadherins with antibodies disrupts cell–cell adhesion in monolayer cell cultures (Yoshida-Noro et al 1984). In order to investigate if cadherins are also essential for maintaining the three-dimensional tissue structures, we tested

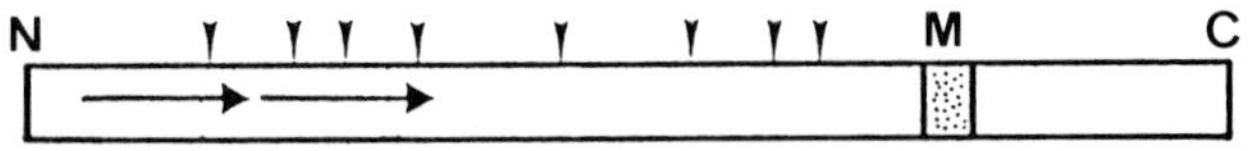

FIG. 1. Structure of the chicken N-cadherin deduced from the cDNA sequence. The extracellular domain contains homologous repeats, of which the major repeats are shown by arrows. Arrow heads represent the N-linked glycosylation sites. N, the amino terminus; M, membrane-spanning region; C, carboxy terminus. All other cadherins show a similar primary structure, although the glycosylation sites vary between different cadherin types.

the effect of antibodies against cadherins on the morphogenesis of various embryonic tissues using organ cultures.

The nervous system expresses N-cadherin. We studied the role of N-cadherin in the histogenesis of chicken neural retina (Matsunaga et al 1988a). In the retina of early embryos, N-cadherin is distributed almost evenly throughout the tissue. As differentiation proceeds, N-cadherin becomes localized gradually in particular cell layers. The amount of N-cadherin is first reduced in the nuclear layers; it then eventually disappears from most parts of the retina, except the outer limiting membrane. When the undifferentiated retinas of early embryos were incubated with Fab fragments of a polyclonal antibody to N-cadherin, they were dissociated into cell clusters. Retinas from older embryos were not dissociated by the Fab fragments, but their morphogenesis was severely affected, for example, the photoreceptor layer was transformed into a rosette structure. Thus, N-cadherin is essential for maintaining the overall structure of early retina and for stabilizing local regions of older retina.

We also tested the effect of antibodies against cadherins on the morphogenesis of lung (Hirai et al 1989). Epithelial cells of the mouse embryonic lung express both E- and P-cadherin, although the amount of the latter diminishes by the new-born stage. We explanted lung primordia onto a membrane filter and looked at the effects of monoclonal antibodies against E- and P-cadherins on their morphogenesis. In a control medium, the lung explants formed typical bronchial trees. In the presence of anti-E-cadherin, they developed into bronchial trees, but their morphology was significantly altered. Epithelial lobules at the tips of bronchial trees had an open luminal space in the control explants, whereas in the presence of anti-E-cadherin, the lobules showed a crushed appearance, that is, they were devoid of a luminal space. This morphogenetic anomaly was enhanced by incubating the explants in the presence of antibodies against both E- and P-cadherins. These results indicate that cadherins are crucial for maintaining the three-dimensional structure of tissues.

Cadherin-mediated specific cell adhesion

We previously found that cells expressing different types of cadherin segregate from each other when mixed (Takeichi et al 1981). This kind of observation led us to propose that different types of cadherin have distinct binding specificities, thus conferring specific adhesive properties on cells (Takeichi et al 1985). The pattern of expression of cadherins in an embryo suggests that these molecules also play a role in cell allocation. When an embryonic cell layer is to be morphogenetically divided into multiple layers, different cadherins are expressed in the different layers. On the other hand, when cells derived from different lineages are to be connected, they express identical cadherins. The expression of cadherins thus always correlates with morphogenetic events such as the segregation or association of cell groups. These observations support the notion that cadherins may be involved in specific cell binding.

In order to examine if each cadherin type has a unique binding specificity, we used cDNA transfection. We isolated L cells transfected with cDNA encoding E-cadherin or P-cadherin, and tested whether these cells with different cadherins can adhere to each other in mixed suspensions (Nose et al 1988). The result was very clear—cells preferentially adhered to the cells expressing the identical

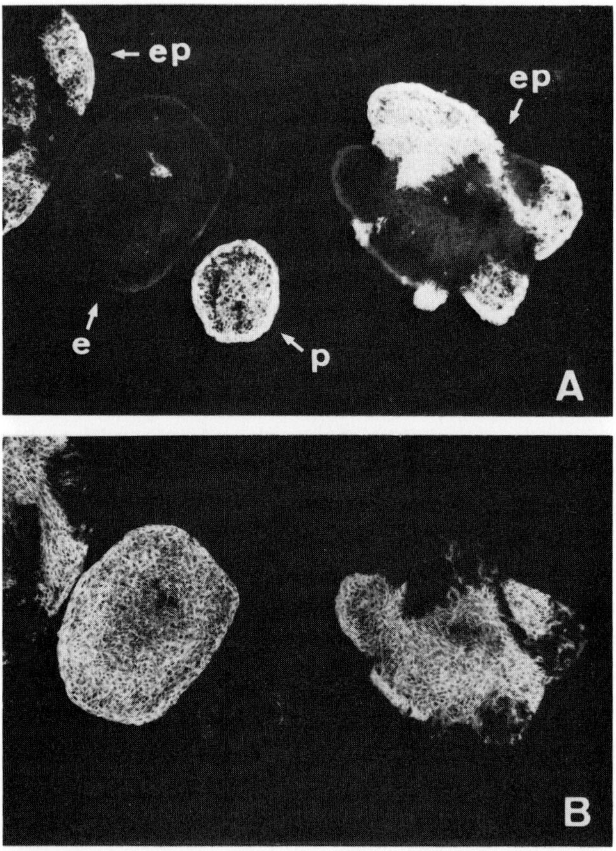

FIG. 2. Segregation of L cells transfected with P-cadherin cDNA from those transfected with E-cadherin cDNA. These two cell lines were singly dispersed and mixed, and then incubated overnight on a gyratory shaker. Aggregates formed were fixed, sectioned and double-stained for P- and E-cadherin. (A) and (B) show the same field, in which (A) represents staining for P-cadherin and (B) for E-cadherin. Note that one aggregate contains only the cells with P-cadherin (p), and another aggregate consists almost purely of the cells with E-cadherin (e). There are two chimaeric aggregates which contain both the P- and E-cadherin transfected cells (ep). In these aggregates, the two cell lines show complementary distribution.

cadherins (Fig. 2). These experiments provide conclusive evidence that E- and P-cadherins have distinct binding specificities and preferentially interact in a homophilic manner. We recently found that N-cadherin also has a unique binding specificity (S. Miyatani & M. Takeichi, in preparation). It is thus concluded that the specificities of cadherins are important for the mechanism of cell sorting.

We extended the above methodology to other biological systems to study the role of cadherins in cell behaviour during morphogenesis. The mouse embryonic lung is composed of epithelial cells expressing E-cadherin and mesenchymal cells without this cadherin. If a suspension of lung cells is incubated to allow aggregation, these two cell types are sorted out from each other during aggregate formation, resulting in the construction of an original lung-like structure. In order to know whether E-cadherin plays any role in this cell sorting system, we added L cells either transfected with E-cadherin cDNA or not to the suspension of lung cells, and observed their behaviour. We found that the untransfected L cells were predominantly incorporated into the mesenchymal region of lung cell aggregates, whereas the cells transfected with E-cadherin were attached to the epithelial tubules. Thus, the L cells originally have an adhesive preference for the mesenchymal cells, but this is altered to a preference for the epithelial cells by the acquisition of E-cadherin expression. These results suggest that E-cadherin is at least partly essential for the segregation of epithelial cells from mesenchymal cells in the lung tissue.

This line of approach was also used to investigate the role of cadherins in axonal guidance. Neurites of various neurons have N-cadherin, for example, the optic axons express this molecule in early embryonic stages. To test if N-cadherin is used for the migration of neurites on the surface of other cells, the following experiments were carried out. We prepared a neuroblastoma cell line, Neuro 2a, and transfected this with N-cadherin cDNA. While the original Neuro 2a cells show little cadherin activity, the transfectants expressed functional N-cadherin molecules derived from the exogenous cDNA. Onto the monolayer cultures of these cell lines, we explanted small pieces of the chicken neural retina and examined the outgrowth of optic axons from the explants (Matsunaga et al 1988b). On the control Neuro 2a cells, very little neurite outgrowth was observed. In contrast, on the transfected cells, a vigorous outgrowth of axons occurred (Fig. 3). The optic axons thus selectively attach to and migrate on the cells expressing N-cadherin. We therefore postulate that N-cadherin molecules are an important factor for mediating the migration of axons on the surface of other cells.

Conclusion

Since the phenomenon that a given cell preferentially adheres to selected cell types was discovered, the molecular mechanism of the specific adhesion has

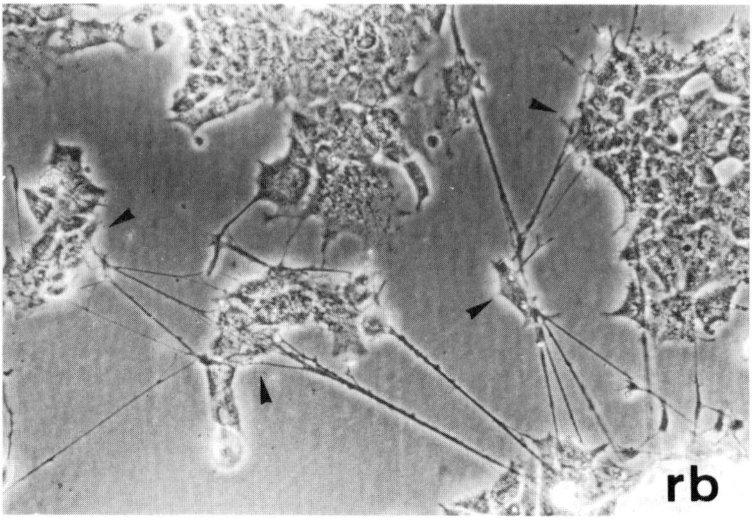

FIG. 3. Migration of optic axons on neuroblastoma Neuro 2a cells transfected with N-cadherin cDNA. A small piece of neural retina of a chick embryo was explanted onto a monolayer culture of the neuroblastoma cells transfected with N-cadherin cDNA and incubated for two days. Neurites extend from the body of the retina (rb) and attach to the surface of the transfected neuroblastoma cells (arrow heads). The migration of the retinal neurites does not occur when untransfected neuroblastoma cells are used. Under the experimental conditions, the retinal neurites do not attach to and migrate on the free surface of the culture dish.

been pursued extensively. The experiments using recombinant cadherins described here demonstrated for the first time that particular molecules are responsible for the specific adhesiveness of cells. Townes & Holtfreter (1955) proposed a model to explain how the neural tube separates from the overlying ectoderm. According to the model, the undifferentiated ectodermal cells are connected to each other by one type of binding molecule. As neurulation proceeds, cells forming the neural tube and those remaining in the overlying ectoderm begin to express distinct binding molecules, thereby inducing the separation of the two cell layers. This model is consistent in part with observations of cadherin expression during neural tube formation—the overlying ectoderm expresses E-cadherin and the neural tube expresses N-cadherin. The experimental evidence for binding specificities of different cadherins, presented here, strongly suggests that the differential expression of these molecules is responsible for the separation of the neural tube from the overlying ectoderm. Probably, cadherins are crucial for many other morphogenetic processes which involve selective cell adhesiveness.

Acknowledgements

This work was supported by research grants from the Ministry of Education, Science and Culture of Japan, and the Special Coordination Funds of the Science and Technology Agency of the Japanese Government.

References

Cunningham BA, Hemperly JJ, Murray BA, Prediger EA, Brackenbury R, Edelman GM 1987 Neural cell adhesion molecule: Structure, immunoglobulin-like domains, cell surface modulation, and alternative RNA splicing. Science (Wash DC) 236:799–806

Gallin WJ, Sorkin BC, Edelman GM, Cunningham BA 1987 Sequence analysis of a cDNA clone encoding the liver cell adhesion molecule, L-CAM. Proc Natl Acad Sci USA 84:2808–2812

Hatta K, Nose A, Nagafuchi A, Takeichi M 1988 Cloning and expression of cDNA encoding a neural calcium-dependent cell adhesion molecule: its identity in the cadherin gene family. J Cell Biol 106:873–881

Hirai Y, Nose A, Kobayashi S, Takeichi M 1989 Expression and role of E- and P-cadherin adhesion molecules in embryonic histogenesis. Development 105:263–270

Hynes RO 1987 Integrin: a family of cell surface receptors. Cell 48:549–554

Matsunaga M, Hatta K, Takeichi M 1988a Role of N-cadherin adhesion molecules in the histogenesis of neural retina. Neuron 1:289–295

Matsunaga M, Hatta K, Nagafuchi A, Takeichi M 1988b Guidance of optic nerve fibers by N-cadherin adhesion molecule. Nature (Lond) 334:62–64

Moscona A, Moscona H 1952 Dissociation and aggregation of cells from organ rudiments of the early chick embryos. J Anat 86:287–301

Nagafuchi A, Shirayoshi Y, Okazaki K, Yasuda K, Takeichi M 1987 Transformation of cell adhesion properties by exogenously introduced E-cadherin cDNA. Nature (Lond) 329:341–343

Nose A, Nagafuchi A, Takeichi M 1987 Isolation of placental cadherin cDNA: identification of a novel gene family for cell–cell adhesion molecules. EMBO (Eur Mol Biol Organ) J 6:3655–3661

Nose A, Nagafuchi A, Takeichi M 1988 Expressed recombinant cadherins mediate cell sorting in model systems. Cell 54:993–1001

Ringwald M, Schuh R, Vestweber D et al 1987 The structure of cell adhesion molecule uvomorulin. Insights into the molecular mechanism of Ca^{2+}-dependent cell adhesion. EMBO (Eur Mol Biol Organ) J 6:3647–3653

Roth SA, Weston JA 1967 The measurement of intercellular adhesion. Proc Natl Acad Sci USA 58:974–980

Takeichi M, Atsumi T, Yoshida C, Uno K, Okada TS 1981 Selective adhesion of embryonal carcinoma cells and differentiated cells by Ca^{2+}-dependent sites. Dev Biol 87:340–350

Takeichi M, Hatta K, Nagafuchi A 1985 Selective cell adhesion mechanisms: role of the calcium-dependent cell adhesion system. In: Edelman GM (ed) Molecular determinants of animal form. Alan R Liss Inc, p 223–233

Takeichi M 1988 The cadherins: cell–cell adhesion molecules controlling animal morphogenesis. Development 102:639–655

Townes PL, Holtfreter J 1955 Directed movements and selective adhesion of embryonic amphibian cells. J Exp Zool 128:53–120

Yoshida-Noro C, Suzuki N, Takeichi M 1984 Molecular nature of the calcium-dependent cell–cell adhesion system in mouse teratocarcinoma and embryonic cells studied with a monoclonal antibody. Dev Biol 101:19–27

DISCUSSION

Kornberg: How do you define cell adhesion? When cells contact each other, is there something more substantial than just touching?

Takeichi: The simplest assay for cell adhesion is to make a single cell suspension culture. If the cells are adhesive, they reaggregate. We are measuring mechanical connections of cells.

Lawrence: Do you have any way of assaying how important these cadherins are in a particular process? Can you, for instance, block them specifically and see if that process is also blocked?

Takeichi: Antibodies are not good tools for that kind of experiment. The best way is to make transgenic mice using a gene linked to an inducible promoter; or to inject antisense mRNA, again connected to some specific promoter to allow it to be expressed in specific tissues. For example, the neural tube loses E-cadherin, so what would happen if we forced the neural tube to express E-cadherin?

Jessell: It's true you have to be careful about the use of antibodies but I think you are being a little modest about the role of cadherins. In most experiments that you and other people have done using antibodies to cadherins and comparing those with equivalent antibodies to N-CAM or any of the other putative cell adhesion molecules, it is the cadherins that play the major role in keeping cells together. That makes it unclear what the function of N-CAM is on cells. N-CAM and cadherins tend to co-distribute but the cadherins seem to be functionally much more important than N-CAM.

Takeichi: Cadherins are important for mechanical connections of cells, at least. N-CAM could be important for other, unknown mechanisms, not necessarily related to adhesion.

Davidson: What are the relative numbers of molecules of N-CAM and cadherin present on the membranes?

Jessell: For N-CAM, it is estimated that there are approximately 10^5 molecules per cell.

Takeichi: The amount of cadherin is probably about the same.

Davidson: How many genes for each of these cadherin types that you mentioned are there?

Takeichi: Each cadherin is encoded by a single gene.

Davidson: Do most cells express only one cadherin?

Takeichi: Not necessarily, many cells express two or even three types of cadherin simultaneously, but it depends on the cell type. For example, the epidermal basal layer expresses both E-and P-cadherin, but the middle layer of the epidermis expresses only E-cadherin.

Wolpert: If you transfect a single L cell with both cadherins, will it express both at the same time?

Takeichi: I haven't tried that.

Wolpert: There may be some exclusion principle. If it expressed both types, would it adhere with both types of other cells containing the homologous cadherin?

Takeichi: I guess it would.

Davidson: Do cells express different cadherins on apical and basal surfaces?

Takeichi: We have the impression that different cadherins show different distribution patterns in cells of certain tissues that express both E-and P-cadherin, but this needs to be confirmed by use of electron microscopy. On the other hand, in some cell lines, we found that multiple cadherin types were co-localized.

Garcia-Bellido: In your homogeneous tissue, do you find quantitative differences in the level of expression along the tissue? That would account for the graded effects in cell sorting.

Takeichi: Yes, there is a gradient in the expression of cadherin in some tissues. For example, in the developing otic vesicle, the distribution of P-cadherin forms a gradient.

Garcia-Bellido: When the cells lose cadherin because they are leaving the packing stage, do they lose the cadherin simultaneously or in a gradient again?

Takeichi: We would need careful observation to answer that. Generally, all cells in a population lose cadherin simultaneously.

Weisblat: During the nerve cell migration, the growth cones express cadherin and yet they can migrate over a substrate that expresses the same cadherin. Isn't that surprising? Because you said that normally migrating cells, like neural crest, don't express any cadherins.

Takeichi: I didn't say that migrating cells have no cadherin. The migrating neural crest cells may express some unidentified cadherins. On the other hand, cells do not use cadherin when moving on non-cellular substrates.

Wolpert: There is a firm prediction from your work, which is surprising. If you take cells from different tissues that express the same cadherin and mix them together, they should not sort out. Is that the case?

Takeichi: In principle, they should not sort out. I would predict that cells derived from different tissues can be connected to each other, if they express identical cadherins, irrespective of their tissue of origin. However, we cannot rule out the possibility that factors other than cadherins may also affect the adhesive behaviour of cells, and those other factors might induce cell sorting, even though the cadherin expression is identical in the mixed cells.

Cabrera: Can you speculate on where and how the system would resolve the problem of the high specificity needed for cell–cell recognition? Adhesion molecules seem to use quite simple interactions, like homophilic interactions. But in a biological system where so many cell–cell interactions are occurring, specific intercellular interaction must happen—how would you resolve this problem with this kind of molecule? Would you compare it, for instance, to the histocompatibility system, which biologically has the potential for great complexity?

Wolpert: You are asking whether these are just the simple glues, and whether the subtle adhesions which cells use for guidance are based on something else?

Takeichi: Different kinds of cell adhesion molecules are co-expressed on cells: the combinations of these molecules could create complexity.

Weisblat: Don't forget that we are not trying to build an animal from a Waring blender mixture of cells; cells have very particular histories during embryogenesis that will constrain their fates.

Cabrera: Yes, but look at cell targeting in the nervous system where similar molecules to these are being used. How do they recognize their targets using such simple structures?

Jessell: One of the striking things about cadherins is their calcium-dependency. Have the sequences given you any information as to why these molecules are so calcium dependent? Secondly, do you think there are any situations in which regulation of extracellular calcium actually regulates the function of cadherins?

Takeichi: We have not detected any homology with other calcium-binding proteins. We don't know which region is the calcium-binding domain, although we do have a candidate for that.

It is hard to measure the concentration of calcium in the microenvironment of the cell, so we cannot tell whether the function of cadherins is regulated by calcium. If cells could control the concentration of calcium at their surface, this ion could regulate the connection of cells mediated by cadherins.

Kimble: Is the calcium binding site external? If the internal concentration of calcium changes, is the function of the cadherins affected?

Takeichi: The external region of cadherin has the ability to bind calcium. We don't know if the internal region also reacts with this ion.

Wieschaus: You mentioned the possibility of unidentified cadherins. Have you any idea how many more there might be?

Takeichi: There should be more because there are a few cell types which do not react with any currently available antibodies, for example neural crest cells, but they do have cadherin-like activity. I don't think there are a large number of cadherins, rather that the precise regulation of the expression of a small number of cadherins is more important.

Garcia-Bellido: Has anyone tried adding large amounts of the extracellular domains of the cadherins to a cell suspension? Would that prevent aggregation?

Takeichi: We have not done that, but there is a report demonstrating that an extracellular fragment of cadherin interferes with cell–cell adhesion when added to cell cultures (Wheelock et al 1987). Therefore, the extracellular domain itself seems to have some reactivity without the cytoplasmic domain. For the complete intercellular binding function, however, the cytoplasmic domain is also required.

Gardner: Could you say more about the relationship with actin? I don't know the valency of these molecules, but is it possible that their distribution within the membrane is important for their function? In other words, if they have to be aggregated rather than dispersed in order to form stable contacts between cells, the cytoskeleton might play a role in holding them together.

Takeichi: Individual cadherin molecules could have only a weak affinity of association; thus it might be necessary for the molecules to be concentrated in the adhesion sites for binding of cells to occur. The cytoskeleton could be involved in such a process.

Martinez-Arias: Is anything known about post-translational modification of these proteins? Are they glycosylated?

Takeichi: They are glycosylated. N-cadherin in different tissues has different molecular weights. The molecular weight is also modified during development.

Cabrera: The carbohydrate modification of N-CAM has been studied (for review, see Jessell 1988). As the carbohydrate moiety is removed, the stickiness of the molecule increases. This is a quantitative effect more than a specific qualitative change.

Meinhardt: Is anything known about the half-life of the molecules on the cell surface? Is there any internalization?

Takeichi: Not much is known about the turnover of cadherin, except that cadherin is restored within a few hours of removal from the cell surface.

Meinhardt: That is a more artificial situation.

Takeichi: Yes, I don't know what happens in the normal situation. We can detect extracellular fragments of cadherin in the blood, so degraded cadherin molecules may be shed from the cell surface.

Kimble: How early in development can cadherins be detected?

Takeichi: In mouse, the fertilized egg has E-cadherin activity.

Struhl: Is anything known about cadherins in insect systems? In insect systems, there is evidence for graded affinities across, for example, an epithelial sheet in a wing disc. There is also evidence suggesting that there are discrete discontinuities, for example posterior compartment cells expressing *engrailed* don't seem to mix with anterior compartment cells which do not express *engrailed*. So you have both possible types of mechanisms, each of which might be mediated by some kind of differential affinity between cells.

Takeichi: Positional segregation between different cell types does occur in insects; for example, neuroblasts segregate from the epidermis in neural development. This type of phenomenon suggests that adhesive recognition is also necessary in insects. An integrin-type molecule and an N-CAM-like molecule have been identified in insects, so why not cadherins?

Struhl: This is not really differentiation of different cell types, but cadherins could be involved in differences within a common cell layer.

Kimble: An obvious way of examining the function of these molecules is to isolate mutations in their genes. Are there cadherins in *C. elegans* or *Drosophila*?

Takeichi: People are looking for invertebrate cadherins, but they haven't found anything with a high homology to vertebrate cadherins yet.

Cabrera: Fasciclin II is a member of the immunoglobulin gene family, like N-CAM (Harrelson & Goodman 1988). When you perturb expression of fasciclin II, the system seems to have great inertia, like redundancy, and you don't get very clear effects.

Meinhardt: Are these types of molecules only used for sticking cells together or are they used as a signalling mechanism? From their structure, it is tempting to speculate in this direction; that the molecules are also the receptors. You have modified internal parts of the molecule and abolished its function. That means that the cell gets feedback, there is an internal representation of what has happened outside.

Takeichi: It is an attractive hypothesis, but I have no evidence to support it, unfortunately.

References

Harrelson AL, Goodman CS 1988 Growth cone guidance in insects: fasciclin II is a member of the immunoglobulin superfamily. Science (Wash DC) 242:700–708

Jessell TM 1988 Adhesion molecules and the hierarchy of neural development. Neuron 1:1–13

Wheelock MJ, Buck CA, Bechtol KB, Damsky CH 1987 Soluble 80-kd fragment of cell-CAM 120/80 disrupts cell–cell adhesion. J Cell Biochem 34:187–202

Polarity and patterning in the neural tube: the origin and function of the floor plate

T. M. Jessell*†, P. Bovolenta*+, M. Placzek*+, M. Tessier-Lavigne*† and J. Dodd*+

Center for Neurobiology and Behavior, Howard Hughes Medical Institute, and Department of Biochemistry and Molecular Biophysics†, and Department of Physiology and Cellular Biophysics+, Columbia University, College of Physicians and Surgeons, New York, NY 10032 USA*

Abstract. Little is known about the cellular and molecular mechanisms that determine neuronal cell fate and the patterning of neuronal connections in the vertebrate central nervous system. In this paper we summarize evidence which indicates that some aspects of neuronal differentiation and axon guidance are regulated by specialized epithelial cells that occupy the medial region of the neural plate and, later, the ventral midline of the spinal cord. This cell group, termed the notoplate/floor plate appears to constitute a distinct compartment within the neural plate that is more closely related in lineage and perhaps also in function to axial mesodermal cells of the underlying notochord than to other neural plate cells. Cells of the notoplate exhibit specialized mechanical and adhesive properties that may contribute to neurulation. At later stages of development, the floor plate appears to guide developing axons in the embryonic spinal cord by releasing a diffusible chemoattractant factor and by virtue of its specialized cell surface properties. The floor plate may also play a role in the determination of cell identity and patterning at earlier stages of neural tube development.

1989 Cellular basis of morphogenesis. Wiley, Chichester (Ciba Foundation Symposium 144) p 255–280

The principles and mechanisms underlying the patterning of neural tissues in vertebrate embryos are poorly understood. Cellular studies of neural development have, however, defined some of the steps that contribute to the generation of patterned neuronal networks. First, neural cells appear to acquire specific identities and differentiate into distinct neuronal classes as a result of their embryonic lineage and as a consequence of local cell interactions (see LeDouarin 1982, Turner & Cepko 1987). Once the identity of individual neurons has been determined, the assembly of precise and stereotyped neuronal connections is achieved by the extension of axons along defined pathways and by the formation of selective synaptic contacts (see Dodd & Jessell 1988).

The rudiments of this neural pattern are laid down soon after primary induction of the neural plate. The analysis of regional differentiation in amphibian neural ectoderm that has been rotated experimentally has revealed that the specification of cell fate and tissue pattern occurs independently along the two axes of the neural plate and that the determination of polarity along the anteroposterior axis precedes that along the mediolateral (ML) axis (Adelmann 1930, Hutchinson 1936, Jacobson 1964, Roach 1945). During

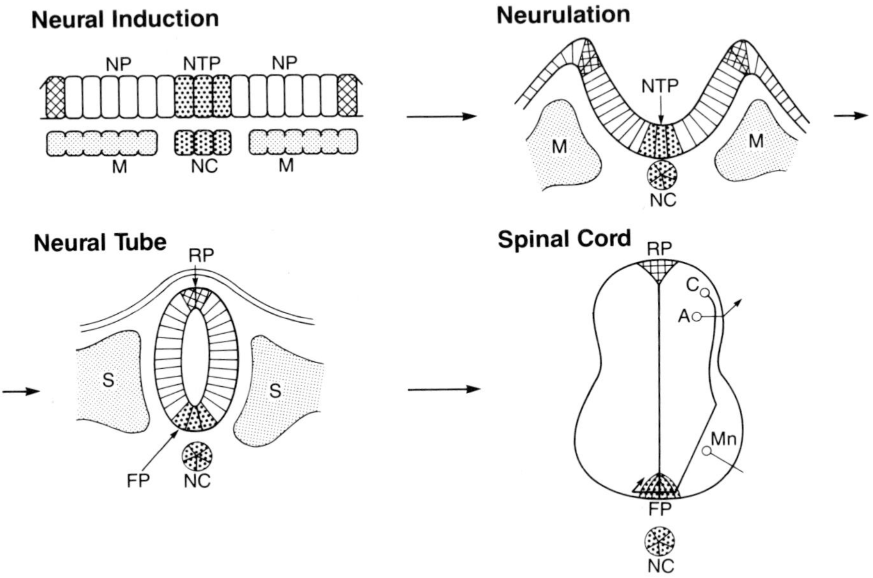

FIG. 1. Stages in the development of the vertebrate neural tube and spinal cord. Neural Plate: At the time of the neural induction, the neural plate (NP) exists as an epithelial layer that can be subdivided into three cell types. The medial region of the neural plate consists of a specialized cell group termed the notoplate (NTP), which overlies the prospective notochord (NC). More lateral regions of the neural plate overlie the prospective somitic mesoderm (M). The lateral margins of the neural plate (hatched cells) may also constitute a specialized cell group at this stage, destined to become the roof plate of the neural tube (see text). Neurulation: The close association between the notoplate and notochord is retained during neurulation and neural tube closure. Neural Tube: After closure, the neural tube consists of a pseudostratified neuroepithelial layer interposed in which are these two specialized midline cell groups. At the ventral midline the notoplate has become the floor plate (FP), while roof plate cells (RP) occupy the dorsal midline. The neural tube is surrounded by newly formed somites (S). Spinal Cord: After the onset of neuronal differentiation, the caudal neural tube gives rise to the spinal cord. The first three classes of neurons to differentiate in the spinal cord are motoneurons (Mn), commissural (C) and association (A) neurons. The axons of each of these three neuronal classes exhibit markedly different initial trajectories.

neurulation, i.e. neural tube formation, the ML axis of the neural plate becomes the dorsoventral (DV) axis of the neural tube (Fig. 1) One of the earliest manifestations of DV polarity is the differentiation of distinct neuronal classes at fixed locations along the DV axis of the spinal cord (Fig. 1).

The mechanisms that determine polarity along the ML and DV axes are not well understood. The regional specification of the neural plate may be imposed, in part, by the underlying axial mesoderm (Spemann 1938). However, there may also be cells within the neural plate itself that exert polarizing influences on neural cell differentiation. One cell group that may play such a role is a set of epithelial cells located at the axial midline of the neural plate, immediately overlying the notochord (Fig. 1). At the neural plate stage in amphibia, this cell population is called the notoplate (Fig. 2A,B) (Jacobson 1981). After neurulation, the notoplate gives rise to the floor plate at the ventral midline of the neural tube and spinal cord (Fig. 1) (Baker 1927, Kingsbury 1930).

In this article we review evidence that suggests that this structure, which we term the notoplate/floor plate, comprises a distinct compartment within the neural epithelium that may be closely related in lineage and properties to axial mesodermal cells of the underlying notochord. In addition, the notoplate/floor plate seems to play a role in neural cell organization and axonal patterning along the ML and DV axes of the developing neural tube and spinal cord. Cells of the notoplate exhibit specialized mechanical and adhesive properties which may contribute to the structural changes that occur in the neural plate and neural tube during neurulation. At later stages of development, the floor plate appears to guide developing axons in the embryonic spinal cord by releasing a diffusible chemotropic factor and by virtue of its specialized cell surface properties.

The notoplate/floor plate and notochord have a common origin

Compilation of fate maps of developing amphibian embryos has shown that cells in the notoplate and floor plate and in the underlying axial mesodermal structure, the notochord, derive from the same blastomere in early cleavage stage embryos. Evidence for this comes from blastomere transplantation and the injection of intracellular markers into individual blastomeres of *Xenopus* embryos at the 32-cell stage (Gimlich 1986, Dale & Slack 1987). At the onset of gastrulation, cells that derive from the dorsal tier blastomere Bl (using the terminology of Dale and Slack) are located in a region overlying the lip of the blastopore called the dorsal marginal zone. After neurulation, labelled cells appear in the notochord and adjacent paraxial mesoderm, but are also present in the floor plate of the neural tube (Fig. 3A). Labelled cells are less frequently found in more lateral regions of the neural tube. Injection of fluorescein-dextran into single progenitor cells at the 128-cell stage of zebrafish embryos labels the notochord and floor plate of the spinal cord in later

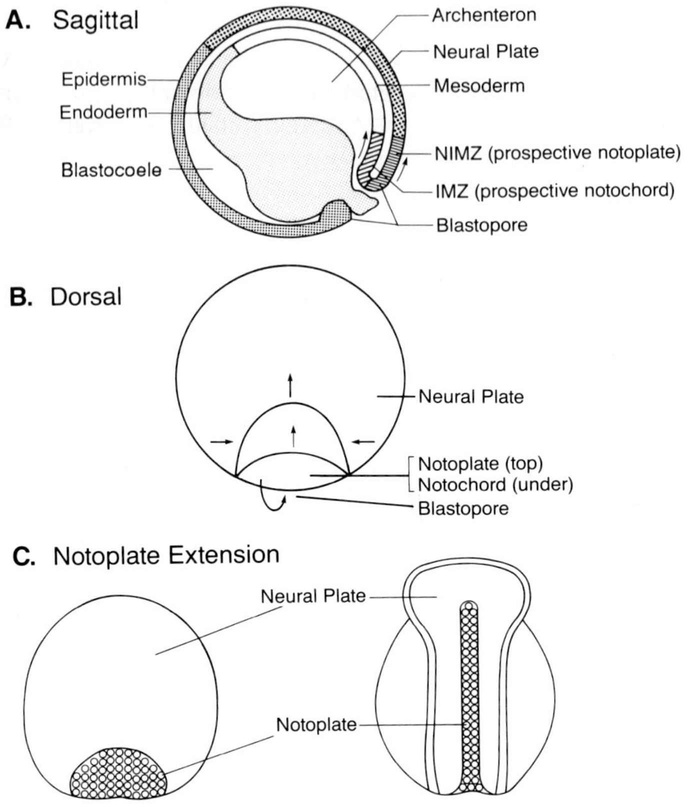

FIG. 2. The relationship of the notoplate and notochord during gastrulation and neurulation in amphibia. A) Saggital view of an amphibian embyro at the middle to late stage of gastrulation, emphasizing the relationship between the involuting marginal zone (IMZ) and non-involuting marginal zone (NIMZ). Arrows indicate the direction of extension of cells of the IMZ and NIMZ after involution of IMZ cells. B) Dorsal view of an amphibian embryo at the same stage as in A). Arrows depict the cell movements that occur during convergent extension of prospective notoplate and notochord cells. C) Diagrammatic representation of the extension of the notoplate during neurulation in amphibian embryos. The notoplate elongates along the midline of the neural plate without a marked change in the number of notoplate cells. This figure has been adapted from the work of Jacobson, Gordon and Keller (for details see Gordon & Brodland 1987, Jacobson & Gordon 1976, Gordon & Jacobson 1978, Keller et al 1985, Keller 1985).

embryos (see Fig. 2d in Kimmel & Warga 1988). These fate maps indicate that cells of the notochord and notoplate/floor plate derive from a common blastomere in early cleavage stage embryos. This common lineage does not restrict the developmental potential of cells in the blastula (Jacobson & Xu 1989), but indicates the lack of intermixing of cells in prospective axial regions before gastrulation (Wetts & Fraser 1989).

Analysis of the cell movements that occur in *Xenopus* embryos has, however, established that, by the onset of gastrulation, cells of the prospective notoplate and notochord are spatially segregated within the dorsal marginal zone and differ in some of their properties (Fig. 2A) (Keller et al 1985, Keller 1985). At the early gastrula stage, the dorsal marginal zone comprises two domains; an involuting marginal zone (IMZ) and a non-involuting marginal zone (NIMZ) (Fig. 2A) (Keller 1985). As implied, cells of the IMZ involute at the blastopore during gastrulation, whereas cells of the NIMZ remain in the ectodermal layer. Before involution, cells that give rise to each domain converge towards the midline, and are spread laterally across the dorsal lip of the blastopore (Fig. 2B). Following the involution of IMZ cells, both domains extend axially, away from the blastopore and towards the prospective anterior region of the embryo (Fig. 2B) (Keller 1985). As a result of this process, termed convergent extension (Keller 1985), the IMZ is transformed into the notochord and cells of the NIMZ form the notoplate. Although both the IMZ and NIMZ undergo axial extension, they are nonetheless functionally distinct cell groups by the onset of gastrulation, since cells transplanted from the NIMZ into the IMZ fail to involute (Keller 1985, R. E. Keller, personal communication).

Appreciation of the lineage relationship between the amphibian notoplate and notochord cells and their axial extension helps to explain the chimaeric composition of neural tubes induced by transplantation of the 'organizer', i.e. the dorsal marginal zone, in the classical experiments on neural induction. Spemann & Mangold (1924) showed that in neural tubes induced by many organizer grafts obtained from late gastrulae, the notochord and floor plate were derived from donor cells, whereas the remainder of the neural tube was derived from host cells (Fig. 3C). In these late gastrulae much of the IMZ had presumably involuted, and thus the graft probably contained NIMZ cells which retained the ability to extend axially in their new location (see Keller 1985) and intercalated themselves between host cells of the secondary, induced neural plate. In experiments in which the graft was obtained from earlier gastrulae and presumably contained only the IMZ, the neural tube did not contain cells of the donor transplant (Fig. 3D). Although this point was not made explicitly by Spemann & Mangold (1924), it appears from their figures that in experiments using earlier gastrulae, the neural tube did not have a morphologically recognizable floor plate (see e.g. Fig. 3D). Thus, the occurrence of a floor plate in an induced neural tube appears to be correlated with the presence of donor cells at the ventral midline of the neural tube, consistent with the idea that the notoplate/floor plate derives exclusively from NIMZ cells in the dorsal marginal zone.

Although less is known about cell movements that occur during gastrulation in other vertebrates, it seems likely that the lineage relationships between the notochord, notoplate and remainder of the neural plate are similar to those

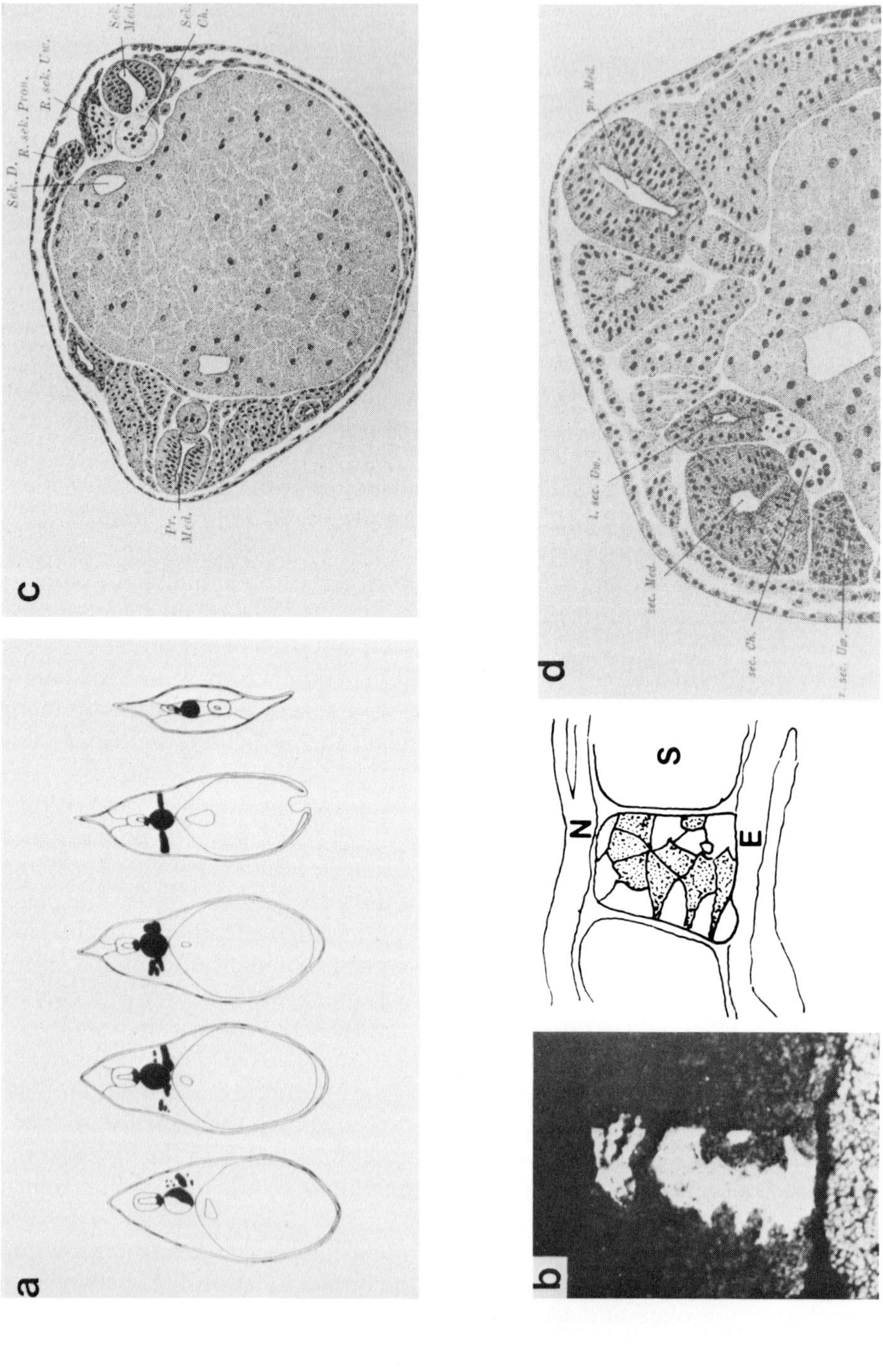

c
Sek. D.
R. sek. Pron.
R. sek. Uw.
Sek. Med.
Sek. Ch.
Pr. Med.
a
d
pr. Med.
l. sec. Uw.
sec. Med.
sec. Ch.
r. sec. Uw.
b
S
N
E

FIG. 3. Relationship between the notoplate/floor plate and notochord revealed by fate mapping and transplantation studies. A) Diagram of cells labelled in floor plate and notochord of *Xenopus* embryos after injection of fluorescent dextran into blastomere B1 of 32-cell stage embryos. (From Dale & Slack 1987.) B) Fluorescence micrograph and diagram of labelled cells in the notochord and overlying notoplate of neural plate stage *Xenopus* embryos. Dorsal marginal zone cells labelled with fluorescein-dextran were grafted to the corresponding region of an unlabelled embryo. The segregation of labelled notoplate cells within the neural plate (N) is apparent. S, somites; E, endoderm. (From Keller 1985.) C) Induction of a second neural axis after organizer grafts from middle-to-late *Triturus* (newt) gastrulae. The induced (right) neural tube is composed of pigmented (*T. taeniatus*) cells with the exception of the floor plate which consists of unpigmented (*T. cristatus*) cells, as does the underlying notochord. The morphological features of the floor plate and ventral neural tube of the induced axis are similar to those of the primary host neural tube. (From Spemann & Mangold 1924.). D) In another graft from the same series of experiments, the induced (left) neural tube does not contain unpigmented donor (*T. cristatus*) cells, although the notochord clearly derived from graft cells. In this embryo, the induced neural tube has a thickened basal mass and no morphological features characteristic of the presence of a floor plate. The correlation between the presence of graft cells at the ventral midline of induced neural tubes and the morphological features of the floor plate is maintained in many other experiments of this class, performed by Spemann, Mangold, Holtfreter and others reviewed in Spemann (1938).

outlined above for amphibia. In avian embryos the notochord derives from Hensen's Node, which lies at the anterior limit of the primitive streak (Ooi et al 1986). After transplantation of Hensen's Node, marked with [^{3}H]thymidine, to an identical location in host embryos, labelled cells are found throughout the length of the notochord and also in the floor plate (Rosenquist 1966, Nicolet 1970). Hensen's Node therefore gives rise to both the notochord and the notoplate/floor plate but not to other regions of the neural plate. This presumably results from the simultaneous deposition of cells in the notochord and notoplate/floor plate as Hensen's Node regresses posteriorly along with the primitive streak during later stages of gastrulation. It is not clear whether cells that give rise to the notochord and notoplate/floor plate are segregated within Hensen's Node as they are within the dorsal marginal zone of amphibian gastrulae or whether they constitute a uniform cell population at this stage.

The notoplate/floor plate forms a discrete compartment in the neural plate and neural tube

The reconstruction by electron microscopy of cell arrangements at the axial midline of the neural plate during neurulation in *Xenopus* indicates that the notoplate is recognizable as a distinct cell population within the neural plate from the time of neural induction (Schroeder 1970). Cells of the notoplate remain spatially segregated from other cells in the neural plate, even after experimental manipulation. This has been demonstrated by transplantation of small patches of the dorsal marginal zone labelled with fluorescein-dextran to equivalent sites in unlabelled *Xenopus* embryos (Keller et al 1985, Keller 1985, Keller & Tibbetts 1989). The labelled NIMZ cells that extend anteriorly within the prospective neural ectoderm give rise to a discrete cell population that coincides with the region defined as the notoplate (Fig. 3B). A boundary between notoplate and other neural plate cells has also been revealed by monitoring cell movements during neurulation in amphibia (Jacobson & Gordon 1976), when the notoplate is transformed from a broad sheet at early stages into a long column of cells stretched along the midline by the onset of neural plate folding. The rapid elongation of the notoplate has been proposed to contribute to the folding of the neural tube (Jacobson & Gordon 1976). However, neural tube formation can occur in embryos lacking a notoplate and notochord (Youn & Malacinski 1981), indicating that other mechanisms contribute to neural tube formation (Burnside & Jacobson 1968, Schoenwolf et al 1988).

Time-lapse monitoring of the movements of individual notoplate and neural plate cells in newt embryos indicates that elongation of the notoplate is achieved by re-positioning cells from the interior of the notoplate region along the notoplate/neural plate boundary, without a marked change in the number of notoplate cells (Jacobson 1981, Jacobson et al 1986, see also Schoenwolf &

Sheard 1989). Extension of the boundary may occur by the preferential adhesion of notoplate and neural plate cells as they make contact at the boundary of these two regions (Jacobson et al 1986). These observations have led to a model that attempts to account for cell movements within the notoplate on the basis of current theories of cortical flow in animal cells (Jacobson et al 1986). In the present context, the most pertinent feature of this model is its emphasis on the notoplate as a structure distinct from the remainder of the neural plate and on the importance of the notoplate/neural plate boundary.

Cells of the notoplate, therefore, derive from precursor cells that are distinct from those that give rise to the remainder of the neural plate. Moreover, these cells respect the boundary between the notoplate and neural plate, despite extensive cell intermingling within each of these regions. It seems appropriate, therefore, to consider the notoplate as a separate compartment within the neural plate, comparable in some ways to compartments that exist in insect development (Crick & Lawrence 1975, Garcia-Bellido et al 1979).

The contribution of induction to notoplate/floor plate differentiation

The classical studies of Spemann & Mangold (1924) and numerous subsequent experiments (reviewed in Spemann 1938) established that the neural ectoderm is induced by the underlying axial mesoderm. In *Xenopus* embryos the notoplate forms a discrete compartment arising from a group of cells that appear to be committed at the onset of gastrulation, before neural induction occurs. During normal development then, it seems that the notoplate/floor plate is not induced by the notochord. Cells of the NIMZ may be specified as the result of an earlier inductive influence of the IMZ or by a graded inductive signal from the vegetal region of the embryo (Jacobson & Sater 1988).

The notochord does seem to be important for the maintenance and later development of the notoplate/floor plate. The axial extension of the notoplate *in vitro* depends on the presence of underlying chordamesodermal cells (Keller 1985). Moreover, extirpation of the prospective or definitive notochord in amphibia prior to neural tube closure results in the absence of a floor plate in the overlying neural tube (Lehmann 1928, Hörstadius 1944, Kitchin 1949, van Straaten & Drukker 1987). In contrast, removal of the notochord at later stages does not appear to affect the development of the floor plate (Kitchin 1949, van Straaten & Drukker 1987; see Watterson et al 1954 for a detailed discussion). Notoplate cells adhere tightly to the underlying notochord cells during neurulation (Schroeder 1970, Keller 1985), suggesting that the apparent dependence of the floor plate on the notochord is contact mediated, although it is difficult to exclude the possibility that ablation of the notochord damages the floor plate directly.

Although in normal development the notochord apparently influences the development of only pre-committed notoplate cells, it may also be capable of inducing a floor plate from uncommitted neural ectoderm. Recently, van Straaten and colleagues (van Straaten et al 1985a,b, van Straaten & Drukker 1987, 1988) have provided some evidence that this can occur. Implantation of an additional chick notochord in a lateral location adjacent to a host neural tube produced a change in morphology of the adjacent neural tube that was interpreted to reflect the presence of an additional floor plate. This interpretation is based solely on morphology and there is as yet no direct evidence that the region of neural tube near to the implanted notochord assumes the molecular and functional properties of the notoplate/floor plate (which are described in the following sections). van Straaten et al (1985a,b) have also cited as evidence in support of floor plate induction by the notochord observations of mutant avian embryos in which duplicated notochords adjacent to a single neural tube had floor plates co-extensive with each notochord (Watterson et al 1954, Limborgh 1956). However, in light of the mechanism of formation of avian floor plate and notochord from Hensen's Node described above, an alternative and perhaps more likely explanation is that in these embryos Hensen's Node became divided into two parts before its regression, such that two notochords and two floor plates were laid down, even though a single zone of neural ectoderm was induced.

It should be possible to determine the contribution of induction to differentiation of the floor plate in avian embryos. Hensen's Node grafts have long been known to induce a secondary neural axis in avian embryos in a manner analogous to the organizer in amphibia (Waddington 1940), and the fate of transplanted Hensen's Node cells can be followed using chick–quail xenografts (McCallion & Shinde 1973, Cuevas & Orts Lorca 1974, Vakaet 1981, 1984). When a quail Hensen's Node is transplanted to lateral sites in primitive streak stage chick embryos, quail cells are observed only rarely in the induced neural plate. This presumably occurs because the grafts are inserted underneath the epiblast rather than into the ectodermal layer, as in the amphibian transplants, so that prospective quail notoplate cells are not in a position to extend with the induced ectoderm. However, embryos in which a secondary neural plate has been induced by transplantation of Hensen's Node have always been analysed before neural tube closure, so that it has not yet been determined whether quail Hensen's Node cells can induce a floor plate in chick ectoderm. In other experiments, rotation of the early chick neural tube by 90° or 180° did not appear to result in an additional floor plate in areas adjacent to the notochord (Jacob et al 1976, although see Steding 1962), which argues against induction of the floor plate by the notochord in avian species, at least at stages after neural tube closure.

Direct evidence that the notochord induces notoplate/floor plate differentiation is, therefore, exiguous at present. Moreover, the original experiments of Spemann & Mangold (1924) provide some evidence against the induction of the notoplate/floor plate by the notochord in amphibia. As discussed above, many of the induced neural tubes in these experiments did not contain donor graft cells (see Fig. 3D), although the secondary notochord clearly derived from the graft. In these embryos, the induced neural tube did not exhibit any morphological features characteristic of a floor plate. It is also worth noting that the differentiation of neural ectoderm in the absence of a notoplate in these and other experiments indicates that the notoplate probably does not (in contrast to the suggestion of Gordon & Brodland 1987) play an essential role in neural induction after involution of the axial mesoderm. It is possible that the NIMZ, before gastrulation, plays a role in conferring neural properties on the adjacent ectoderm (Sharpe et al 1987, Kintner & Melton 1987).

The notoplate and floor plate may influence neuroepithelial cell proliferation

A variety of experimental or genetic perturbations of vertebrate embryos which result in the absence of the notochord and notoplate/floor plate suggest that these two structures may influence neuroepithelial cell proliferation. For example, UV irradiation of *Xenopus* embryos soon after fertilization results in embryos that develop without a notochord and without a floor plate in the neural tube (Youn & Malacinski 1981, Malacinski & Youn 1982). The ventral neural tube in these embryos has a thickened basal mass. Similarly, UV irradiation or surgical removal of Hensen's Node in chick embryos results in the absence of a notochord and in the formation of a thickened neural tube that lacks a floor plate (Wolff 1936, Grabowski 1956). Analysis of mouse mutations that affect the development of the primitive streak suggests that the same relationship holds for mammalian embryos. Two such mutations, *Danforth short tail (Sd)* and *truncate (tc)*, are characterized by the absence or premature degeneration of the notochord (Gruneberg 1958, Theiler 1959). At segmental levels in which the notochord is absent, the floor plate is also missing and the somitic mesoderm fuses at the ventral midline. The ventral neural tube at affected segmental levels exhibits a characteristic thickening similar to that observed in amphibian and avian embryos in which the prospective notochord or Hensen's Node has been eliminated (Holtfreter & Hamburger 1955). Holtfreter & Hamburger suggested that the ventral mesoderm causes neuroepithelial cell proliferation and, as a consequence, a thickened ventral neural tube. It is also possible that the floor plate and/or notochord normally exerts an inhibitory effect on the proliferation of adjacent neuroepithelial cells (see van Straaten & Drukker 1987). Elimination of the notochord or floor plate would remove this inhibitory influence and in this way lead to neuroepithelial cell proliferation.

Involvement of the floor plate in cartilage induction

Many of the experimental manipulations, discussed above, that revealed the common behaviour and interdependence of the notochord and notoplate during gastrulation and neurulation raise the possibility that the floor plate may regulate mesodermal differentiation at stages after neural tube closure. Removal of the notochord and neural ectoderm from amphibian embryos at or before the neural plate stage results in the perturbation of axial cartilage formation (Hörstadius 1944, Kitchin 1949, Holtzer & Detwiler 1953). The notochord and neural tube function independently *in situ*, with each cell group inducing the cartilage that eventually surrounds it (Watterson et al 1954). It is possible to induce cartilage *in vitro* across porous membrane filters; these experiments have shown that the inductive properties of the neural tube are confined mainly to its ventral half (Grobstein & Holtzer 1955, Flower & Grobstein 1967). Moreover, the grafting of additional chick neural tubes to ectopic sites in host embyros leads to cartilage formation adjacent to the ventral midline of the neural tube (see Fig. 30 in Watterson et al 1954), suggesting that the floor plate may be primarily responsible for the ability of the neural tube to induce cartilage. At present, the nature of the cartilage-inducing agent in the ventral spinal cord has not been determined, although the trans-filter experiments suggest that it is a diffusible factor (see Hall 1977).

Chemotropic action of the floor plate

Soon after neurulation, the neural tube consists of a relatively simple pseudostratified epithelium, punctuated by floor plate cells at the ventral midline and a second specialized cell group, termed the roof plate, that occupies the dorsal midline (Fig. 1C). Distinct classes of neurons differentiate in the intervening neural epithelium at defined positions along the dorsoventral axis of the neural tube. Motoneurons differentiate near to the floor plate, followed soon after by two sets of sensory relay neurons, termed commissural (*C*) and association (*A*) neurons, in the dorsal neural tube adjacent to the roof plate (Fig. 1D). The axons of these three sets of neurons exhibit markedly distinct initial trajectories (Fig. 1D), suggesting that they respond differentially to guidance cues in their local environment within the spinal cord. Several observations indicate that the guidance of *C* axons may be influenced by floor plate cells.

The initial projection of the axons of *C* neurons occurs ventrally through the neural epithelium near to, but not in direct contact with, the external limiting membrane of the spinal cord. At this stage *C* axons do not fasciculate, so that extension of the first axons occurs almost exclusively in contact with the surface of neural epithelial cells and not other axons. Recent studies have shown that the prospective neuroepithelial pathway of *C* axons in the dorsal spinal cord expresses a laminin-heparan sulphate proteoglycan complex that may contribute to the initial extension of *C* axons (Letourneau et al 1988; J. Dodd et al, in preparation). Once *C* axons enter the ventral half of the spinal cord, they break away from the lateral edges of the spinal cord and project ventromedially towards the floor plate (Fig. 1D) (Holley & Silver 1987, Dodd et al 1988). This change in trajectory, which takes *C* axons directly through the developing motor column to the floor plate, combined with the specialized properties of the floor plate cells discussed in previous sections, raises the possibility that the floor plate guides *C* axons by releasing a diffusible chemoattractant factor. The directed growth of axons to the floor plate after mechanical perturbation of the developing neural tube in chick embryos had prompted earlier speculation on the role of the floor plate in axon guidance (Weber 1938).

It has been shown that the floor plate releases a chemotropic factor that guides *C* axons by co-culturing explants of E11 rat floor plate and dorsal neural tube in a collagen gel matrix capable of stabilizing gradients of diffusible factors (Fig. 4) (Tessier-Lavigne et al 1988). Dorsal spinal cord explants grown in the absence of a floor plate target exhibit little or no axon outgrowth after 40–44 hours *in vitro*. In contrast, in the presence of a floor plate there is profuse and oriented *C* axon outgrowth from the dorsal explant, towards the floor plate (Fig. 4a, b). The axons that extend from dorsal explants appear to derive from *C* neurons, since they express the cell surface glycoprotein TAG-1 that is present on *C* but not *A* axons *in situ* (Fig. 4c) (Dodd et al 1988). Although in these experiments *C* axons do not extend into collagen gels in the absence of the floor plate, many TAG-1 immunoreactive axons can be detected within the dorsal explants, providing evidence that the floor plate affects axon extension without a detectable effect on the survival or differentiation of *C* neurons (Tessier-Lavigne et al 1988).

The effect of the floor plate is selective for *C* axons, since it has no apparent influence on the pattern of growth of *A*, motor or sensory axons (Tessier-Lavigne et al 1988). Moreover, the induction of *C* axon outgrowth from dorsal explants is highly specific to the floor plate. Other neural and non-neural embryonic tissues, including the remainder of the E11 neural tube and notochord, do not mimic the effect of the floor plate. The activity of the floor plate is not mimicked by nerve growth factor, laminin or several other neurotrophic and mesodermally derived growth factors (Placzek et al 1988). The absence of activity in the notochord provides evidence that by this stage of development the notochord and floor plate are no longer equivalent. This observation also suggests that the factor responsible for guidance of *C* axons is distinct from that involved in cartilage differentiation, since the notochord and ventral neural tube retain the ability to induce cartilage formation at this stage. The identity of the floor plate-derived tropic factor is not known, although preliminary studies indicate that it is a protein with a molecular weight in excess of 30 kDa (Placzek et al 1988). These observations provide evidence for the specialized properties of floor plate cells and represent the first evidence that the axons of vertebrate central neurons are guided by chemotropism.

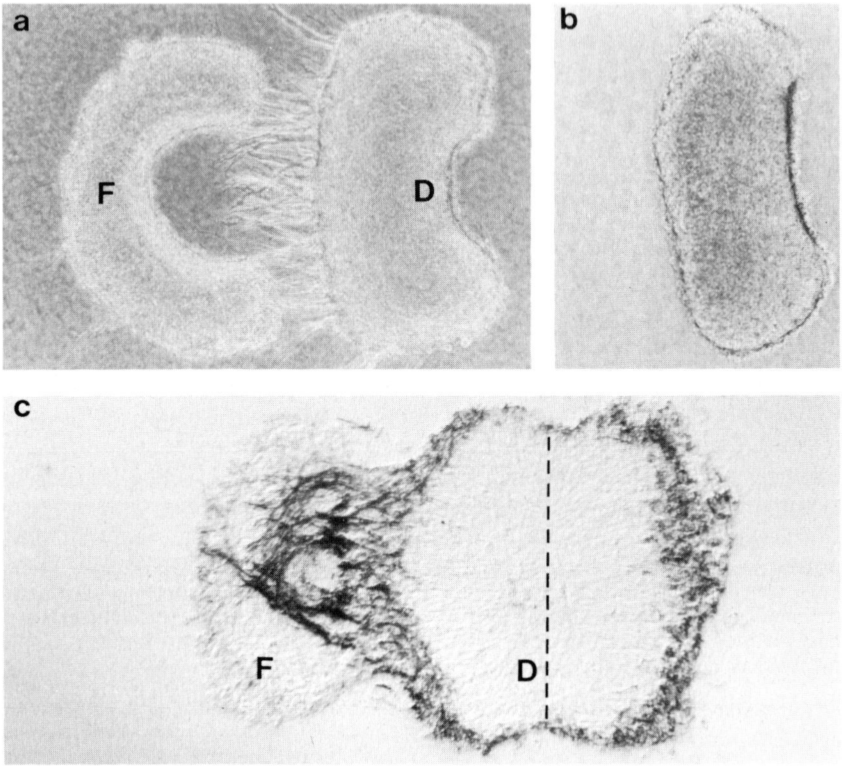

FIG. 4. The floor plate secretes a chemoattractant molecule that guides axons in the developing mammalian nervous system. a) An explant of the dorsal spinal cord of an E11 rat neural tube (D) grown in a 3-dimensional collagen gel matrix in the presence of a strip of explanted floor plate (F). After 40 hours in culture, numerous axon bundles can be seen projecting from the dorsal explant towards the floor plate. b) A similar dorsal explant grown in culture for 40 hours without a floor plate explant. No axons emerged from the explant. c) The two sides of an E11 dorsal explant (D) are opened at the dorsal midline (dashed line) in an 'open book configuration' and a floor plate explant (F) placed opposite one side of the dorsal explant. Commissural axons are identified by immunoperoxidase localization of the TAG-1 glycoprotein. TAG-1^+ axons project from the dorsal explant on the side facing the floor plate explant and project into the floor plate tissue. On the side away from the floor plate, no TAG-1^+ axons project from the dorsal explant, although numerous TAG-1^+ cell bodies and axons can be observed within the explant itself, indicating that commissural neurons are viable in the absence of a floor plate target. Scale bar = 100 μm. For details see Tessier-Lavigne et al (1988).

Axon guidance at the floor plate

Once *C* axons arrive at the ipsilateral boundary between the neuroepithelium and floor plate, they continue to project in the transverse plane and cross the ventral midline through the floor plate (Dodd et al 1988). However, as *C* growth cones emerge on the contralateral side of the floor plate, they undergo a marked

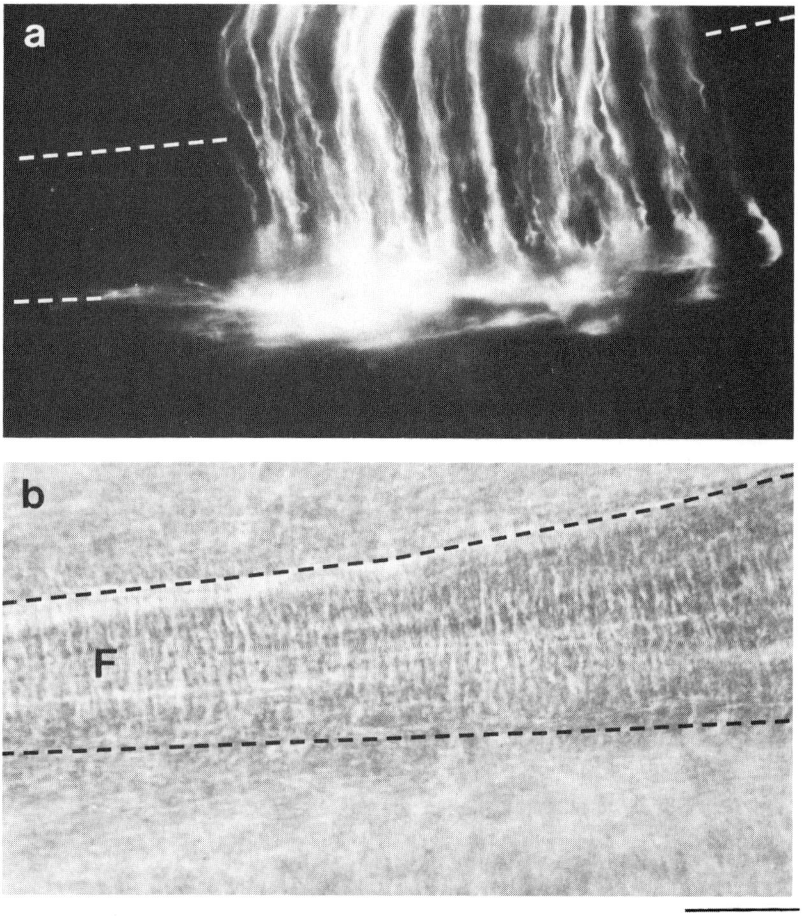

FIG. 5. Developing commissural axons alter their trajectory at the floor plate. A group of commissural axons in E12 rat spinal cord labelled with the fluorescent carbocyanine dye, Di-I, project towards and across the floor plate (a) (viewed from above). The lateral borders of the floor plate (F) are marked with dashed lines and viewed in phase contrast (b). At the contralateral edge of the floor plate, commissural axons make an abrupt right-angled turn and project in the longitudinal plane adjacent to the lateral surface of the floor plate. Scale bar = 50 μm.

change in axonal trajectory, turning orthogonally to project longitudinally in the ventral funiculus. The transition in the plane of axon extension occurs precisely as growth cones encounter the contralateral edge of the floor plate (Fig. 5) (P. Bovolenta & J. Dodd, in preparation), providing additional evidence for specialized properties of cells at this boundary region. Given the bilateral organization of the spinal cord, the behaviour of *C* growth cones at the ventral midline raises the issue of why *C* axons ignore ipsilateral boundary cues, since such cues can clearly be recognized by contralateral *C* axons approaching the same boundary at the same time. One possible explanation is that the properties of *C* axons are altered as a function of passage through the floor plate, such that they recognize the floor plate/neural epithelium boundary only after such modification.

There is evidence that modification of the surface properties of *C* axons does occur during passage through the floor plate. The TAG-1 glycoprotein is expressed by *C* axons during their ipsilateral growth in the transverse plane, but expression of this glycoprotein ceases as *C* axons emerge on the contralateral side of the floor plate (Dodd et al 1988). Conversely, the L1 axonal glycoprotein is not detectable on *C* axons during their growth in the transverse plane and is first expressed only as growth cones pass through the floor plate, reaching high levels on the contralateral segment of axons as they fasciculate in the ventral funiculus. The onset of fasciculation may be a consequence of the expression of L1 (Dodd et al 1988). These observations suggest that the modification of surface molecules may contribute to the ability of *C* growth cones to respond selectively to the contralateral floor plate/neural epithelium boundary.

The nature of the boundary cues that initiate the change in trajectory of *C* axons is not known. One possibility, consistent with observations of notoplate cell movements discussed above, is that the floor plate and lateral neural epithelium have markedly different adhesive properties. The orthogonal turn of *C* axons may then reflect their preferential adhesion to the contralateral face of the floor plate. There is some direct evidence for the specialized adhesive properties of the floor plate. At the stage at which *C* axons first traverse the midline, the floor plate expresses high levels of the adhesion molecule N-CAM, whereas the surrounding neural epithelium expresses very low levels (Dodd et al, in preparation). In addition, floor plate cells express several cell surface antigens not present on other cells of the neural tube, which may confer specialized adhesion or recognition properties to the floor plate (Dodd & Jessell 1988).

A final line of evidence that implicates the floor plate in the guidance of *C* axons derives from the analysis of axon pathfinding in *Danforth short tail* mice, which lack a floor plate. *C* axons in affected embryos exhibit marked perturbations in axon trajectory at the ventral midline in the region normally occupied by the floor plate (Bovolenta et al 1988). In many such embryos, *C* axons project out of the spinal cord, forming a supernumerary nerve that projects ventrally into the underlying mesenchyme. One factor contributing to this misrouting of axons may be the breakdown of the basal lamina at the ventral midline. *Danforth short tail* mutant embryos exhibit defects in extracellular matrix deposition in the region of the floor plate and notochord (Paavola et al 1980, Center et al 1982). Moreover, antigens selectively expressed at the basal surface of floor plate cells and not at other regions of the neural tube basal lamina are absent in embryos that lack a floor plate (Bovolenta et al 1988).

Collectively, these observations provide additional evidence for the specialized properties of the floor plate and indicate that the guidance of at least one set of axons in the developing spinal cord is influenced by the floor plate. Floor plate cells thus exert multiple influences on developing axons, by the release of diffusible chemotropic factors, by their specialized cell surface adhesive properties, and perhaps by the modification of axonal surface properties and the deposition of specialized extracellular matrix components.

The floor plate and the establishment of polarity in the developing neural tube

This article has emphasized the specialized properties of the notoplate/floor plate and its role in neurulation and axon extension. This set of axial cells derives from the region of vertebrate embryos that corresponds to Spemann's organizer, and shares several behavioural and molecular properties with the chordamesodermal cells that are responsible for primary neural induction. As mentioned in the introduction, little is known about the mechanisms by which polarity is established along the mediolateral axis of the developing neural plate, although these mechanisms are likely to differ from those operating along the anteroposterior axis. The presence of notoplate cells at the midline of the neural plate suggests that these cells may play additional roles in the morphogenesis of the developing central nervous system.

One intriguing possibility is that notoplate cells, and a second specialized set of cells located at the lateral border of the neural plate destined to form the roof plate of the neural tube, constitute cell groups that release diffusible morphogens whose function is to specify the programme of differentiation of intervening neural plate cells. Although there is no direct evidence that the notoplate secretes morphogens, one set of grafting experiments is of potential interest in defining such molecules. Hensen's Node, which gives rise to the notochord and notoplate/floor plate, mimics the action of the zone of polarizing activity (ZPA) in that it can specify additional digits when grafted to anterior regions of the chick limb bud (Hornbruch & Wolpert 1986). The notochord exhibits similar properties. The action of the ZPA can also be mimicked by local sources of retinoic acid (Tickle et al 1982) and there is recent evidence for a gradient of endogenous retinoids in the chick limb bud (Thaller & Eichele

1987), suggesting that retinoic acid may function as an endogenous morphogen. Since both the notoplate and notochord derive from Hensen's Node, it is possible that these two structures represent sources of endogenous retinoids that function as morphogens in neural tube development.

There is also evidence that within the neural plate prospective roof plate cells synthesize diffusible molecules that may function as morphogens. Prospective and definitive roof plate cells express mRNA encoding the proto-oncogene *int-1* (Wilkinson et al 1987), which has been proposed to function as a diffusible morphogen on the basis of its structural homology with the TGF-β growth factor family (Sporn & Roberts 1988) and with the *Drosophila* segment polarity gene, *wingless* (Rijsewijk et al 1987). The identification of putative morphogens such as retinoic acid and *int-1* may open the way for a molecular analysis of the role of the notoplate/floor plate and the roof plate in the determination of cell identity and differentiation at early stages of neural tube development.

Acknowledgements

This work was supported by grants from the NIH, The McKnight Foundation and The Muscular Dystrophy Association. TMJ is an investigator of the Howard Hughes Medical Institute. We thank Gary Struhl, Lorna Role and Andrew Lumsden for critical comments on the manuscript, Ray Keller for helpful discussions and R. Lenertz for help in preparation of the manuscript.

References

Adelmann H 1930 Experimental studies on the development of the eye II. The eye forming potencies of the median portions of the urodelean neural plate. J Exp Zool 54:291–317

Baker RC 1927 The early development of the ventral part of the neural plate of amblystoma. J Comp Neurol 44:1–27

Bovolenta P, Jessell TM, Dodd J 1988 Disruption of commissural axon guidance in the absence of the midline floor plate. Soc Neurosci Abstr, in press

Burnside MB, Jacobson AG 1968 Analysis of morphogenetic movements in the neural plate of the newt *Taricha torosa*. Dev Biol 18:537–552

Center EM, Spigelman SS, Wilson DB 1982 Perinotochordal sheath of heterozygous and homozygous Danforth's short-tail mice. J Hered 73:299–300

Crick FHC, Lawrence PA 1975 Compartments and polyclones in insect development. Science (Wash DC) 189:340–347

Cuevas P, Orts Llorca F 1974 Induction primaire chez le jeune blastoderm de poulet (Gallus Gallus) pare greffe du noeud de Hensen de caille (Coturnix coturnix janonica). Acta Anat 89:423–430

Dale L, Slack JMW 1987 Fate map for the 32-cell stage of the *Xenopus laevis*. Development 99:527–551

Dodd J, Jessell TM 1988 Axon guidance and the patterning of neural projections in vertebrates. Science (Wash DC) 242:692–699

Dodd J, Morton SB, Karagogeos D, Yamamoto M, Jessell TM 1988 Spatial regulation of axonal glycoprotein expression on subsets of embryonic spinal neurons. Neuron 1:105–116

Flower M, Grobstein C 1967 Interconvertibility of induced morphogenetic responses of mouse embryonic somites to notochord and ventral spinal cord. Dev Biol 15: 193–205

Garcia-Bellido A, Lawrence PA, Morata G 1979 Compartments in animal development. Sci Am 241:102–110

Gimlich RL 1986 Acquisition of developmental autonomy in the equatorial region of the *Xenopus* embryo. Dev Biol 115:340–352

Gordon R, Brodland GW 1987 The cytoskeletal mechanics of brain morphogenesis. 1. Cell state splitters cause primary neural induction. Cell Biophys 11:177–238

Gordon R, Jacobson A 1978 The shaping of tissues in embryos. Sci Am 238:106–113

Grabowski CT 1956 The effects of the excision of Hensen's Node on the early development of the chick embryo. J Exp Zool 133:301–343

Grobstein C, Holtzer H 1955 In vitro studies of cartilage induction in mouse somite mesoderm. J Exp Zool 128:333–358

Gruneberg H 1958 Genetical studies on the skeleton of the mouse XXII. The development of Danforth's short tail. J Embryol Exp Morphol 6:424–443

Hall BK 1977 Chondrogenesis of the somitic mesoderm. Adv Anat Embryol Cell Biol 53:7–47

Holley JA, Silver J 1987 Growth pattern of pioneering chick spinal cord axons. Dev Biol 123:375–388

Holtfreter J, Hamburger V 1955 Embryogenesis: progressive differentiation. In: Willier BH et al (eds) Foundations of Experimental Embryology. Saunders Press, Philadelphia, p 230–296

Holtzer H, Detwiler SR 1953 An experimental analysis of the development of the spinal column. III. Induction of skeletogenous cells. J Exp Zool 123:335–366

Hornbruch A, Wolpert L 1986 Positional signalling by Hensen's node when grafted to the chick limb bud. J Embryol Exp Morphol 94:257–265

Hörstadius S 1944 Uber die Folgen von Chorda exstirpation an spaten Gastrube and Neurulae von Amblystoma punctatum. Acta Zool 25:75–87

Hutchinson C 1936 Reconstitution in the nervous system following unilateral reversal of the dorsoventral axis in part of the spinal cord of amblystoma punctatum. J Comp Neurol 63:465–487

Jacob HJ, Christ B, Jacob M, Ahistrom P 1976 Differenzierung des neuralrohres. Acta Anat 94:204–220

Jacobson AG 1981 Morphogenesis of the Neural Plate and Tube. In: Connelly TG et al (eds) Morphogenesis and Pattern Formation. Raven Press, New York, p 233–263

Jacobson AG, Gordon R 1976 Changes in the shape of the developing vertebrate nervous system analyzed experimentally, mathematically and by computer simulation. J Exp Zool 197:191–246

Jacobson AG, Oster GF, Odell GM, Cheng LY 1986 Neurulation and the cortical tractor model for epithelial folding. J Embryol Exp Morphol 96:19–49

Jacobson AG, Sater AK 1988 Features of embryonic induction. Development 104:341–359

Jacobson CO 1964 Motor nuclei, cranial nerve roots and fiber pattern in the medulla oblongata after reversal experiments on the neural plate of axolotl larvae. I. Bilateral operations. Zool Bidrag Uppsala 36:7–160

Jacobson M, Xu W 1989 States of determination of single cells transplanted between 512-cell Xenopus embryos. Dev Biol 131:119–125

Keller RC, Danilchik M, Gimlich R, Shih J 1985 The function and mechanism of convergent extension during gastrulation of *Xenopus laevis*. J Embryol Exp Morphol (Suppl) 89:185–209
Keller RE 1985 The cellular basis of amphibian gastrulation. In: Browder LW (ed) Developmental Biology. Plenum, New York, vol 2:241–327
Keller R, Tibbetts P 1989 Mediolateral cell intercalation in the dorsal axial mesoderm of Xenopus laevis. Dev Biol 131:539–549
Kimmel CB, Warga RM 1988 Cell lineage and developmental potential of cells in the zebrafish embryo. Trends Genet 4:68–74
Kingsbury BF 1930 The developmental significance of the floor-plate of the brain and spinal cord. J Comp Neurol 50:177–207
Kintner CR, Melton DA 1987 Expression of Xenopus NCAM RNA in ectoderm is an early response to neural induction. Development 99:311–325
Kitchin IC 1949 The effects of notochordectomy in ablystoma mexicanum. J Exp Zool 112:393–415
LeDouarin N 1982 The neural crest. Cambridge University Press, Cambridge
Lehmann FE 1928 Die Bedeutung der Unterlagerung fur de Entwicklung der Medullar platte von Triton. Arch F Entwick Mech D Organ 113:123–171
Letourneau P, Madsen AM, Palm SM, Furcht LT 1988 Immunoreactivity for laminin in the developing longitudinal pathway of the brain. Dev Biol 124:135–144
Limborgh van J 1956 The influence of adjacent mesodermal structures upon the shape of the neural tube and plate in bird embryos. Acta Morphol Neerl Scand 1:155–166
Malacinski GM, Youn BW 1982 The structure of the anuran amphibian notochord and a re-evaluation of its presumed role in early embryogenesis. Differentiation 21: 13–21
McCallion DJ, Shinde VA 1973 Induction in the chick by quail Hensen's Node. Experientia (Basel) 29:321–322
Nicolet PG 1970 Analyse autoradiographique de la localisation des differentes ebauches presomptives dans la ligne primitive de l'embryon de Poulet. J Embryol Exp Morphol 23:79–108
Ooi VEC, Sanders EJ, Bellairs R 1986 The contribution of the primitive streak to the somites in the avian embryo. J Embryol Exp Morphol 92:193–206
Paavola LG, Wilson DB, Center EM 1980 Histochemistry of the developing notochord, perichordal sheath and vertebrae in Danforth's short-tail (Sd) and normal C57BL/6 mice. J Embryol Exp Morphol 55:227–245
Placzek M, Tessier-Lavigne M, Jessell TM, Dodd J 1988 Tissue specificity and properties of a floor plate-derived factor that influences commissural axon outgrowth. Soc Neurosci Abstr 14:595
Rijsewijk F, Schuermann M, Wagenaar E et al 1987 The *Drosophila* homolog of the mouse mammary oncogene Int-1 is identical to the segment polarity gene wingless. Cell 50:649–657
Roach FC 1945 Differentiation of the central nervous system after axial reversals of the medullary plate of amblystoma. J Exp Zool 99:53–77
Rosenquist GC 1966 A radioautographic study of labelled grafts in the chick blastoderm. Contrib Embryol Carnegie Inst 38:71–110
Schoenwolf GC, Sheard P 1989 Shaping and bending of the avian neural plate as analysed with a fluorescent-histochemical marker. Development 105:17–25
Schoenwolf GC, Folsom D, Moe A 1988 A re-examination of the role of microfilaments in neurulation in the chick embryo. Anat Rec 220:87–102
Schroeder TE 1970 Neurulation in *Xenopus laevis*. An analysis and model based upon light and electron microscopy. J Embryol Exp Morphol 23:427–462

Sharpe CR, Fritz A, DeRobertis EM, Gurdon JB 1987 A homeobox-containing marker of posterior neural differentiation shows the importance of pre-determination in neural induction. Cell 50:749–758

Spemann H 1938 Embryonic development and induction. Yale University Press, New Haven

Spemann H, Mangold H 1924 Uber induktion von embryonanlagen durch implantation artfremder organisatoren. Wilhelm Roux' Arch Entwicklungsmech Org 100:599–638

Sporn MB, Roberts AB 1988 Peptide growth factors are multifunctional. Nature (Lond) 332:217–219

Steding GV 1962 Experimente zur morphogenese des Ruckenmarkes. Acta Anat 49:199–231

Tessier-Lavigne M, Placzek M, Lumsden AGS, Dodd J, Jessell TM 1988 Chemotropic guidance of developing axons in the mammalian central nervous system. Nature (Lond) 336:775–778

Thaller C, Eichele G 1987 Identification and spatial distribution of retinoids in the developing chick limb bud. Nature (Lond) 327:625–628

Theiler K 1959 Anatomy and development of the "truncate" (boneless) mutation in the mouse. Am J Anat 104:319–343

Tickle C, Alberts B, Wolpert L, Lee J 1982 Local application of Retinoic acid to the limb bud mimics the action of the polarizing region. Nature (Lond) 296:564–566

Turner DL, Cepko C 1987 A common progenitor for neurons and glia persists in rat retina late in development. Nature (Lond) 328:131–136

Vakaet L 1981 Neurale inductie bij vogels. Verh K Acad Geneeskd Belg 43:79–103

Vakaet L 1984 Early development of birds. In: LeDouarin N, McLaren A (eds) Chimaeras in Developmental Biology. Academic Press, London, p 71–88

van Straaten HWM, Hekking JWM, Wiertz-Hoessels EJLM, Thors F, Drukker J 1988 Effect of the notochord on the differentiation of a floor plate area in the neural tube of the chick embryo. Anat Embryol 177:317–324

van Straaten HWM, Drukker J 1987 Influence of the notochord on the morphogenesis of the neural tube. In: Wolff JR et al (eds) Mesenchymal-epithelian interactions in neural development. NATO ASI (Adv Sci Inst) Ser Vol H5, Springer-Verlag, Berlin, p 153–162

van Straaten HWM, Hekking JWM, Thors F, Wiertz-Hoessel ELMJ, Drukker J 1985a Induction of an additional floor plate in the neural tube. Acta Morphol Neerl Scand 23:91–97

van Straaten HWM, Thors F, Wiertz-Hoessels L, Hekking J, Drukker J 1985b Effect of a notochordal implant on the early morphogenesis of the neural tube and neuroblasts: Histometrical and histological results. Dev Biol 110:247–254

Waddington CH 1940 Organizers and genes. Cambridge University Press, Cambridge

Watterson RL, Fowler I, Fowler BJ 1954 The role of the neural tube and notochord in development of the axial skeleton of the chick. Am J Anat 95:337–399

Watterson RL, Goodheart CR, Lindberg G 1956 The influence of adjacent structures upon the shape of the neural tube and neural plate of chick embryos. Anat Rec 122:539–553

Weber A 1938 Croissance de fibres nerveuses commissurales lors de lesion de la moelle epiniere chez de jeunes embryons de poulet. Biomorphosis 1:30–35

Wetts R, Fraser SE 1989 Slow intermixing of cells during Xenopus embryogenesis contributes to the consistency of the blastomere fate map. Development 105:9–15

Wilkinson DG, Bailes JA, McMahon AP 1987 Expression of the proto oncogene int-1 is restricted to specific neural cells in the developing mouse embryo. Cell 58:79–88

Wolff E 1936 Les bases de la teratogenese experimentale des vertebres amniotes d'apres les resultats de methods directes. Arch Anat Histol Embryol 22:1–375
Youn BW, Malacinski GM 1981 Axial structure development in ultraviolet-irradiated (notochord-defective) amphibian embryos. Dev Biol 83:339–352

DISCUSSION

McKeown: The results from the mutant mouse also say that the floor plate is not necessary for those commissural neurons to find the ventral surface.

Jessell: We think we understand why these axons get to the midline in the absence of the floor plate. This early phase of their trajectory appears to be dependent on cell–substrate interaction. Laminin forms a precise corridor that delineates the prospective pathway of these axons. So this first phase is probably independent of the effect of the floor plate. Laminin is expressed in this dorsal region and its expression ceases in the ventral half of the spinal cord as motoneurons differentiate. In the absence of motoneurons, laminin expression persists ventrally, with the result that there is a continual corridor of laminin that extends to the ventral midline. So in the mutant, we think these axons reach the midline in the absence of the floor plate target because they continue to migrate on this corridor of laminin. This is, in effect, a default pathway.

Wolpert: If there is a chemotactic factor, why do the axons cross the floor plate? One would expect them to stop there.

Jessell: Presumably, the concentration of this factor is roughly uniform inside the floor plate. It would attract the axons towards the floor plate, at which stage a different mechanism of guidance may come into operation. We know there is highly specialized expression of adhesion molecules on floor plate cells.

Kimble: Some years ago the idea of pioneer neurons was very popular. Do you have any comments about their possible role in this system?

Jessell: There are neurons that get to the midline first. The differentiation of these commissural neurons is relatively protracted, in the rat it lasts for about 48 hours. It seems that newly differentiated cells, as judged by antigen expression, migrate through early neural epithelium and don't use other axon scaffolds. That's certainly true of the first axons that get there. We think that there are pioneering neurons, at least as defined by the first class that migrate.

Cooke: You described what is thought about the derivation of different cells and this, to my mind, is the closest approach we have seen to the sort of compartmentation that people working on *Drosophila*, notably Peter Lawrence, have been telling us to look for in vertebrate embryos. This floor plate and spinal cord look like an integrated morphological structure, composed of neatly non-intermingling cells that come from adjacent parts of a sheet where the pattern formation took place earlier on. I find it interesting that they share molecular characteristics with notochord cells which lie underneath them. Are they really more like notochord cells than like cells of the nervous system? Don't they also show various neural markers? There are several monoclonal antibodies that identify unknown proteins in the nervous system.

Jessell: The properties of the floor plate are not identical to those of the notochord—we can distinguish them in later stages. But by many criteria, cells of the floor plate more closely resemble notochord cells. Many of the general neural markers are also expressed by the notochord. N-CAM, for example, is expressed at high levels in the neural tube and notochord. There are several antigens that within the neural epithelium label only the floor plate cells and many of these also label the notochord. I don't want to give the impression that these cells at all times are identical to those of the notochord. From Ray Keller's experiments, it appears that by the onset of gastrulation, cells that give rise to the notoplate and notochord are functionally distinct in that even if you transplant prospective notoplate cells to the site occupied by involuting cells destined to give rise to the notochord, they will never involute.

Cooke: As you said, there are various ways in which you can create embryos that have intermittent patches of notochord missing in the final situation when the axis is differentiated and that correlates quite strongly with floor plate structure being omitted in the neural tube. It is hard to see how that would come about if this population of cells isn't locally and continuously dependent on, or induced by, a notochord underneath it.

Jessell: The point is that these cells have a different origin from the rest of the neural epithelium. There is also evidence of the sort that you describe, indicating that the survival of these cells within the neural epithelium is in some way dependent on the continued presence of notochord.

Cooke: Do you mean survival, or does lack of underlying notochord allow transformation into some other type of cell?

Jessell: There is not enough information. At early stages the notochord adheres tightly to these cells, certainly in some amphibian species. The axial extensions of these two cell groups are probably mutually dependent. Ray Keller has some evidence in support of this. There is some evidence that the survival or at least the continued differentiation of these cells is dependent on the underlying notochord. The observations that you describe could be due to such interdependence at later stages of development.

Davidson: Does this observation require that as these commissural neurons cross the floor plate they detect something, send a signal back to their genomes, turn on the L1 gene, send the gene product down the axon and secrete or display it at the end of the axon? Are all of these functions necessary?

Jessell: We don't know. After the onset of expression of L1, the protein continues to be expressed only on the distal contralateral segment of the axon. One presumes that the protein is synthesized in the cell body and transported down the axon, although there is no direct evidence for this. You could think of scenarios where stable expression of L1 is regulated by the environment of the neuron. After crossing, commissural axons tend to fasciculate, with the result that the predominant environment of any given axon after crossing the midline is another axon, whereas before crossing the midline most axons contact neural epithelial cells.

Davidson: Protein synthesis in these cells presumably occurs in the cell body, far up on the dorsal side?

Jessell: I would presume that. We have cloned L1 and are currently looking by *in situ* hybridization at its site of synthesis. There are examples in polarized neural cells, for example oligodendrocytes, where proteins such as myelin basic protein are not synthesized in the cell body but out in the processes.

Davidson: Then this would be a strange translational control mechanism.

Jessell: We really don't yet understand how the spatial regulation of L1 expression is brought about.

Kimble: You said that TAG-1 and L1 were distinct molecules. Does that mean that they are distinct immunologically or are they distinct proteins?

Jessell: We think they are different proteins. Melitta Schachner's group in Heidelberg have cloned L1 from mouse brain. Polyclonal antibodies that recognize deglycosylated forms of L1 and TAG-1 don't cross-react. There are many molecules in this size range. I don't think it is a post-translational modification, I think they really are distinct proteins.

Gardner: What is the fate of the floor plate? Is it a transitory structure that's present during the development of this system and then disappears or does it persist?

Jessell: It is thought, and you can follow this with antigen expression, that these cells maintain their luminal and their external connections. If you look at a later developing spinal cord, the midline is invaginated. We think that floor plate cells continue to be present but form a raphé at the midline which plays some structural role. At later stages of development the midline is refractory to the passage of axons, so the specialized properties of these cells which permit axons to cross are a transient feature. The cells probably persist at least until post-natal development. The antigens that we have studied turn off at later stages, so we can't follow the fate of those cells at much later times.

Gardner: Kothary et al (1988) have a transgenic mouse line which may carry an insertional mutation at the *dystonia* locus. The expression of the β-galactosidase transgene is almost wholly confined to the floor plate and not seen in the notochord.

Jessell: In their pictures and from talking to Janet Rossant it's not clear yet whether the notochord labels. They generated an insertional transgene which gave a phenotype that was allelic to a known mouse mutation, *Dystonia musculorum*. The phenotype of the mutation involves sensory neuron degeneration, but the β-galactosidase expression pattern indicates expression is selectively in the floor plate of the neural tube. This doesn't make a lot of sense. It is possible that the expression pattern that they see is not the appropriate or complete expression pattern of the normal gene. It is unclear to me how perturbation of a gene normally expressed in the floor plate would lead to the phenotype they observe.

Gardner: They did see a little expression in the dorsal root ganglia.

Jessell: It was very sporadic. From their pictures, the β-galactosidase expression is predominantly in the floor plate.

Melton: Are there experiments which rule out that the right and left commissural neurons know which side of the body they should grow down. Can you cut a growing neural tube longitudinally and rotate it through 180° to see what happens?

Jessell: It is unlikely that there is a left/right sidedness to it. A way to address that is in chick embryos. We are working with Andrew Lumsden in London, who is trying exactly that sort of experiment to see what happens when you switch. It is easier to explain it on the basis of some sort of switch, since we know that such an event occurs, but it is possible that there is that asymmetry. We also don't understand why axons project rostrally instead of caudally.

Meinhardt: Would it be possible to transplant floor plate cells into a position halfway between their normal position and the origin of the nerves, i.e. at the side of the neural tube? In that case, after passing through the transplanted floor plate cells, the growth cone would still be on the same side of the spinal cord. That would show whether a left-right or right-left transition is required to trigger the change in the direction of the neural outgrowth.

Jessell: Those are experiments that we are trying to do in chick with Andrew Lumsden. The eventual way to understand this is to identify what the floor plate is secreting and then perhaps alter expression of this factor and see whether we can alter axon guidance.

Lawrence: This business of axons recognizing midline is fascinating. If you make a nuclear transplant mosaic in *Drosophila*, and put donor cells which stain the axons into only one side of a host in which no axons stain and then look at the central nervous system, you see an absolutely straight line down the midline. On the ipsilateral side of the donor there is very dark staining of many axons going right up to that straight line. On the contralateral side there is pale staining where some, but only a fraction of, axons have crossed that line. That line is very sharp, probably of the order of about 10–20 μm, and is recognized by those axons as they grow in the tangled neuropile.

Jessell: In *Drosophila* there is clearly something specialized about these midline cells. Paulson originally described them as mesoectodermal cells. From work in Cory Goodman's lab and others, it is clear that the molecular properties of those cells are quite distinct from those of other cells.

Lawrence: But I am talking about the neuropile, far from the cells—the axons growing through there appear to know where they are with respect to the midline.

Struhl: Are you suggesting that in insect systems there could be diffusible substances, much like the one that you define, which tell the first axons to reach the midline to do that.

Jessell: It is possible—I don't think there is any evidence for or against it. Most of the neurobiological approaches in *Drosophila* have not been looking for that. Even in vertebrate systems, it is only with these *in vitro* methods that you can get strong evidence that there is a diffusible molecule.

Struhl: You could perhaps genetically manipulate these midline cells or create midline cells in other places.

Jessell: The more interesting gene in terms of possible diffusible molecules is the *slit* gene that Spiros Artvanis Tsakonas and others are working on, which has EGF-like repeats. By *in situ* hybridization the gene is expressed selectively in the midline. The protein is expressed on the midline cells and it diffuses away onto axon tracts. So it is conceivably a diffusible molecule that is synthesized by the midline cells and diffuses away.

Cabrera: Schaller found that the homologue of the differentiation factor in the head of Hydra in *Drosophila* is expressed exclusively in the nervous system. It is a small peptide that is likely to diffuse. Its role in *Drosophila* is not yet clear.

Reference

Kothary R, Clapoff S, Brown A, Campbell R, Peterson A, Rossant J 1988 A transgene containing lac z inserted into the dystonia locus is expressed in neural tissue. Nature (Lond) 335:435–437

Cellular communication in the developing *Drosophila* eye

Andrew Tomlinson

MRC Laboratory of Molecular Biology, Hills Road, Cambridge CB1 2QH, UK

Abstract. Undetermined cells are directed to their fate in the *Drosophila* eye by positional cues which are probably presented by adjacent differentiating cells. The *sevenless* protein is thought to function as a signal-receiving protein on the plasma membrane of the cell destined to form the seventh photoreceptor, R7. Localization of the protein indicates the interactions between undetermined and differentiating cells and suggests the existence of other receptor proteins similar to *sevenless.*

1989 Cellular basis of morphogenesis. Wiley, Chichester (Ciba Foundation Symposium 144) p 281–289

Pattern formation in the *Drosophila* eye is lineage independent and is directed by positional cues (Ready et al 1976, Lawrence & Green 1979). As a developmental system it is well suited to analysing the cellular interactions which specify cell fate. The *Drosophila* compound eye is made from many hundred identical ommatidia and during development these are assembled in a precise and invariant manner. This assembly occurs in two distinct phases. First, a group of five cells (the presumptive photoreceptors R2, R3, R4, R5 and R8) are organized into a structured cellular unit (termed the precluster) as surrounding cells undergo division. The second phase follows as the cluster progressively recruits surrounding cells and the organized unit expands radially (Ready et al 1976, Tomlinson 1985).

The pattern formation described occurs in the eye imaginal disc, in which all cells extend the depth of the epithelium. Cluster cell nuclei lie in the apical half of the epithelium and those of undetermined cells are located basally. When a cell joins the growing cluster, it first achieves, in the apical region of the epithelium, a precise set of contacts with other cluster cells. These contacts are subsequently extended deeper into the tissue, so that cells initially adjacent in only the apical region eventually contact each other along their whole apical/basal extent. As the apical contacts extend deeper into the tissue, the nucleus of each newly incorporated cell rises from the basal regions (Tomlinson 1985). Differentiation such as the expression of neural antigens coincides roughly with the rise of the nucleus (Tomlinson & Ready 1987a). Hence, there is a series of events which can be identified in the process of recruitment into the ommatidium. Apical contacts are the first evidence of a cell joining the unit. Imposition of these contacts deeper into the epithelium follows and the rise of the cell's nucleus and differentiation then occur.

The fate of a cell entering the growing cluster appears to be controlled by the position it occupies within the unit. This and other observations have led to the proposal that the fate of an undetermined cell is controlled by the combination of cell types in the cluster that it contacts (Tomlinson & Ready 1987a). Recent work on the *sevenless* gene has cast some light upon the nature of this mechanism and I wish to speculate about pattern formation in the ommatidium in relation to these and other results.

sevenless is a mutation that causes each ommatidium specifically to lack photoreceptor 7 (R7) (Harris et al 1976). During ommatidial development in *sevenless* flies, a cell occupies the cluster position normally held by the presumptive R7 and achieves precisely the same cellular contacts but fails to differentiate as R7, becoming instead a lens-secreting cell type (Tomlinson & Ready 1986). For an ommatidium to generate the R7 photoreceptor, *sevenless* gene function is required only in the presumptive R7 (Harris et al 1976, Campos-Ortega et al 1979, Tomlinson & Ready 1987b). The nucleotide sequence of the *sevenless* gene suggests that it encodes a transmembrane protein related in general structure to several peptide hormone receptors such as the insulin and EGF receptors, with an intracellular tyrosine kinase domain and a large extracellular domain (Hafen et al 1987). It was therefore proposed that the *sevenless* gene encodes a protein which receives and transduces information required for the R7 developmental pathway by binding of ligand to the extracellular domain and consequent modulation of its tyrosine kinase activity (Hafen et al 1987).

There are two descriptions of *sevenless* protein localization (Banerjee et al 1987, Tomlinson et al 1987). The two accounts concur that many cells express the *sevenless* protein but disagree about in which cells and where in those cells this occurs. The following discussion is based upon the descriptions of Tomlinson et al (1987) and is accordingly subjective in assessing the available results, but many interesting indications of the cellular interactions occurring in the developing ommatidium follow from this.

sevenless protein is detected in the apical plasma membrane of a specific subset of ommatidial cells in a precise sequence (Fig. 1). Cells express the protein for only a few hours. In the precluster, *sevenless* protein is expressed in R3 and R4 and two other cells associated with the unit called mystery cells (cells associated with the developing precluster yet not destined to be part of it, see Tomlinson et al 1987). Initially, many cells surrounding the precluster show indications of low level expression of the protein; then R7 and the anterior and posterior cone cells begin to express high levels of *sevenless* protein, followed

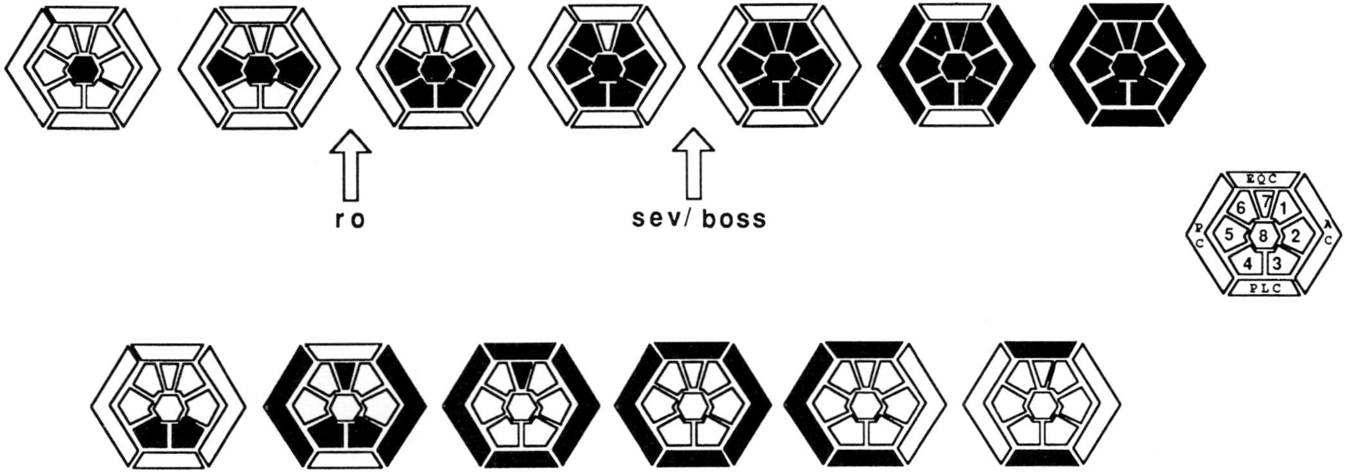

FIG. 1. Diagram depicting the cellular sequence of differentiation (upper) and expression of *sevenless* protein (lower) in the inner twelve cell unit of the developing ommatidium. These occur on different time scales and are not intended to be directly compared. Cells are shown in the position they come to occupy rather than the position they occupy at the differentiation stage shown. Upper. The twelve cells begin to differentiate in a precise sequence (derived from Tomlinson & Ready 1987a). By examining mutants in which the pattern formation breaks down at particular points, genes involved at the different stages can be identified. *sevenless* (*sev*) and *rough* (*ro*) develop normally up to the point shown, then deviate from wild-type pattern formation. *boss* is tentatively included, but pattern formation in developing *boss* ommatidia has not yet been investigated. Lower. The sequence of turning on and off *sevenless* protein expression: the mystery cells have been omitted (derived from Tomlinson et al 1987). The key to the right indicates the cell identities. Photoreceptors are numbered; AC, anterior cone cell; PC, posterior cone cell: EQC, equatorial cone cell; PLC, polar cone cell.

by the equatorial and polar core cells (Fig. 1). Cells that synthesize *sevenless* protein and contact R8 (R3, R4 and R7) accumulate protein at the point of contact with R8, suggesting that R8 expresses a ligand for the protein. Hormone receptor proteins similar in structure to *sevenless* are thought to be activated by ligand binding to the extracellular domain which modifies intracellular kinase activity; this transmits signals to intracellular components. The inferred ligand binding in R3, R4 and R7 suggests that the *sevenless* protein is activated in these cells. However, *sevenless* protein is required only in R7. This begs the question of the reason for superfluous expression. Further to this, some cells indulge in this superfluous expression and others do not.

Since cells are directed to their fate by positional cues then prior determination, these cells are expected to be pluripotent; a cell should be able to develop into one of many different types, the type being dictated by the position in the cluster it comes to occupy. *sevenless* as a protein used in receiving the positional information for the R7 developmental pathway is required only in the presumptive R7, but other cells should be able to differentiate as R7. In this regard expression of the *sevenless* protein in these cells is expected, not superfluous but indicative of the cells' pluripotency. *sevenless* protein as a receptor for positional cues can be viewed as a sensory protein used by the cells to assess whether they are receiving cues for the R7 developmental pathway. The cells' pluripotency argues that they will also be expressing similar sensory proteins required for the specification of other cell types which may or may not eventually be superfluous to them. Expression patterns of other sensory proteins may well appear similar to that of *sevenless*.

In the simplest model of the *sevenless* system, activation of the protein would be sufficient to trigger R7 development and all cells would express the *sevenless* protein, but the ligand would be presented only to the cell in the R7 position. However, from the description given above this is not the case. Some cells express the protein and others do not and there are strong indications that activation of the protein alone is insufficient for R7 development to proceed. The cells of the ommatidium develop in a clear order, which is respected by the sequence of cells that express *sevenless* protein (Fig. 1). The question now arises of why the cells express in sequence and why some fail to participate. Because the sequence of cells expressing the protein mimics the incorporation of these cells into the growing unit, a two-step process of recruitment can be envisaged. A cell is inducted into the cluster and cued to express its sensory proteins. Using this sensory capability, the cell then determines its fate and differentiates into a fully fledged member of the unit. This cell in turn expresses signals peculiar to its type for the use of other cells subsequently entering the cluster. Evidence exists from the *sevenless* mutant to support this notion. In the mutant, a cell occupies the R7 position, behaves as the presumptive R7 for many hours (achieving the same cellular contacts and displaying the same nuclear migration), but never shows any indications of differentiating as that cell type and subsequently adopts a completely different fate. The mutant cell occupying the R7 position may be expressing its sensory proteins but because it bears a defective *sevenless* protein it is incapable of receiving a necessary part of the R7 cue and sits in the position until directed to another fate.

The sequential nature of the recruitment of cells into the developing ommatidium (Fig. 1) may explain why some cells fail to accumulate *sevenless* protein. From being determined to a particular type, a delay is expected before a cell will express signals which other cells can read. This is indicated by the behaviour of R1, R6 and R7. These cells appear to be incorporated almost simultaneously into the precluster, but R1 and R6 differentiate many hours before R7 (Tomlinson 1985, Tomlinson & Ready 1987a). If all the positional cues directing R1 and R6 to their fate are already in place on the precluster cells, then they may be rapidly informed of their fate and remain in the undetermined state expressing sensory proteins for only a short time before beginning differentiation. Hence, although these cells do not stain with the antibody, they may well express *sevenless* protein as part of their sensory battery, but it fails to reach detectable levels. The presumptive R7 contacts R8 and R1 and R6. If it is the combination of these cell types that cues the R7 pathway, then R7 will need to wait for R1 and R6 to express their signals, during which time the sensory battery builds up in this cell and *sevenless* expression is high. A similar situation can be envisaged in the precluster. R3 and R4 are the last two cells incorporated and these may show high levels of *sevenless* expression while they wait for the other precluster cells to display the required signals. From this perspective, expression of *sevenless* protein indicates cells in the undetermined condition attempting to establish their appropriate fate.

In the simple model given above, the cell expressing *sevenless* protein and achieving ligand binding would be the one cued to be R7. However, the indication is that the *sevenless* protein expressed by R3 and R4 also binds ligand and clearly these cells do not become the R7 type; neither do they need the *sevenless* gene function for their development. This may reflect the combinatorial basis of the code specifying cell type. The presumptive R7 may require a combination of signals, one of which is mediated through the *sevenless* protein. The most likely candidates for the source of these other signals would be R1 and R6. R3 and R4, although receiving part of the signal required for the R7 development, differentiate as different types and may be directed by signals which override the *sevenless* activation. If a combinatorial code is in operation in the developing ommatidium, then predictions can be made. If the presumptive R7 requires developmental cues from R8 and R1 and R6, mutations causing a *sevenless*-like phenotype, which define genes encoding products required in these cells for R7 to develop, should be recoverable. In a screen for other mutations causing the absence of R7 from the ommatidia, *bride of sevenless* (*boss*) has been identified (Reinke et al 1988). The *boss* gene product is required only in R8 for an ommatidium to generate R7 and provides a clear example of non-cell autonomous gene function, indicative of communicative signals between cells (Reinke et al 1988). If R7 is specified by the combined interactions with R8, and R1 and R6, then gene products similar to *boss* can be predicted that will be required only in R1 and R6 for R7 to develop.

In the eyes of *sevenless* and *boss* flies, the absence of R7 is the only major defect in the ommatidia, indicating that R7 is a dispensible pattern element and any signals expressed by the developing R7 cell are not used by other cells joining the cluster. Mutations affecting cell fate in other cells are not expected to have a 'clean' phenotype, since errors in these cells will lead to downstream effects in other cells and highly disrupted ommatidia may result. In searching for genes of the cellular communication process in cells other than R7, the phenotype of mutant adult eyes helps little, it is the developmental phenotype that provides the major clues. If a gene is involved in the determination mechanism of a particular cell class, in a fly mutant for this gene the ommatidia should develop normally up to the point when the gene is used then deviate. In the mutant *rough*, the initial stages of ommatidial development appear as in wild-type—R8, R2 and R5 begin to develop normally, but the subsequent addition of R3 and R4 does not occur correctly and ommatidial development following this point is aberrant (Fig. 1). The gene product is probably required only in R2 and R5 for a normal ommatidium to arise and the nucleotide sequence of the gene suggests it encodes a homeobox-containing protein (Saint et al 1988, Tomlinson et al 1988). The homeobox sequence implies a possible role for the *rough* protein in transcriptional regulation in R2 and R5, but it is R3 and R4 which are affected in the mutant. Hence the *rough* gene product may regulate the expression of signals in R2 and R5 which are required by R3 and R4.

References

Banerjee U, Renfranz PJ, Hinton DR, Rabin BA, Benzer S 1987 The sevenless protein is expressed apically in cell membranes of developing *Drosophila* retina; it is not restricted to cell R7. Cell 51:151–158

Campos-Ortega JA, Jurgens G, Hofbauer A 1979 Cell clones and pattern formation: studies on *sevenless*, a mutant of *Drosophila melanogaster*. Wilhelm Roux's Arch Dev Biol 186:27–50

Harris WA, Stark WS, Walker JA 1976 Genetic dissection of the photoreceptor system in the compound eye of *Drosophila melanogaster*. J Physiol (Lond) 256:415–439

Hafen E, Basler K, Edstrom JE, Rubin GM 1987 sevenless, a cell-specific homeotic gene of *Drosophila*, encodes a putative transmembrane receptor with a tyrosine kinase domain. Science (Wash DC) 236:55–63

Lawrence PA, Green SM 1979 Cell lineage in the developing retina of *Drosophila*. Dev Biol 71:142–152

Ready DF, Hanson TE, Benzer S 1976 Development of the *Drosophila* retina, a neurocrystalline lattice. Dev Biol 53:217–240
Reinke R, Zipursky SL 1988 Cell–cell interaction in the Drosophila retina: the *bride of sevenless* gene is in photoreceptor cell R8 for R7 cell development. Cell 55:321–330
Saint R, Kalionis B, Lockett TJ, Elizur A 1988 Pattern formation in the developing eye of *Drosophila* is regulated by the homeo-box gene, *rough*. Nature (Lond) 334:151–154
Tomlinson A, Ready DF 1986 *Sevenless*: A cell specific homeotic mutation of the *Drosophila* eye. Science (Wash DC) 231:400–402
Tomlinson A, Ready DF 1987a Neuronal differentiation in the *Drosophila* ommatidium. Dev Biol 120:366–376
Tomlinson A, Ready DF 1987b Cell fate in the *Drosophila* ommatidium. Dev Biol 123:264–275
Tomlinson A, Bowtell DDL, Hafen E, Rubin GM 1987 Localization of the sevenless protein, a putative receptor for positional information in the eye imaginal disc of *Drosophila*. Cell 51:143–150
Tomlinson A 1985 The cellular dynamics of pattern formation in the eye of Drosophila. J Embryol Exp Morphol 89:313–331
Tomlinson A, Kimmel BE, Rubin GM 1988 A cellular and molecular analysis of rough, a Drosophila homeobox gene required in photoreceptors R2 and R5 for inductive interactions in the developing eye. Cell 55:771–784

DISCUSSION

Gardner: What is in between those units (ommatidia)? If you have a continuum of cells, how does one of those cells on the periphery relate to that particular group rather than to another? The cells have particular neighbourhood relations but presumably outside the group you highlight there are other cells which must relate to one group rather than to another.

Tomlinson: It appears as though a cell is drawn into one ommatidium or a neighbouring ommatidium according to the local environment. Initially, the eye is essentially a sea of cells in which nucleation centres are forming. The wave of induction spreads from each centre and the cells, by virtue of their position, are incorporated into an ommatidium either early or late.

Kimble: Do you have any speculations about how the morphogenetic furrow establishes the identity of cell 8?

Tomlinson: Cobalt sulphide staining of eye discs selectively highlights the very apical tips of the cells of the epithelium. In the morphogenetic furrow, lines of cells thrown into periodic undulations are evident. Groups of cells corresponding to each of the undulations appear to rearrange and form circular five cell units. These are the preclusters containing photoreceptors 2, 3, 4, 5 and 8.

Wieschaus: Are the cells in that undulating layer divided up into these five cell groups?

Tomlinson: Not all of them, but it's difficult to observe precisely.

Wolpert: So there is a spacing pattern mechanism rather like bristles?

Tomlinson: Yes.

Cooke: Has it been possible to investigate whether the morphogenetic furrow is moving across tissue because it is propagating, by intercellular interactions that are close in time to what you see, or whether the thing has been set up beforehand. In other words, is a wave of new activity moving across that tissue by jumping from cell to cell, or is there a wave-like appearance because cells at one end were programmed to do it earlier than cells at the other end? The obvious experiment would be a barrier or surgical interruption of the field.

Tomlinson: These experiments have been done in mosquitoes. The results are ambiguous. Probably, there is a meristematic region in the anterior of the eye which generates tissue to the posterior. We may be looking at a time lapse, so although it appears as though the furrow moves anteriorly, it may be that the furrow is a restriction which occurs in cells of a certain age.

Weisblat: What is the timing of the expression of *bride of sevenless*?

Tomlinson: The data on *bride of sevenless* are based on mosaics. It is required in cell 8 for cell 7 to develop.

Lawrence: Do you know the expression pattern?

Tomlinson: It has not been cloned and we don't know the developmental phenotype.

Kimble: Is *sevenless* in ommatidia later, or is it developmentally specific?

Tomlinson: It is developmentally specific. There is evidence for some expression in the adult but it hasn't been localized yet. Essentially a cell only expresses it for about 2–3 hours.

Wieschaus: You mentioned that not all cells express *sevenless*. Does this mean that cells which do not express *sevenless* cannot become a cell 7 because they don't have the receptor? Is there some pre-selection among potential R7 cells?

Tomlinson: Banerjee et al (1987) reported that they saw *sevenless* expressed by all cells in the developing ommatidium. The pattern we see is that cells 8, 2 and 5 do not express *sevenless*. I am not sure whether cells 1 and 6 do express it, although we don't see it in those cells. If *sevenless* is a sensory protein used by a cell in an undetermined condition to establish what its fate is, cells 1 and 6 may be informed of their fate very quickly because they are cued from precluster cells which have been around for many hours. They know who they are very quickly and the protein is not built up to a level that we can detect.

Wieschaus: I meant for the cells that are surrounding the cluster.

Tomlinson: The cells surrounding the cluster show *sevenless* expression at a low level. As the cell is brought into the ommatidium, it expresses extremely high levels of *sevenless* and then stops.

Kornberg: So the example that you showed where cells 3, 4 and 7 were *sevenless* was just a temporal window.

Martinez-Arias: Is it in bees that there is another cell 7?

Tomlinson: In the bee there is what we think is an extra cell 7. It does appear to be a duplicated cell 7.

Jessell: In *sevenless*, the cell 7 becomes a cone cell. What happens in the cone lineage, do you displace a cone cell?

Tomlinson: There is no lineage. We never see in *sevenless* an indication of that cell which would normally have become the cone cell. Presumably, because the cell in the R7 position is already that cone cell. The way to view the system is that the cells sit around the developing unit in a neutral condition until they are drawn in and told to do something. Because in *sevenless* cell 7 is switched into this other cell type, there is no reason to recruit another cell to do that.

Garcia-Bellido: The number of cells in this field is finite; the ones that are not recruited to be a particular cell type simply die. What are their centres of reference for the recruiting reaction? It would be nice to know if there are different periodicities for different types of cells or if the periodicity applies to groups of cells which are later increasingly specified in smaller groups. You have chosen to study cell 7 but there are lots of other ordered cells in the system.

Wolpert: There must be some grander order, because all the ommatidia are in register. There is a well defined anteroposterior polarity and a well defined left–right polarity.

Tomlinson: I described the assembly of the central parts of an individual ommatidium. There are definitely factors which we do not know anything about that organize the field of ommatidia into an eye.

Garcia-Bellido: So the question of priority, who comes first as a reference, cannot be answered yet.

Struhl: I would like to relate this to the discussion we had on the cell state model versus morphological cues (Lawrence, this volume). There are two types of properties involved, there are global field properties and there are the very local interactions defining the final pattern. In the formation of the segment pattern of *Oncopeltus*, there could be field properties and there could be local interactions, the two things must exist together. You cannot explain everything by local interactions.

Reference

Banerjee U, Refranz PJ, Hinton DR, Rabin BA, Benzer S 1987 The sevenless protein is expressed apically in cell membranes of developing *Drosophila* retina; it is not restricted to cell R7. Cell 51:151–158

Lawrence PA 1989 Cell lineage and cell states in the *Drosophila* embryo. In: Cellular basis of morphogenesis. Wiley, Chichester (Ciba Found Symp 144) p 131–149

General discussion IV

Davidson: The regulatory molecular biology that is emerging suggests that in different metameres homeotic genes are expressed differentially, which means there must be different regulatory factors active in these different places. To what extent can we name the genes that produce these spatially localized factors?

Garcia-Bellido: We know from segmentation that genes like *Ubx* have limits to their realm of expression. Both genetic and molecular evidence suggest that these borders are defined by other genes which act hierarchically earlier. This defines fields or groups of cells with no further quality. Then comes the question of how pattern is defined within these areas. This is going to be controlled by the homeotic selector genes which specify compartments.

Davidson: But you can't explain this just with homeotic genes.

Garcia-Bellido: That's right. The internal landscape within compartments becomes the commander of expression of the selector genes. This landscape is defined by genes involved in the generation of the space that the selector genes are controlling. In the wing of *Drosophila* there are mutations that affect wing formation, possibly with periodicity functions, which in turn control the expression of *Ubx*. The selector genes high in the hierarchy may be just members of a general class of periodicity genes. They talk to each other, via homeoboxes or transcription factors, and have been doing that for a long time in evolutionarily simpler forms. Selector genes that evolved later, a subset of homeobox-containing genes, became co-ordinators. They are still operationally intercalated between the periodicity genes (segmentation and morphogenetic genes, or in our terminology activator and realizator genes).

Davidson: Do we have any evidence, either direct genetic evidence or molecular evidence, as to the identity of the possible genes?

Lawrence: You have to keep cellular spatial aspects in the fore and ask which genes define the set of cells in which *Ubx* is expressed. In Lawrence et al (1987), we argued that the cells of parasegment 6 lie between the anterior boundary established by *ftz* and the posterior boundary established by *eve*. Thus, these boundaries together delimit the set of cells that will express *Ubx*.

Martinez-Arias: We had already shown (Ingham & Martinez-Arias 1986) that absence of *ftz* results in the decreased expression of several homeotic genes in even numbered parasegments. Returning to the question raised by Eric, when you observe expression of homeotic genes during development in the wild-type or in mutants, a gene comes on in a very broad domain, then its expression becomes modulated within that domain. We have identified some genes that affect this modulation. What came as a bit of a surprise was that when analysing homeotic gene expression in different mutants, absence of one of the elements at any stage does not preclude the others from modulating later (Irish et al 1989). For example, you might fail to initiate *Ubx* expression because the gene that defines the initial domain is absent. However, later, when in the wild-type the genes that would be modulating that expression are turned on, you start seeing *Ubx* expression in very bizarre patterns. That says you don't need a step for the next one to follow, the steps are independent.

Wolpert: You have this set of segments or parasegments, and something happens in one of these which is different to the adjacent ones. We want to know the nature of the 'something'.

Davidson: What you are saying doesn't distinguish between an upstream effect and something which follows from that spatial specification which then in turn might activate another gene that produced a DNA-binding protein. How far is the phenomenon that you are describing from the effect of this regulator on that piece of DNA? Are there ways of addressing this question?

Martinez-Arias: There is a list. Each of us here would contribute to that list in a different way—is it worth it?

Lawrence: I only gave part of the answer which is how you define a particular set of cells. I think that's the important part of it. Why *Scr* is switched on in parasegment 2 rather than in parasegment 3 is another problem.

Nüsslein-Volhard: But you don't know what is first and what is second.

Lawrence: We need both parts of the answer.

Nüsslein-Volhard: The second part of the answer, the regional specificity, is probably the gap genes. Of the known genes the gap genes are the obvious candidates. They are supposed to code for DNA-binding proteins and they could control this region specificity, perhaps not in a completely direct manner but they are certainly close.

Scott: There is pretty good evidence that certain members of each of the three classes of segmentation genes, ignoring the maternal ones, that Janni and Eric defined in their screen do affect the expression of homeotic genes. There might be cases where the gap genes that act earliest control pair-rule genes in some way and, *in addition*, 'short circuit' things and control homeotic genes directly.

Integration of inputs from gap, pair-rule and segment polarity genes by the homeotic genes gives rise to the complex patterns of homeotic gene expression. There are some segments where homeotic genes are expressed in the unit, the parasegment, fairly uniformly in certain tissues at certain stages. As Gary was emphasizing yesterday, in other segments or other tissues there can be very fine patterning within a parasegment or segmental unit of the expression of these homeotic genes. How does this intrasegmental (or intra-parasegmental) patterning occur? Some of it may be due to the only genes known to subdivide those segments and parasegments in terms of their expression patterns, which would be pair-rule genes and segment polarity genes, many of which are not even cloned yet. The pair-rule and segment polarity genes may set up complicated patterns of expression of the homeotic genes. These segmentation gene products, many of which we now presume are transcription factors, are differentially expressed in very fine patterns within segments, and it seems likely that they are responsible for precise control of homeotic gene expression when it occurs. In cases where homeotic gene expression appears to be uniform, the effects of pair-rule and segment polarity genes (and genes controlling dorsoventral differentiation) may be to affect how cells *respond* to a uniformly distributed homeotic product. Cells within a segment or parasegment may have very different histories that predispose them to respond to a uniformly expressed homeotic gene in that segment differently. So the earlier acting segmentation genes, we may speculate, could control both expression of homeotic genes and the effects of homeotic gene products.

Davidson: These are all very reasonable possibilities. My question was, can one get beyond this with respect to molecular identification of *cis*-regulatory sites. What do we know is differentially expressed in a way that parallels the pattern of expression, anterior to posterior, of a gene like *Ubx*?

Scott: The targets of action of homeotic genes are mostly unknown, and I would expect those targets to integrate inputs from, for example, *Ubx* and other regulators. So I would not expect a target gene to be expressed in a *Ubx*-like pattern. The *cis*-regulatory sites of homeotic genes themselves are being studied in many labs at the moment. Anne Boulet in my lab has analysed regions upstream of one of the *Antennapedia* promoters and she finds discrete regions responsible for specific parts of the expression pattern. She is mapping, for example, DNA sequences that respond to other homeotic genes that impinge on that promoter.

Davidson: It is easier to get at the *cis* regions than the *trans* entities. Because I am interested in the network, I want to know about *trans* regulation.

Scott: Many of them are probably unknown. They may be tissue-specific factors which generally have not come out of these studies, at least not yet. In the case I mentioned, we believe that the *trans*-acting factors are *Ubx*, *abdA* and *AbdB* acting on an upstream region of the *Antennapedia* promoter. Where *Krüppel* might be acting on an *Antennapedia* promoter is not yet known.

Struhl: We have many putative *trans*-acting factors and many putative *cis*-acting responses. Janni gave a very nice example of the *bicoid* protein which sees a target on the *hunchback* gene. I am sure that in five years time we will know which actions are direct.

Nüsslein-Volhard: The gap genes are very good candidates for the controllers of this region specificity. It has not previously been possible to do these binding studies *in vitro*, but this is underway.

Kornberg: Do you then distinguish between what happens in the establishment of the initial pattern—when these genes are expressed and present in the embryo—from what happens later when we know the pattern is so stable that you can add large quantities of these proteins without the pattern changing very much? The later expression cannot be due to the same products, so is there something universally present which has not been identified?

Struhl: The whole beauty of the system is that there are initial transient signals which lead to an activation event. After that activation event has occurred, those signals are no longer required and they disappear. Even if you reintroduce them via heat shock experiments, they have no effect. After that you have a system for making it heritable and that involves other systems of genes, many of which have been identified. It is a two step process.

Cooke: For the benefit of people who work on other systems: it is reasonable to assume that a lot of the genes are now known, even though only a restricted sub-set have been characterized. Is it turning out that all the levels of this hierarchy that regionalizes the blastoderm, even right back to the earliest levels, seem largely to be DNA-binding proteins? Developmental biologists had expected that genes earlier in the hierarchy than the gap genes would encode some other kind of molecule, on the grounds that functions in early development are expected to be conserved in evolution, and *Drosophila* is unusual in being a syncytium when early pattern formation is occurring.

Nüsslein-Volhard: The later level is only discussed on the level of DNA-binding proteins, the gap genes; pair-rule genes and segment polarity genes, however, certainly do not all code for DNA-binding proteins. There are probably others not yet identified.

Scott: At the moment, it looks as though the switch to the sorts of cell interaction or signal transduction systems you might guess must be relevant may be occurring predominantly as the segment polarity genes come into action. Some of the segment polarity genes almost certainly do not encode DNA-binding proteins. Very few of the segment polarity genes or gene products have been looked at so far, there are probably quite a few more.

Cooke: It seems striking to me that grading things on small scales requires proteins that are intercellular signals, whereas right at the beginning, when the spatial organization is being determined on a very large scale, it is done by DNA-binding proteins.

Wieschaus: Some of these may be nuclear but not necessarily DNA-binding proteins. The way that cells talk to each other prior to cellularization is by direct competition for DNA binding sites. Different nuclei might make different RNAs and the resultant proteins could diffuse between nuclei and compete with each other. Any nuclear communication that occurs prior to cellularization would occur at that level. Once cell membranes are made, cell communication may occur only in the cytoplasm or at the membrane.

Nüsslein-Volhard: Competition is not necessarily involved.

Struhl: There is one other question which relates to how proteins, such as *Ubx*, *abd-A* and *Abd-B*, which all seem to be able to bind similar or identical sites, produce very different qualitative patterns of differentiation. Gines Morata has done a very nice experiment which suggests that the specificity lies outside the homeobox.

Morata: It concerns the C1 mutation which is a deletion that breaks in an intron of *Ubx* and in another intron of *abd-A*. Both genes are transcribed in the same direction. This deletion removes the 5′ end of *Ubx* and the 3′ end of *abd-A*, and fuses together the two parts that are left. The breaks are in introns, so all the coding is in phase. This was described by Rowe & Akam (1988), they did the molecular studies and we did the developmental characterization (Casanova et al 1988). The protein is part *Ubx* and part *abd-A*, but the homeobox is from *Ubx*. It also contains the whole of the *abx cis*-regulatory region of *Ubx*. We eliminated the rest of the *bithorax* complex, to make a chromosome containing exclusively that hybrid gene function. This gene behaves like a conventional homeotic gene, with properties between those of *Ubx* and *abd-A*, although the expression is that of *Ubx* because of the *abx* regulatory element. Parasegment 5, which typically has *Ubx* identity, is essentially normal in this mutant. The amount of *Ubx* in the fusion protein is the homeobox plus 11 or 13 amino acids, the rest is from *abd-A*. The fact that the hybrid gene promotes *abd-A* patterns indicates that at least part of the specificity lies outside the homeobox.

Scott: I would like to comment on that in case everyone gets the impression that there is one class of homeoboxes and one class of homeodomains and they all bind to the same things. John Tamkun, George Hartzell and I recently tried to compile a list of the known homeodomain sequences; we have 87 and I know that's far from complete. You can group them in about 10 classes plus there are about 14 homeoboxes that aren't in any of these classes. The classes are based on distinctive amino acids at certain positions and look quite different from each other. It is possible that in some cases the proteins of one class will bind to different arrays of target DNA sequences than proteins in another class, or at least have different preferences. Some classes, such as the *Antennapedia* class or the *engrailed* class, which have been studied a lot already, may have related binding sites or related target sites—we don't know. Some of the other classes look much more different from each other, including the homeodomain of the protein encoded by the *rough* gene that Andrew Tomlinson talked about (Tomlinson, this volume), so those may have very different arrays of targets or different affinities for targets. I certainly agree with what Gary said, that a lot of these proteins will have specific actions according to the protein sequence outside of the homeodomain, but between classes there may also be differences in what these proteins bind to.

Kimble: Are these 87 homeoboxes all from *Drosophila*?

Scott: No, they are from many species. 21 of them are from *Drosophila*, which is only a fraction of what exists in *Drosophila*, I am sure.

References

Casanova J, Sanchez-Herrero E, Morata G 1988 Developmental analysis of a hybrid gene composed of parts of the *Ubx* and *abd-A* genes of *Drosophila*. EMBO (Eur Mol Biol Organ) J 7:1097–1105

Ingham P, Martinez-Arias A 1986 The correct activation of *Antennapedia* and *bithorax* complex genes requires the *fushi tarazu* gene. Nature (Lond) 324:592–597

Irish V, Martinez-Arias A, Akam M 1989 Spatial regulation of the *Antennapedia* and *Ultrabithorax* homeotic genes during Drosophila early development. EMBO (Eur Mol Biol Organ) J, submitted

Lawrence PA, Johnston P, Macdonald P, Struhl G 1987 Borders of parasegments in *Drosophila* embryos are delimited by the *fushi tarazu* and *even-skipped* genes. Nature (Lond) 328:440–442

Rowe A, Akam M 1988 The structure and expression of a hybrid homeotic gene. EMBO (Eur Mol Biol Organ) J 7:1107–1114

Tomlinson A 1989 Cellular communication in the developing *Drosophila* eye. In: Cellular basis of morphogenesis. Wiley, Chichester (Ciba Found Symp 144) p 281–289

Summary

Lewis Wolpert

Department of Anatomy & Developmental Biology, University College & Middlesex School of Medicine, Windeyer Building, Cleveland Street, London W1P 6DB

Compared to the 1974 meeting on Cell Patterning, we know an enormous amount more and discussion is quite different. It has become much more difficult because there is more information and models are so much more constrained. Molecular biology has not solved the problems yet. The simple models have gone and there is a sense of knowing more and understanding less.

If you look in terms of cell-to-cell interactions, the range over which those seem to be occurring have shrunk. It seems that a large number of cell interactions occur over short distances. Also, the important events in determination are occurring earlier. So events have become both more local and earlier. While emphasizing how important local interactions of the kind Andrew Tomlinson was talking about are, one has to reconcile them with more global properties. The global properties are manifested by the evidence for gradients, and the evidence for size regulation. One is going to have to reconcile somehow what seem to be the important local interactions with these more global properties and how they relate to the more long range interactions, which undoubtedly also take place.

One will also have to face the question of axes and polarities. I feel we haven't come to grips with this problem, largely because we don't have sufficient to say about them. Whether one looks at the sea urchin or the amphibian embryo, their axes really are Cartesian. It is so clear for example, that the anteroposterior axis and the dorsoventral axis of the *Drosophila* embryo are at right angles. If we look at the *Drosophila* eye, there again there is global orthogonality.

Expressing my own prejudices, I think we are going to have to deal with the problems of localized boundaries. Peter Lawrence touched on this issue of whether there are special properties at boundary regions which control the pattern in adjacent regions.

[1]1989 Cellular basis of morphogenesis. Wiley, Chichester (Ciba Foundation Symposium 144) p 296–297

I still am not persuaded that there isn't a different mechanism for spacing patterns and local patterns. For example, the process that spaces the segments or ommatidia of the *Drosophila* eye may be somewhat different to that which specifies the interactions within the segments or the ommatidium. These processes may turn out to be related but at the moment I have a feeling that they may be different. In vertebrates, for example, I think there may be one mechanism that specifies the spacing of feather follicles in birds, distinct from that which affects the particular feathers they are going to develop into.

Then there is the question of universality. On the one hand, it is very encouraging that those working on amphibians can use genes from *Drosophila* and do exciting things with them! There clearly are powerful similarities in some of these processes, but just how similar they are it's very hard to know. In another context, how many signalling molecules are we going to find? If we take all the different parts of the fly, is each group of cells going to have its own set of signals or is there going to be a common set of signals to which the response will be determined by the history of the cell? Judith Kimble said it rather nicely; she thought there would be lots of different molecules, but they would fall into natural classes and that will bring some coherence to the field. When one talks about all these different molecules, it is still not clear to me which is more complex, embryo or cell. The interactions between molecules within the cell are undoubtedly very complex. Cell biologists trace a response to a signal to a phosphorylated protein or a rise in calcium ions and then one comes up against a brick wall. If there really are a thousand and one kinases, as Tony Hunter (1987) has said, and many are involved in patterning, it is going to be quite hard to unravel. It may well be that the interactions between molecules within the cells will turn out to be more complex than the interactions between cells within the embryo. My guess is that the real problem for the future will not be the interactions between cells but what's going on inside cells.

There has been some suspicion of mathematical models. I think that's very unfortunate. Only if we have well defined, concrete models, can we test our ideas. Inevitably these models must be mathematical and we need to know how they behave, which requires analytical solutions or computer simulation. The days are gone when ill-defined models can be entertained.

Finally, there is the problem of evolution. I think it is probably a little too early to make progress, but at some stage we will have to give it some attention. How have all these remarkable developmental systems evolved? The whole of the animal world as we know it depends not so much on the evolution of animals—of course that's important—but the evolution of development itself.

Thank you all for a very nice meeting!

Reference

Hunter T 1987 A thousand and one protein kinases. Cell 50:823–829

Index of contributors

Non-participating co-authors are indicated by asterisks. Entries in bold type indicate papers; other entries refer to discussion contributions

Indexes compiled by Liza Weinkove

*Astrow, S. H., **113**
*Austin, J., **212**

*Bovolenta, P., **255**
*Britten, R. J., **156**
*Busturia, A., **227**

Cabrera, C. V., 57, 125, 142, 223-225, 251-254, 280
*Casanova, J., **227**
Cooke, J., 10, 14, 32, 33, 34, 57, 89, 93, 111, 112, 142, 153, 166, 167, **187**, 201-206, 208, 209, 210, 276, 277, 288, 293

Davidson, E. H., 11, 15, 29, 31, 55, 56, 97, 98, 126, 128, 140, **156**, 166-171, 182, 183, 204, 208, 209, 210, 222, 225, 239, 250, 251, 277, 278, 290, 291, 292
*Dodd, J., **255**

*Franks, R. R., **156**

Garcia-Bellido, A., **5**, 10-15, 32, 61, 62, 63, 92, 93, 97, 112, 125, 126, 127, 144, 151, 152, 153, 170, 210, 221, 222, 239, 240, 251, 252, 289, 290
Gardner, R. L., 13, 93, 110, 127, 128, 169, 171, **172**, 181-185, 203, 204, 209, 210, 253, 278, 287

*Hatta, K., **243**
Heasman, J., 30, 35, **99**, 109-112, 124, 166, 224

Jessell, T. M., 32, 110, 183, 202, 250, 252, **255**, 275-280, 289

Kimble, J., 129, **212**, 221-226, 252, 253, 254, 276, 278, 287, 288, 295
Kornberg, T., 61, 88, 150, 151, 155, 224, 250, 289, 293

Lawrence, P. A., 13, 31, 92, 93, 97, 98, 109, 110, 127, **131**, 140-144, 146, 147, 148, 155, 181, 206, 209, 222, 240, 241, 242, 250, 279, 288, 290, 291

Martinez-Arias, A., 12, 13, 31, 36, 56, 59, 94, 97, 141, 144, 147, 148, 238, 239, 241, 253, 289, 290, 291
*Matsunaga, M., **243**
McKeown, M., 30, 31, 58, 62, 145, 223, 275
Meinhardt, H., 12, 14, 29, 32, 56, 86, 90, 95, 96, 129, 142, 143, 169, 171, 221, 253, 254, 279
Melton, D. A., **16**, 29-36, 59, 61, 87, 109, 128, 140, 166, 202, 205, 279
Morata, G., 60, 61, 63, 142, 144, 225, **227**, 239, 240, 241, 294

*Nagafuchi, A., **243**
*Nose, A., **243**
Nüsslein-Volhard, C., **37**, 55-63, 88, 89, 90, 92, 94, 95, 96, 143, 144, 148, 152, 153, 221, 223, 225, 291-294

*Placzek, M., **255**

*Roth, S., **37**
*Ruiz i Altaba, A., **16**

*Sánchez-Herrero, E., **227**
Sander, K., 58, 59, 60, 96, 151, 152, 154, 167

Scott, M. P., 35, 62, 63, 98, 146, 147, 150, 291, 292, 293, 294, 295
*Snape, A., **99**
*Sokol, S., **16**
Struhl, G., 33, 34, 58, 59, 62, 63, **65**, 86–96, 98, 112, 128, 145, 148, 151, 154, 205, 209, 210, 211, 222, 223, 239, 240, 253, 279, 280, 289, 292, 293, 294

Takeichi, M., 111, 184, **243**, 250–254
*Tessier-Lavigne, M., **255**
Tomlinson, A., 124, **281**, 287, 288, 289
*Turner, A., **99**

Weisblat, D. A., 30, 93, 109, **113**, 124–129, 167, 171, 184, 185, 201, 202, 206, 221, 251, 252, 288
Wieschaus, E., 11, 34, 58, 86, 94, 96, 126, 141, 148, 153, 167, 168, 185, 210, 222, 223, 225, 252, 287, 288, 293
Wolpert, L., **1**, 13, 15, 31, 36, 58, 91–95, 97, 98, 110, 112, 127, 129, 143, 145, 147, 153, 154, 168, 169, 170, 185, 204, 205, 206, 208, 209, 211, 224, 240, 250, 251, 252, 276, 288, 289, 291, **296**
*Wylie, C. C., **99**

*Yisraeli, J., **16**

Subject index

abdominal-A (*abd-A*) gene, 66–67, 230, 292, 294
  functional characteristics, 235–236, 238–239
  genetic analysis, 232–234
Abdominal-B (*Abd-B*) gene, 66–67, 230, 292, 294
  functional characteristics, 235–236, 239
  genetic analysis, 234–235
Actin gene, CyIIIa cytoskeletal *See CyIIIa cytoskeletal actin gene*
Actin microfilaments, teloplasm formation and, 119
Adhesion, cell
  axonal guidance and, 270, 276
  calcium-dependent, 108, 110–111, 244, 252
  calcium-independent, 111
  Xenopus embryos, 107–108, 110–111
  See also Cadherins
Aggregates, cell
  mouse lung cells, 247
  Xenopus blastula cells, 102–108, 109–111
Anchor cell (AC), *C. elegans*, 213, 214, 222
Animal–vegetal polarity
  leech eggs, 115–116
  sea urchin eggs, 164, 167–171
  Xenopus eggs, 16–22, 27, 29–31, 188–189, 191–192
Annelids, factors specifying cell lineages, 113–129
Antennapedia (*Ant*) gene, 63, 66–67, 239–241, 292
Anterior body patterning, insect embryos, 38–40, 73–74, 82–83
Anteroposterior axis
  insect embryos, 30, 37–47, 55–60, 133–135
    genetic analysis, 39–40
    gradient mechanisms, 51–53, 55–57, 58–59, 65–84, 86–91
    maternal signals and, 40–47, 68–69
    organizing centres, 38–39
  Xenopus embryos
    organization before gastrulation, 191–192, 194–197, 206
    role of *Xhox3*, 22–27, 31–36
Anti-cadherin antibodies, 244–245
Antisense *Vg1* RNA, 17, 18
Antisense *Xhox3* DNA, 23–26
armadillo gene, 153–154
Axes, 296
  insect embryos, 37–53, 55–64, 65–84, 86–91
  mammalian embryos, 184
  Xenopus embryos, 16–27, 31–36
  See also Animal–vegetal polarity; Anteroposterior axis; Dorsoventral axis; Mediolateral axis; Oral–aboral axis; Polarity
Axonal guidance, 257
  to floor plate/notoplate, 266–268, 275–276
  pioneer neurons, 276
  recognition of ventral midline, 279–280
  role of cadherins, 247, 248, 251
  within floor plate/notochord, 269–271

bFGF, basic FGF *See Fibroblast growth factor*
bicaudal embryos, 52, 53, 76–78, 94
bicoid (*bcd*) gene, 39
  control of anterior body patterning, 40, 73–74, 76–79, 133
  mutant phenotypes, 39–43, 60
  protein, 35
    activation of target genes, 46–47, 55–57, 63, 71–72
    caudal interactions, 69, 74, 87–88
    graded distribution, 44–46, 52, 66–67, 69

RNA, localization, 44, 45, 61
Bithorax complex (BX-C) genes, 227–236, 238–242
developmental role, 229
functional characteristics, 235–236
number, 229–230
structure, 230–235
See also abdominal-A gene; Abdominal-B gene; Ultrabithorax gene
Blast cells (leech), 116
supernumery, 117, 122–123, 125, 126
Blastocyst, mammalian
differentiation, 172–174
implantation, 184
Blastoderm, insect, 35, 38, 68, 293
cell states, 135, 141–142
Blastula, *Xenopus*
mesoderm pre-organization, 189–190, 192–200, 201–202, 204–207
regional determination, 99–108, 109–112
Body plan formation
insects, 37–53, 55–64, 65–84, 86–91, 132–135
Xenopus, 16–27, 29–36, 187–200, 201–208
boss gene *See bride of sevenless gene*
Boundaries, lineage, 296
Drosophila embryos, 92–93, 132, 133, 136, 138
neural tube, 262–263
bride of sevenless (*boss*) gene, 283, 285–286, 288

cactus gene, 48, 50, 51
Cadherins, 107–108, 243–248, 250–254
histogenesis and, 244–245
invertebrate, 253–254
specificity of cell adhesion, 245–247
structure, 244
Caenorhabditis elegans, cell interactions in development, 125, 212–219, 221–226
Calcium-dependent adhesion, 108, 110–111, 244, 252
See also Cadherins
Cartilage induction, 266
caudal (*cad*) protein
bicoid interactions, 69, 74, 87–88
graded distribution, 52, 66–67, 71
cDNA transfection, cadherins, 244, 246–248, 250–251
Cell adhesion *See Adhesion, cell*
Cell cycles
asynchrony, 110, 177
variability, 183–184
Cell interactions *See Intercellular interactions*
Cell lineages, 1–2, 6, 12
Caenorhabditis elegans, 224–225
and cell states in *Drosophila*, 131–138, 140–148
leech, factors specifying, 113–129
mouse embryos, 172–185
sea urchins, specific gene activation, 156–171
See also Commitment of cells
Cell sorting, 251
mouse lung cells, 247
Xenopus blastula cells, 102–108, 109–111
Cell states, 12, 132, 289
definitions, 135, 141, 146–147
mesoderm, 137–138
models, 135–137, 141–148
somatic sex determination, 145–146
Cell surface
differences in *Xenopus* blastula cells, 104, 107–111
modification in growing axons, 270
See also Adhesion, cell
Cell transplantation, *Xenopus* embryos, 100, 102–104
Centrifugation
leech teloplasm, 117–118, 124, 126, 127, 128
sea urchin eggs, 170
Chemotropic activity, floor plate/notoplate, 266–268, 276
Chloramphenicol acetyltransferase (CAT) gene, 158–164
cis-Regulation, 6
bithorax complex genes, 231, 232, 233, 235, 292
sea urchin embryos, 156–165
Colchicine, effects on teloplasm formation, 119

Commitment of cells
  sea urchin embryos, 169
  Xenopus embryos
    germ layer-specific, 100, 101, 110
    in vitro cell sorting and, 104–108, 109–111
    regional, within endoderm, 102, 111–112
  See also Cell lineages
Compartments, 2
  Drosophila embryos, 132, 135, 136, 141–142
  mammalian embryos, 181–182
  neural tube, 262–263
  versus nematode equivalence groups, 225–226
Cooperativity, gene, 92, 93
Coordinate genes, anteroposterior axis formation, 39
CyIIIa cytoskeletal actin gene
  expression, 157–158
  in vivo competition, 158–162
  lineage-specific regulation, 162–164
Cytochalasin, effects on teloplasm formation, 119
Cytoplasmic determinants, leech teloblasts, 116–118, 121–123
Cytoplasmic rearrangements, annelid embryos, 115, 118, 119, 124
Cytoplasmic transplantation, insect embryos, 40, 48, 50, 74–75
Cytoskeleton
  association of *Vg1* mRNA with, 21–22, 29–30
  cell–cell adhesion and, 253
  polarity of sea urchin eggs and, 169, 170
  teloplasm formation and, 119–121, 124

Diachronic analysis, 7
Differentiation centre, Siedel's, 151–152, 153
Distal tip cell, 212–215, 219
DNA-binding proteins, 62–64, 166, 293–295
  See also bicoid gene protein; Bithorax complex genes; CyIIIa cytoskeletal actin gene; Homeobox-containing genes; Homeotic selector genes
dorsal-group genes, 47–51, 52
Dorsoventral (DV) axis, 27
  insect embryos, 38, 47–53, 59, 61
  mammalian embryos, 184
  vertebrate neural tube, 257, 266

E-cadherin, 244–248, 250–251, 253
Ectoderm
  differentiation in mouse embryos, 175–178, 182
  induction by mesoderm in *Xenopus*, 22–27, 31–35
Embryonal carcinoma (EC) cells, differentiation, 176–177, 179
Endoderm
  mouse
    differentiation into visceral and parietal, 178–179
    specification, 175–178, 182
    visceral, trophoblast proliferation and, 174, 184–185
  Xenopus
    cell surface differences, 104–110
    commitment of vegetal cells to, 100, 101
    regional differentiation in gastrula, 102, 111–112
engrailed (*en*) gene, 72, 135, 241
  cell state and, 135, 136, 137, 144, 148
  expression in mesoderm, 138
  patterns of expression, 66–67, 133
  short germ band insects, 150–151
Entelechy, 8, 10
Epidermal growth factor (EGF), 224
Equivalence groups, *C. elegans*, 213, 215, 225–226
even-skipped (*eve*) gene
  function, 92, 134, 143
  homology to *Xhox3*, 23, 31
  patterns of expression, 66–67, 96, 133, 136
Evolution, 297
  conservation of morphogenetic genes, 6, 7, 10–11
  strategies of anterior and posterior body patterning and, 82–83
Extracellular matrix
  sea urchin embryos, 171
  trophoblast proliferation and, 174–175

exuperantia (*exu*) gene, 40, 44
Eye development, *Drosophila*, 132, 281–289

Fasciclin II, 254
Fibroblast growth factor (FGF), 17, 26, 27, 197–198, 202–203
Fibronectin, 109
Floor plate/notoplate, 256, 257
  axon guidance within, 269–271
  cartilage induction by, 266
  chemotropic guidance of axons to, 266–268, 275–276
  as a compartment, 262–263
  differentiation, induction by notochord, 263–265, 277
  fate, 278
  neural tube polarity and, 271–272
  neuroepithelial cell proliferation and, 265
  relationship to notochord, 257–262, 276–277
fushi tarazu (*ftz*) gene, 62, 64, 92, 290
  function, 98, 134, 143
  patterns of expression, 66–67, 93, 96, 133, 134, 136
  regulation of expression, 154–155

Gap genes, 39, 134
  functions, 72, 92, 93, 95, 291–293
  patterns of expression, 45, 46, 66–67
  regulation of expression, 46–47, 71–72
  See also hunchback gene; knirps gene; Krüppel gene; tailless gene
Gastrula, *Xenopus*
  cell commitment, 100–102, 111–112
  cell sorting, 105, 106
Gastrulation, 13
  insect embryos, 38
  mammals, regional pre-organization, 208–209
  neural tube relationships, 257–262
  Xenopus, 22
    effects of disruption, 25
    mesoderm organization preceding, 189–190, 192–202, 204–207
Gene transfer, CyIIIa-CAT fusion gene, 157, 158
Germ cells, mouse, 224
Germline cell regulation, *C. elegans*, 212–215, 218–219, 221
glp-1 gene, 212–215
  functions, 218–219, 221–225
  genetic analysis, 216, 218
  genetic mosaic studies, 215, 216, 221
  mutant phenotypes, 214–215
Glycoprotein
  L1, 270, 277–278
  TAG-1, 267, 270, 278
Gradients, 2, 3
  cadherins, 251
  insect embryos
    anteroposterior patterning, 51–53, 55–59, 61, 65–84, 86–91
    dorsoventral patterning, 48–49, 52–53
    primary, 68–69
    secondary, 69–71
    segment determination and, 71–72, 92–93, 95–98, 143–145
  sea urchin eggs, 168–171
  Xenopus mesodermal organization and, 193–194
  See also Polarity; Positional information

Haemopoietic cells, 192, 203, 224
hairy gene (*h*), 66–67, 96, 98, 134, 154
Hensen's Node, 262, 264–265, 271–272
Homeobox-containing genes
  activation of target genes, 62–64
  classes, 294–295
  vertebrates, 10–11, 23, 31–32, 35–36
Homeotic selector genes, 152, 229
  expression in adults, 240–241
  patterns of expression, 66–67, 133, 138
  regulation of expression, 72, 240, 290–292
  See also Bithorax complex genes
Horseradish peroxidase labelling, 183
hunchback (*hb*) gene, 39, 40, 52, 64
  control of anteroposterior body patterning, 74–83, 86–87
  pattern of expression, 45, 66–67
  regulation of expression, 46–47, 55–56, 63, 69–71, 87
Induction
  cartilage, 266
  ectoderm of *Xenopus*, role of *Xhox3*, 22–27, 31–35

Induction (*cont.*)
floor plate/notoplate differentiation, 263–265, 277
lineage specification in sea urchins, 164–165, 168–169, 171
mesoderm of *Xenopus See Mesoderm induction in Xenopus*
Inner cell mass (ICM)
differentiation of cells within, 175–178
orientation in relation to uterus, 184
role in differentiation of trophectoderm, 174–175
specification, 172–174
int-1 proto-oncogene, 272
Integrin, 109, 243
Intercalation, 8, 11–12, 60
Intercellular interactions, 1-2, 9, 10, 293, 296, 297
Caenorhabditis elegans development, 125, 212–219, 221–226
Drosophila eye, 281–289
mesoderm induction in *Xenopus*, 16–17, 27, 188, 189, 197–204
Notch gene of *Drosophila*, 218, 222–223
See also Adhesion, cell; Induction; Positional information
Intracellular interactions, 9, 297
invected gene, 150

knirps gene (*kni*), 39
pattern of expression, 45, 47, 66–67
posterior body patterning and, 40, 53, 74, 77, 79–82, 86, 94–95
regulation of expression, 46, 71, 90
Krüppel gene (*Kr*), 152, 292
pattern of expression, 45, 66–67, 92–93
posterior body patterning and, 74, 79–82, 86–87, 94–98
regulation of expression, 46, 47, 71, 90

L-CAM (liver cell adhesion molecule), 244
L cells, mouse, cadherin-transfected, 244, 246–247, 250–251
L1 glycoprotein, 270, 277–278
Laminin, 267, 276
Laminin–heparan sulphate proteoglycan complex, 267
lin-12 gene, 212, 214, 224
functions, 222, 223
genetic analysis, 216–218
mutant phenotypes, 214, 215, 221
Lineages, cell *See Cell lineages*
Lithium treatment, 26, 167–169
Lung morphogenesis, mouse, adhesive mechanisms, 245, 247

Maternal signals 131
body patterning in insects, 39, 40–50, 69–71, 75–82, 86–87, 152
germline cells of *C. elegans*, 215, 225
leech teloplasm, 121–123, 128–129
Xenopus eggs, 17–22, 27, 29–31
Mathematical models, 297
Mediolateral (ML) axis, 256–257, 271
Mesoderm
cell states in, 137–138
ectoderm induction in *Xenopus*, 22–27, 31–35
induction in *Xenopus*, 188, 189
role of maternal *Vg1* mRNA, 16–18, 27, 31
XTC-MIF and FGF and, 197–204
pre-organization in *Xenopus* blastula, 189–190, 192–202, 204–207
types, gastrulating *Xenopus*, 197
See also Notochord
Mesoderm-inducing protein (MIF) *See XTC-MIF*
Metamerization *See Segmentation*
Microenvironment, role in cell allocation, 173
Microinjection
antisense *Xhox3* DNA, 23, 25–26
Vg1 RNA, 19–20
Xhox3 mRNA, 23–25, 31
Microtubules, teloplasm formation and, 119–121
Morphogenesis
approaches to study, 5–9
theories, 1–3
Morphogenetic genes, 6–7, 9
evolutionary conservation, 6, 7, 10–11
Morphogens, 2–3
dorsal gene product, 48, 50
gradients *See Gradients*
neural, 271–272
Mosaic theory, 131–132

N-cadherin, 108, 244, 245, 247, 248, 253
N-CAM (neural cell adhesion molecule), 111, 250, 253, 270, 276
naked gene, 136–137, 144, 153
nanos gene (*nos*), 39, 40, 73
activation of gap genes, 46, 71
control of posterior body patterning, 52–53, 69–71, 74–82, 86–87, 94–96
graded distribution of product, 44–46, 66–67, 69
mutant phenotypes, 39–43
Nervous system
adhesive mechanisms, 111, 245, 247, 248, 251
Notch gene of *Drosophila* and, 222–223
Xenopus, early organization, 189
Neural tube, 111, 248
compartments, 262–263
polarity and patterning, 255–272, 275–280
stages of development, 256
See also Floor plate/notoplate
Neurite guidance *See Axonal guidance*
Neuroblastoma cell line, 247, 248
Neuroepithelial cell proliferation, 265
Neurons
differentiation, 256, 266
pioneer, 276
Nocodazole, effects on teloplasm formation, 119–121
nos gene *See nanos gene*
Notch gene, 218, 222–224
Notochord, 33, 256
induction of notoplate/floor plate, 263–265, 277
lineage relationship to floor plate, 257–262, 276–277
Notoplate *See Floor plate/notoplate*
Nurse cells, 30, 44

Ommatidia, 281–289
Optic axons, cadherin-mediated outgrowth, 247, 248
Oral–aboral axis, sea urchins, 164–165

P-cadherin, 244–247, 250–251
Pair-rule genes, 134, 143, 153, 291–293
patterns of expression, 66–67, 96, 147
regulation of expression, 72, 154
See also even-skipped gene; fushi tarazu gene; hairy gene; runt gene
Parasegments, 132, 143
formation, 133–134
grasshopper embryos, 150–151
mesodermal, 137–138
role of homeotic genes, 228–236, 240–242, 290–292
See also Segmentation
patched gene, 136–137, 141, 144
Pattern formation
mechanisms, 10–13, 296–297
role of cell division, 129
See also Axes; Body plan; Segmentation
Periodicity genes, 290
Pharyngeal development, anterior, *C. elegans*, 213, 215, 219, 225
Polarity, 296
animal–vegetal *See Animal–vegetal polarity*
cell states and, 136, 142, 144
cells, role in lineage determination, 129
Drosophila eye, 289
gradients and, 68–69, 89–90, 94–98
mouse blastomeres, 173, 185
neural tube, 256–257, 271–272
See also Axes
Polyadenylated RNA (polyA$^+$ RNA), leech teloplasm, 120, 121
Polyadenylic acid (polyA), induction of supernumerary teloblasts, 122–123, 127–129
Positional information, 2, 3
differentiation of inner cell mass cells and, 175–178
Drosophila eye, 281–286
mesodermal organization in *Xenopus*, 23, 27, 32–34, 194–200
Posterior body patterning, insect embryos, 38–40, 74–84
See also Anteroposterior axis
Pre-pattern model, 2–3
Primitive streak, 184, 208–209
pumilio gene, 46

Regional determination
insect segmental units, 72
mouse trophectoderm, 174–175
neural tissues, 256–257
Xenopus blastula, 99–108, 109–112

Retina, chick neural, adhesion mechanisms, 245, 247, 248
Retinoic acid, 271–272
Retroviral insertion, 182–183
Roof plate, 256, 266, 271, 272
rough gene, 283, 286, 294
runt gene, 66–67, 96, 134, 153

Sea urchin embryos
  cis-regulation of lineage-specific genes, 156–171
  size regulation, 15
Segment polarity genes, 134–135, 143, 291–293
  patterns of expression, 66–67, 136–137, 147
  regulation of expression, 72, 153–154
  See also engrailed gene; wingless gene
Segmentation (metamerization), 227–229
  cell state models, 135–137, 141–144
  differentiation centres and, 151–152, 153
  genetic control, 66–67, 72, 133–135, 150–155, 290–294
  simultaneous versus sequential, 11–12, 150–151, 155
  telson formation and, 60–61
Segmentation genes, 10–11, 134–137, 291–293
  See also Gap genes; Pair-rule genes; Segment polarity genes
Segments, 38, 132
  expression of selector genes, 240–242
Selector genes *See Homeotic selector genes*
sevenless gene, 282–286, 288–289
Sex combs reduced (*Scr*) gene, 66–67, 138, 242, 291
Sex determination in *Drosophila*, 145–146
Sex lethal (*Sxl*) gene, 145–146
Short germ band insects, body patterning, 60, 83, 150–151, 155
Siedel's differentiation centre, 151–152, 153
Size regulation, 13–15, 58
slit gene, 280
snake gene, 50
Spemann's organizer, 26, 33, 151, 259, 271
Strypsin, 183
swallow (*swa*) gene, 40, 44

TAG-1 glycoprotein, 267, 270, 278
tailless gene (*tll*), 39, 40, 53, 56
  pattern of expression, 45, 66–67
  posterior body patterning and, 74, 79, 81, 82, 94–95
  regulation of expression, 46–47, 71
Teloblasts, 116
Teloblasts, leech
  cytoplasmic determinants, 116–118
  supernumerary, 117–118, 122–123, 125–128
Teloplasm
  distribution during cleavage, 115–116, 124, 127
  factors determining cell lineage, 116–118, 121–123
  formation, 114–115, 119–121
  transplantation, 124
Terminal group genes, 39, 40
  See also tailless gene; torso gene
Timing mechanisms, 209–212
  Xenopus embryogenesis, 196–198, 201–202, 205, 208
Toll gene, 48, 50, 51, 61
torso gene (*tor*), 39, 40
  function, 46, 56, 60, 69, 71, 81
  mutant phenotypes, 39–43, 57, 94–95
  pattern of expression, 44, 45, 66–67
trans regulation, 292
Transforming growth factor-β (TGF-β), 17, 27, 202, 272
Transgenic mice, 278
Transgenic sea urchin embryos, CyIIIa-CAT fusion gene, 158–164
Transplantation *See Cell transplantation; Cytoplasmic transplantation*
Trophectoderm
  diversification of cells within, 174–175, 182, 184–185
  specification, 172–174, 183
Trophoblast cells, differentiation, 174–175, 184–185
Tubulazole, effects on teloplasm formation, 119–121
Twinned body patterns, 190
twist gene, 50, 51

Tyrosine kinase, 224
membrane-spanning, 46, 57
sevenless gene and, 282, 284

Ultrabithorax (*Ubx*) gene 63, 230
functional characteristics, 235–236, 238–240, 290, 291, 292, 294
genetic analysis, 230–233
homology to *abd-A*, 232–233
patterns of expression, 66–67, 133, 138
UV-irradiated *Xenopus* embryos, 25, 26, 33, 192, 203, 265

VC1 marker, 112
Vgl gene, 202
Vgl mRNA, 16–22, 29–31
association with cytoskeleton, 21–22, 29–30
localization of synthetic, 19–20
role of localized, 16–18, 27
translocation to vegetal pole, 18–19
Vulval development, *C. elegans*, 214, 222

W gene, 224
Wing development, 2, 132, 231, 290
wingless (*wg*) gene, 272
cell state models and, 136–137, 141–144, 148
function, 137, 140, 153–154
pattern of distribution, 66–67
regulation of expression, 72
x gene, 45, 47, 52, 55–56, 74
Xhox1A gene, 31
Xhox3 gene, 22–27, 31–36, 205
XTC-MIF (XTC factor mesoderm-inducing protein), 17, 27, 202–204
role in mesoderm pre-organization, 197–199
Xhox3 expression and, 26

yellow gene, 135

Zygotic target genes, 39, 40, 50, 51, 53
activation by maternal signals, 46–47
See also Gap genes